Analog Electronics for Scientific Application

DENNIS BARNAAL
Luther College

Prospect Heights, Illinois

For information about this book, write or call:
Waveland Press, Inc.
P.O. Box 400
Prospect Heights, Illinois 60070
(847) 634-0081

1989 reissued by Waveland Press, Inc.

ISBN 0-88133-422-7

Printed in the United States of America

12 11 10 9

Contents

Preface

This book could be subtitled "A short course in electronics for students in physics, chemistry, engineering (other than electronic), biology, medicine, experimental psychology, geology, nursing, computer science...." Such a subtitle would illustrate the wide-ranging influence of electronic instruments and techniques in contemporary science. A laboratory in any area of science uses electronic instruments to study much of the phenomena under investigation. Today, the inside of a hospital's intensive care ward resembles the interior of the physics lab of a few years ago. Indeed electronics is pervasive throughout modern society. It is now routinely found in the home and in industry. Industrial manufacturing, chemical processing, and the like use electronics extensively, and most engineering students will benefit from a general understanding of electronic instrumentation and techniques.

People working in these areas need an acquaintance with electronics for a number of reasons: to make proper use of instruments; to understand instruments including new instruments becoming available; to conceive of new possibilities and techniques and convey them to designers; and on occasions, to save time and keep the operations going by fixing the instrument or building it oneself!

Because of these needs electronics courses have been taught in physics departments for many years. However, they have typically been "scaled-down electrical engineering courses" not well suited to the needs of scientists. For most laboratory applications, we need not concern ourselves with complex amplifier design, particularly now with the advent of marvelous integrated circuit blocks. With the understanding of some basic concepts, we can learn to use these units to perform many tasks.

Undergraduate course programs for science students are large, with little room for additions. There are many desirable and interesting courses in the major, in supplementary fields such as mathematics and computer science, and in the arts and humanities. Therefore, in 1972 a pair of two-credit minicourses was introduced at Luther College to provide flexibility in tight student schedules, and the results have been quite satisfactory.

However, no textbooks suitable for half-semester minicourses were available. Short paperback books on the market were intended primarily for the hobbyist. Furthermore, to pick and choose readings from larger texts is quite unsatisfactory. Consequently, this short book tailored to the needs of a concise course was written.

For the first few years, the minicourse was primarily concerned with the characteristics of discrete devices (tubes, transistors) and their design into various amplifier types. This emphasis left little time, particularly in a minicourse, to consider *overall design of instrument systems.* Today the availability of a sophisticated assortment of linear integrated circuits clearly renders design centered on discrete devices obsolete for all but a few specialized situations. Students in biology, chemistry, and medicine in particular can solve almost all situations with that powerful function block, the operational amplifier. Therefore, this book greatly decreases the attention paid to discrete active devices (for example, hybrid equivalent circuit analysis for transistors). The design of function blocks around integrated circuit devices and their implementation into instrumentation for representative tasks in science and medicine are emphasized. Applications of the various electronic circuits and techniques in the laboratory are frequently outlined; in general we have tried to avoid simply presenting a catalog of circuit types.

A good understanding of Ohm's Law and dc circuits, as well as basic ac circuits, remain important in using the new devices, however. Chapter A1 reviews and occasionally extends these basic areas for the student. Some component characteristics practical to working with electronics are also considered here. Chapter A2 is also rather practical in that it describes some important features of various kinds of measuring equipment and instruments widely used in electronics and the laboratory. Much of this basic material is timeless since it has remained pertinent through the vacuum tube, transistor, and now integrated-circuit generations.

Electronics is useful because of its contact with the physical world outside the electronic box. Contact is made through a transducer, and it is important to learn about a number of the transducer possibilities. Chapter A3 attends to this. Then the basic but useful diode is taken up in Chapter A4. A very common application of diodes makes possible dc power for electronic devices from ac power, and this application deserves examination. Thus, basic power supplies and modern integrated-circuit regulators are considered. Further, a basic discussion of the behavior of semiconductors is given here.

Chapter A5 presents some general features of amplifiers and amplifier use before examining specific cases in the following chapters. This chapter provides a general perspective of amplifiers that in fact remains relevant throughout the electronic generations. Chapters A6 and A7 get to the heart of the course with a survey of the uses of linear integrated circuits, especially the operational amplifier. Some pulse or digital circuit considerations are taken up in Chapter A7, however, since they can be useful in applications and circuits that are primarily analog.

A comment on *analog electronics* versus *digital electronics* is in order here. This book is primarily concerned with the former, while the latter is discussed in my book *Digital and Microprocessor Electronics for Scientific Application*. In the era of the 1980s, almost everyone has an idea of the difference between digital and analog. People are familiar, for example, with the widely used digital watch that now often replaces the "analog watch" on wrists. The term *digital* implies the use of discrete numbers and electronic circuits that are just "on" or "off" like a switch. The term *analog* implies the use of a needle (such as a watch hand) along the range of a scale and electronic circuits whose currents and voltages may be any of a continuous range of values.

Finally, Chapter A8 briefly presents various transistor types as well as some other important semiconductor devices. It is not the intention of the chapter to develop a facility for transistor amplifier design. However, scientists often encounter transistors and even vacuum tubes as separate or discrete elements in older circuitry; furthermore, integrated circuits consist largely of interconnected transistors. Therefore a basic understanding of transistor characteristics and how they accomplish the "magic of electronics" is desirable, for it permits a comprehension of the operation of discrete circuits and facilitates an appreciation of the operating characteristics and limitations of integrated circuits.

As background for the course, the book assumes only that the student has studied electricity in the context of a noncalculus college physics course. However, the first chapter provides a reasonably self-contained review of these topics, and the preliminary version of the book has been used in a general no-prerequisite course at one institution. Higher mathematics such as differential equations, Fourier analysis, and Laplace transforms have been completely avoided; indeed they are not particularly necessary for much of electronics as it is actually practiced in the laboratory. Almost all derivations require only Ohm's Law and some algebra. Although calculus is used at a few places in the book (for the benefit of students with calculus background), students without calculus can skim through these steps to the conclusions derived.

At Luther College, the analog and digital electronics minicourses are generally elected by physics majors and pre-engineering students as a 4-credit course (3 lectures and one 3-hour laboratory each week) in the first semester of their second year. The study of electronics early in the curriculum of physical science permits this important tool to be realistically exploited throughout the later laboratory work. Nonphysics majors typically elect one or both of the minicourses during their third or fourth year in college.

Laboratory work remains crucial to the art of using electronics. Merely reading about it is not enough; a certain intuition needs to be developed through *doing*. To assist the student in developing this intuition, electronic diagrams in the book often include values for components rather than only sterile symbols. Furthermore, we will not be so aloof as to avoid "nitty-gritty" problems and practice of real-world electronics as used in the laboratory (or even in the home).

The laboratory work at Luther College was reorganized at the same time that the preliminary version of this book was developed in 1977 under a Local Course Improvement (LOCI) grant from the National Science Foundation. The laboratory experiments investigate a number of important function blocks that may be accomplished with the operation amplifier and the 555 timer integrated circuits. The concluding experiments for the course involve the merger of these function blocks into instruments that each accomplish some desired task; in a real sense, the laboratory work with these instruments integrates the course.

NOTE TO INSTRUCTORS: Especially when multiblock instruments are constructed, conventional breadboard techniques can yield a rather confusing haystack of wires and components. Therefore, at Luther College modules are used that are rather similar to those sold commercially by Pasco Scientific (San Leandro, California). The modules are used earlier in the laboratory work to study the important function blocks; in the capstone experiments they are interconnected to accomplish the desired instruments. Several (pedagogical) medical instrument designs that students find very interesting were developed by the Curricular Development Laboratory of the Technical Education Research Center (TERC) (Cambridge, Massachusetts), and laboratory booklets for these experiments are available from Pasco Scientific.

Finally, it should be mentioned that a number of my colleagues reviewed the manuscript during the various stages of its development. Thanks for their helpful comments and suggestions are extended to Mark J. Engebretson, Augsburg College; Thomas I. Moran, University of Connecticut; Roy Knispel, University of Wisconsin; and Charles Duke, Grinnell College. A special word of thanks goes to Ed Francis of Breton Publishers who inspired me to develop my original work into a textbook suited to a broad, national audience. And Ellie Connolly of the Breton production staff is to be commended for her many contributions to the final product.

Dennis Barnaal
Luther College
Decorah, Iowa

Analog Electronics for Scientific Application

CHAPTER A1

Passive Components and Networks

The first chapter deals with elements of electronics that do not "amplify"; that is, they do not cause an increase in the power level of the signal. Much of the material is at least touched upon in a first-year physics course (considered a prerequisite to this book). Although the presentation is reasonably self-contained, the first chapter functions primarily as a review. Nonamplifying elements, however, are the foundation of electronics, and any rusty memories must be reoiled. Further, working with electronics is applying Ohm's Law again and again; therefore, the student must understand basic circuits very well.

In the course of the chapter we review electrical concepts such as charge, current, resistance, voltage, and electrical power. We consider how to analyze the behavior of dc circuits and simple ac circuits. The important passive elements are reviewed; these include the resistor, potentiometer, battery, capacitor, inductor, and transformer. However, we do discuss important characteristics of the real components used in electronics, and this may not be review material. Also, the emphasis with respect to ac circuits is on frequency filter circuits. Finally, a remarkable property of electrical circuits called Thevenin's Theorem is discussed. This theorem is the basis of an approach to electronics that is emphasized in this book—that is, the black box model of amplifiers and system components. Although the contents of the boxes may change over the years, this approach promises general and quite timeless utility.

Summary of Basic Concepts

Electricity begins with the Coulomb Law for the force F between point electrical charges Q_1 and Q_2 that are a distance r apart. In MKS or SI units it is

$$F = k\left(\frac{Q_1 Q_2}{r^2}\right) \tag{A1–1}$$

where $k = 8.987 \times 10^9 (\simeq 9 \times 10^9) \mathrm{Nm^2/C^2} = 1/(4\pi\epsilon_0)$

The unit of charge is the *coulomb,* which is defined through current in the MKS system. We have the familiar story that unlike charges attract and like charges repel. We imagine that a charge experiences a force because of the electrical field at that point due to other charges. If F is the force on a charge Q, then the electric field E at that point is

$$\vec{E} = \frac{\vec{F}}{Q} \tag{A1–2}$$

Combining Equations A1–1 and A1–2 quickly shows that the magnitude of electric field due to an isolated point charge Q is

$$E = \frac{kQ}{r^2} \tag{A1–3}$$

where r is the distance from the charge. The total electric field from a group of charges is the vector sum of individual contributions given by Equation A1–3.

If one moves a charge Q between two points A and B in an electric field, *work* W is generally done (a force is exerted through a distance). We speak of the *potential difference* or the *voltage difference* V_{AB} between two points; that is,

$$V_{AB} = \frac{W}{Q} \tag{A1–4}$$

Thus the voltage between two points is the *"work per unit charge"* to move the charge between the two points, and the units are joules/coulomb $\equiv$ volts in the MKS system. Since these are the "practical units" of the electrician, MKS units have become dominant in electronics and, indeed, throughout science. One of the remarkable features of electric fields is that the work done in moving a charge between two points is independent of the path followed. This permits us to speak of a *conservative field* and to use the term *potential.* Kirchhoff's second law follows directly from this fact, as we note shortly.

Charges in motion constitute electric current. The net charge passing through a surface per second is the current in amperes. If ΔQ is the net charge that flows in a time Δt, then the current I is

$$I = \frac{\Delta Q}{\Delta t} \tag{A1–5}$$

It follows that 1 ampere = 1 coulomb/second.

The direction of conventional current flow is the direction that positive charges move (or at least would move if they were there). Negative charge moving in the opposite direction is counted as equal to positive current moving in the conventional direction. Thus, ions of opposite sign in an electrolyte solution move in opposite directions, but these currents add when calculating net current.

Review of dc Circuits

The first section of our review covers circuits and circuit elements for the case when the current is nearly constant and in one direction only—that is, the *direct current* or *dc* case.

Ohm's Law

If a voltage difference exists between two points on a length of wire or other material, the difference is due to the existence of an electric field between the two points. Any charges—that is, *charge carriers*—that are free to move in the material will do so because of the push of the electric field. The charges move at a *drift velocity* that is limited by frictional effects in the material, and a certain current results. The current is proportional to the push of the electric field that in turn is proportional to the voltage difference between the two points. On the other hand, it is inversely proportional to the frictional effects, or *resistance R,* to current flow between the two points. We write this as Ohm's Law:

$$I = \frac{V}{R} \quad \text{or} \quad V = IR \tag{A1–6}$$

The unit of resistance is the *ohm* and Equation A1–6 shows 1 ohm = 1 volt/ampere.

The resistance of a length of wire is proportional to its length L but inversely proportional to its cross-sectional area A. We have

$$R = \frac{\rho L}{A} \tag{A1–7}$$

where the proportionality constant ρ is called the *resistivity* of the material used for the wire. Table A1–1 gives resistivities for several metals, alloys, and insulators at room temperature.

TABLE A1–1. *Some Resistivities*

Material	*Resistivity (ρ) in 10^{-8} ohms × meters*
Silver	1.5
Copper	1.7
Aluminum	2.6
Manganin	44.0
Constantan	50.0
Nichrome	100.0
Carbon	350.0
Silicon (pure)	625.0
Glass	10^{20}

Batteries

Batteries serve as electrical elements that by chemical action maintain a potential difference (voltage) between two terminals whether a current is flowing or not. A single cell consists of two electrodes made from different materials that are inserted in some electrolyte. The potential or *emf* of the cell is determined by the composition of the electrodes and the electrolyte concentrations, not by the physical size of the cell. Strictly speaking, batteries are cells that are connected in series to increase the voltage; however, the term is often used for single cells as well. Figure A1–1 shows the conventional circuit symbols for a cell and a battery.

Batteries have regained popularity in electronics because of the low power requirements of contemporary solid-state circuits. They are necessary for equipment used in remote places and useful when equipment must be isolated from power line (wall socket) noise. It is desirable for the circuit designer and the user to have a basic knowledge of different cell characteristics.

The *lead-acid storage battery* is the common car battery in the United States and has been in use for many years. It has an emf of 2.06 V to 2.14 V when charged, requires a liquid (acid) electrolyte, is heavy because of the lead, and is capable of high currents (up to hundreds of amperes). It is rechargeable by reversing the current through the battery and thus is reusable for a few years. A sealed form is now available.

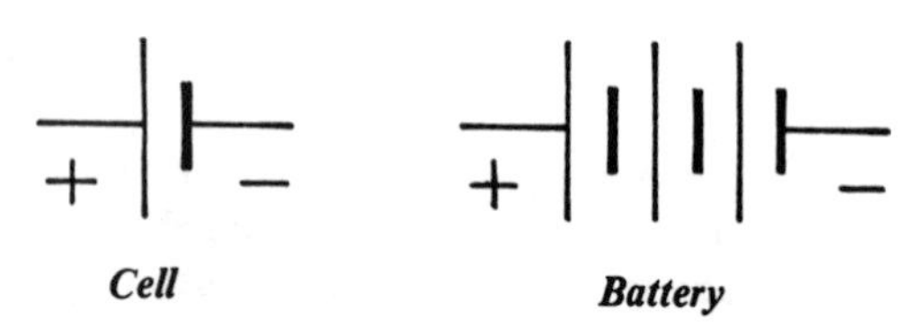

FIGURE A1–1. *Conventional Circuit Symbols for a Cell and a Battery*

The *carbon-zinc battery* is the familiar flashlight battery of many years use. It is called a *dry cell* because of the paste electrolyte used and has a fresh emf of about 1.55 V. It is inexpensive but has poor performance at low temperatures and a shelf life only of a year or so; that is, even if unused (or left on the shelf), it becomes useless after a period of time because of the internal discharge that takes place. The battery can be only partially recharged over a period equaling the shelf life.

The *alkaline-manganese battery* is another paste electrolyte cell with an emf of about 1.5 V. It has a higher current capability and longer shelf life than the carbon-zinc, but at approximately four times the price. It is not rechargeable.

The *nickel-cadmium battery* is a rechargeable battery that has been used in Europe for many years but is perhaps twice as expensive as the lead-acid type. It uses a liquid electrolyte, but is available in sealed form in smaller sizes (for example, for use in calculators). It is capable of high discharge and charge rates and has a long service life. The cell emf varies from 1.4 V at full charge to about 1.25 V for most of the discharge period.

The *mercury battery* is popular in scientific instruments because its voltage remains very constant at 1.35 V over the discharge period of the battery. It uses a paste electrolyte, but is considerably more expensive than other paste types and does not perform well at low temperatures. See the forms of discharge curves for several battery types in Figure A1–2.

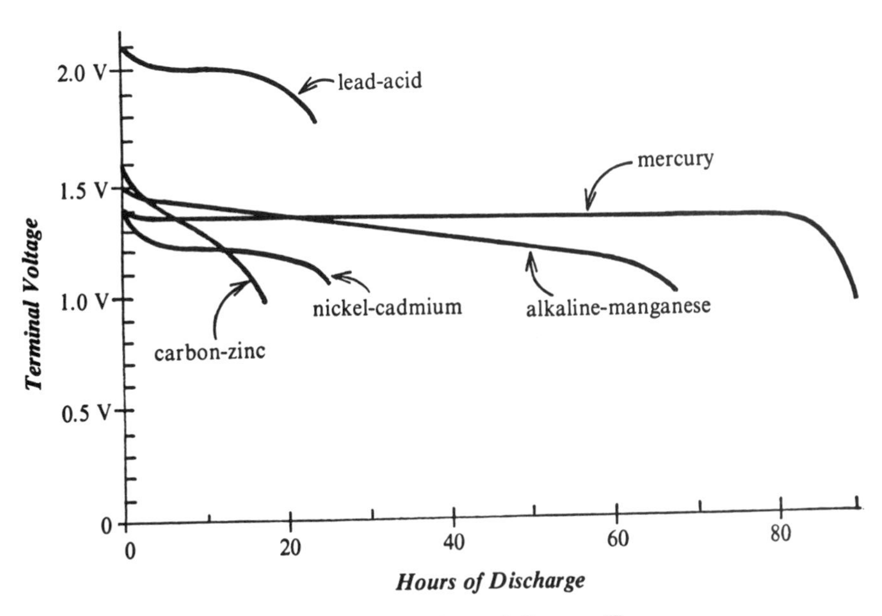

FIGURE A1–2. *Discharge Curves for Several Battery Types*

For all battery types it is important to use a battery that is physically large enough to handle the current drain. Large plate area implies a lower internal resistance and less heat dissipated in the battery, as well as a smaller drop in terminal voltage. Batteries are rated in *ampere-hours,* or in *milli-ampere-hours.* This rating is simply the product of current drain times the discharge period that we may expect from the battery, and is basically a unit of energy. Eveready (Union Carbide) has a substantial handbook listing properties of their large line of dry cells.

Power in Electrical Circuits

The *voltage* of a battery is the work per unit charge to move the charge from one terminal to the other. If positive charge Q moves from the positive terminal to the negative terminal, the battery performs work W equal to VQ. *Power* P is the time rate of doing work. Dividing by the time involved quickly gives (for steady current)

$$P = \frac{W}{t} = V\left(\frac{Q}{t}\right) = VI \qquad P = VI \qquad \textbf{(A1–8a)}$$

as the power delivered when a current flows between a voltage difference V. If the voltage appears across a resistance R, Ohm's Law gives two alternate ways to compute the power (in the form of heat dissipated in the resistor):

$$P = I^2R \qquad P = \frac{V^2}{R} \qquad \textbf{(A1–8b)}$$

When the current is in amperes, the voltage in volts, and resistance in ohms, the resulting power is in *watts.*

Resistors

Electronics makes regular use of circuit elements with definite resistance, called *resistors.* Various types of resistors are made to fit the needs of low cost, small size, high power consumption, high accuracy, low-temperature effects, and so on. It is not possible to meet all of these requirements with one type.

The basic *fixed resistor* (or simply *resistor*) is a large length of wire wound as a coil on a ceramic tube. Material of high resistivity is chosen so that the length is not outlandish. Further, temperature has a small effect on the resistance, and nichrome or manganin wire are often used to meet these two requirements. The circuit symbol for a resistor is a zig-zag line suggested by the side view of a coil of wire (see Figure A1–3).

FIGURE A1–3. *Schematic Symbols for Fixed and Variable Resistors*

If a sliding contact is arranged so the length of wire is effectively variable, we obtain a *variable resistor.* In general there are three terminals available from this device; they are from the two ends of the fixed resistor and from the slider or tap. Only one end terminal and the slider terminal need to be used for a variable resistor. This two-terminal device is called a *rheostat,* although the term rheostat is encountered primarily for large variable resistors capable of dissipating 10 watts or more. The symbol for a two-terminal variable resistor is shown in Figure A1–3 as a zig-zag line with a cross arrow to suggest variability.

The three-terminal variable resistor is commonly called a *potentiometer* because all three terminals are put to use when it is employed in a variable potential divider (or voltage divider). The circuit symbol for a potentiometer is shown in Figure A1–3 as a zig-zag line with an arrow to the edge; this arrow suggests the sliding contact connection. Almost all variable resistors used in electronics are in fact potentiometers, with the electrical connections then made to the slider terminal and to one end terminal.

The *wirewound resistor* is commercially available virtually as already described for the fixed resistor. It can be made with high precision (for example, 1% for $.50 and 0.1% for several dollars each) and is usually put on a ceramic form for high power. Five-watt, 10-watt, 20-watt, and even 50-watt sizes are quite common, with size proportional to the ability to dissipate power as heat. When operating at capacity, it is hot to the touch. Manganin is the alloy usually used to obtain a small temperature effect on the resistance (temperature coefficient). The wire is usually folded back on itself to obtain a *noninductive winding,* but even so, a wirewound resistor does not work well at radio frequencies (100 kilohertz and above) because of its inductance.

Certainly the workhorse resistor of electronics is the *carbon composition resistor* with its hallmark color coding. Because of the high resistivity of carbon, wires may be inserted into opposite ends of a cylinder of carbon to give resistances between about 1 ohm and 22 megohms. This resistor has a low inductance, a fairly small temperature coefficient, and a small size. Power capabilities available are 1/8, 1/4, 1/2, 1, and 2 watts. The commonly used tolerance is 10% (silver band), with price on the order of $.05 per unit, although 5% (gold band) is becoming similarly priced and is often preferred. See Figure A1–4 and Table A1–2 for the color code scheme.

The first to third color bands on the resistor provide the digits that represent the resistance. The first and second bands give the first two digits and the third band is the multiplier that shows how many zeros to add to the first two digits. For example, if the bands were brown, red, and yellow, the

TABLE A1–2. *Carbon Color Code*

Color	*Digit*	*Color*	*Digit*
black	0	green	5
brown	1	blue	6
red	2	violet	7
orange	3	gray	8
yellow	4	white	9

resistance would be 120,000 ohms. The fourth band gives the tolerance. A fifth band, if present, gives the reliability of the resistor. Reliability is used for military specifications.

For precision resistors with low inductance and moderate power capabilities, the *thin film resistor* is a good choice. A thin film of metal, carbon, or metal oxide is deposited on a glass or ceramic cylinder with a manufacturing tolerance of usually 1%. The metal type is the best because it has the lowest noise and a low temperature coefficient, but each metal resistor costs $.50 to $1.00. The value of resistance is usually stamped on the metal film units, although color coding is common for the carbon film resistors.

Figure A1–5 illustrates a number of common resistor types: power wirewound (Figure A1–5a), carbon composition (Figure A1–5b), and metal or carbon film (Figure A1–5c).

The most common *potentiometer*—that is, three-terminal variable resistor—used in electronics has a sliding contact moving over a circular segment of carbon resistive film. The interior construction and exterior appearance are shown in Figure A1–6a. It is typically rated at 2 watts, and the center lug (of the three available) is the sliding contact. For higher resolution *multiturn potentiometers* are available that typically have a linearity of 0.1%. The interior construction of the popular small rectangular *trimmer* type is

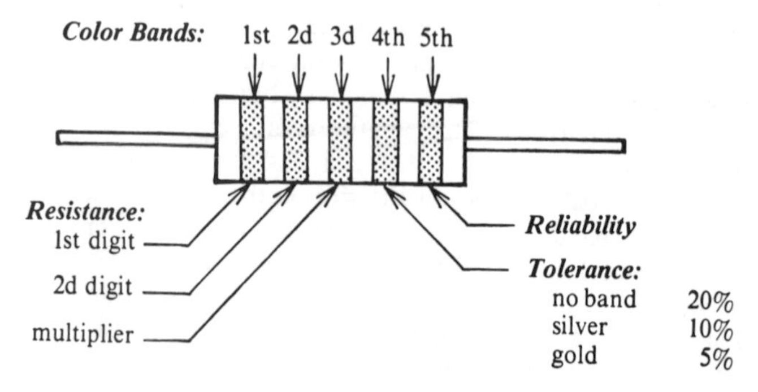

FIGURE A1–4. *Color Code for Carbon Composition Resistors*

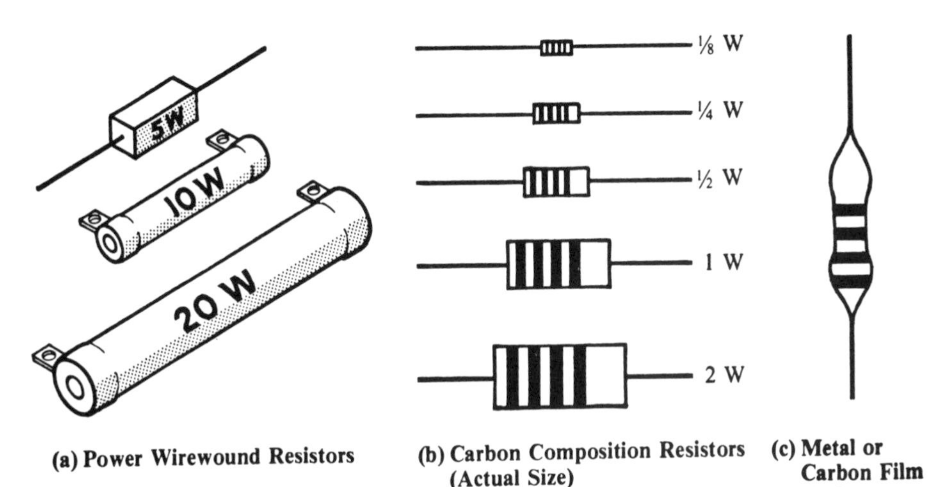

(a) Power Wirewound Resistors **(b) Carbon Composition Resistors (Actual Size)** **(c) Metal or Carbon Film Resistor**

FIGURE A1–5. *A Sample of Types of Common Resistors*

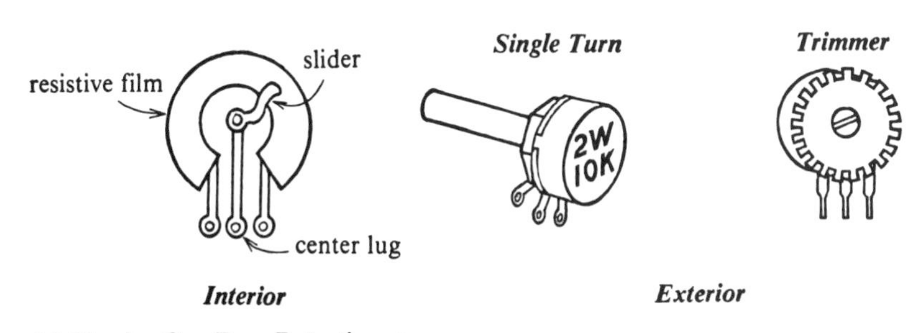

(a) Circular One-Turn Potentiometers

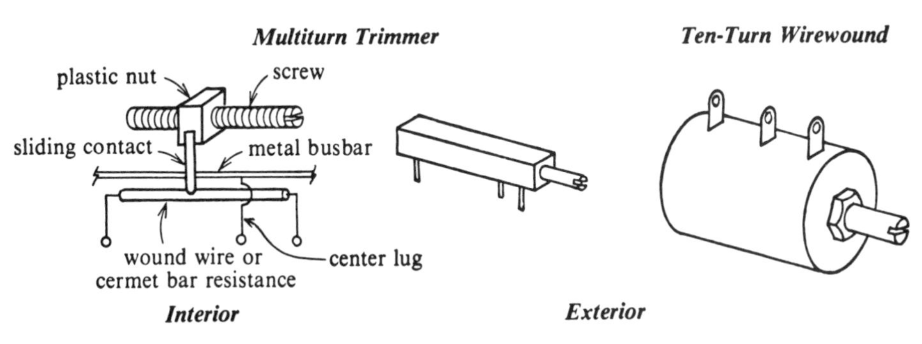

(b) Multiturn Potentiometers

FIGURE A1–6. *Common Variable Resistors or Potentiometers*

illustrated in Figure A1–6b. The resistive element in this type is often a bar of *cermet* material (ceramic and metal) that permits *infinite resolution;* the power dissipation capability is on the order of 0.5 to 1 watt. The larger *tubular* types generally use a coil of wire and can dissipate 3 watts (see the 10-turn wirewound potentiometer in Figure A1–6b). The term *pot* is a common colloquial expression for potentiometers; thus the term *trim pot* refers to a small adjustable potentiometer often placed directly on a circuit board rather than on a control panel.

Series and Parallel Networks and Kirchhoff's Laws

In solving problems where resistors are interconnected in quite general ways to form networks, the first point to remember is the relationship between relative polarity of voltage across a resistor and the direction of current through it. The more positive end is always the end at which current enters, as pictured for resistor R in Figure A1–7.

The second point is that the voltage computed between two points is the same no matter what path is chosen between the points (because electric fields are conservative). This second point in essence is *Kirchhoff's Voltage Law.* It may be stated more concisely as: The sum of rises and falls of potential (or voltage) around any closed path is zero; that is,

$$\sum_{\substack{i \\ \text{closed} \\ \text{loop}}} V_i = 0 \qquad \textbf{(A1–9)}$$

Consider the circuit shown in Figure A1–8. We have three *resistors in series* (because the same current must flow through each) that are connected to a battery of voltage V. The voltage at B compared to A, $(V_B - V_A)$, must

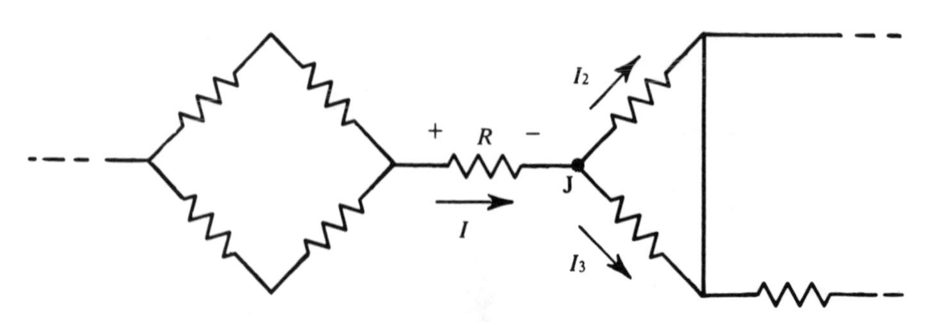

FIGURE A1–7. *Polarity and Current Direction in a Resistor*

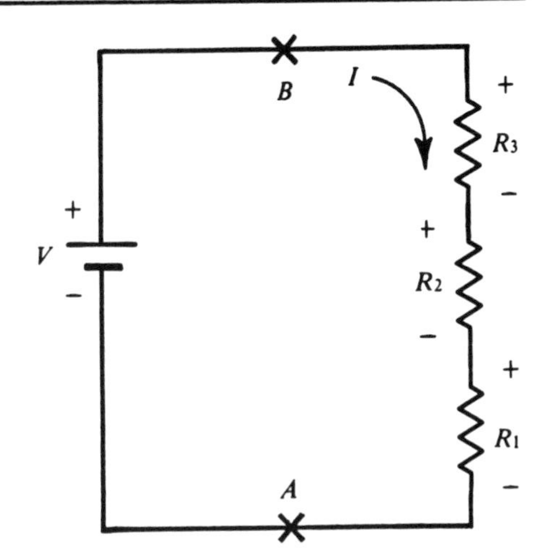

FIGURE A1–8. *Series Resistance Circuit*

be equal to V when considering the left path from A to B. Going up the right side involves successive voltage rises of V_1 across R_1, V_2 across R_2, and V_3 across R_3. The total potential rise is $V_1 + V_2 + V_3$, which must be the same as V. Therefore,

$$V = V_1 + V_2 + V_3 \qquad \textbf{(A1–10)}$$

However, Ohm's Law requires that $V_1 = IR_1$, $V_2 = IR_2$, and $V_3 = IR_3$. Substituting into Equation A1–10 results in

$$V = IR_1 + IR_2 + IR_3$$

Solving for I gives

$$I = \frac{V}{R_1 + R_2 + R_3} = \frac{V}{R_s}$$

if we let

$$R_s = R_1 + R_2 + R_3 \qquad \textbf{(A1–11)}$$

The three resistors in series behave as one equivalent resistance R_s, and Equation A1–11 is the familiar *additive rule for resistors in series.*

Note that direct application of Equation A1–9 results in the same equation as Equation A1–10. Suppose we begin at A and travel counterclockwise around the circuit, adding potential rises and falls. We go up through R_1 an amount V_1 (since we go − to +), again up V_2 and again up V_3, but then down $-V$ for + to − through the battery as we return to A. Then adding these voltage changes gives

$$V_1 + V_2 + V_3 - V = 0$$

which is the same as Equation A1–10 if V is moved to the other side. We may show the same happens if we think of going clockwise from A and back to A (or again, if B is chosen as the starting point).

Before reviewing parallel resistors, the third fundamental point about electric circuits—called *Kirchhoff's Current Law*—should be mentioned. When there is a junction of current paths (a fork in the road) such as J in Figure A1–7, the current into the junction must equal the current out. (Otherwise net charge would accumulate there and literally explode from coulomb repulsion.) Therefore $I = I_2 + I_3$ in Figure A1–7. A neater way of stating the same thing is the algebraic sum of currents into a junction must give zero; that is,

$$\sum_i I_i = 0 \tag{A1–12}$$

Here a current is reckoned positive if going into the junction and negative if leaving it.

Consider now the circuit shown in Figure A1–9. We have three *resistors in parallel* since there are three current paths literally parallel from A to B. Applying Kirchhoff's Current Law to the junction at A gives

$$I = I_1 + I_2 + I_3 \tag{A1–13}$$

The voltage between B and A must be V when the battery path is followed, but it must be $V_1 = I_1 R_1$ or $V_2 = I_2 R_2$ or $V_3 = I_3 R_3$ when moving through the respective resistors from B to A. Solving for the currents in these three statements and substituting into Equation A1–13 gives

$$I = \frac{V_1}{R_1} + \frac{V_2}{R_2} + \frac{V_3}{R_3} = V\left(\frac{1}{R_1} + \frac{1}{R_2} + \frac{1}{R_3}\right)$$

We see that

$$I = V\left(\frac{1}{R_p}\right)$$

where
$$\frac{1}{R_p} = \frac{1}{R_1} + \frac{1}{R_2} + \frac{1}{R_3} \tag{A1–14a}$$

Thus the three resistors in parallel behave like one equivalent resistance R_p, which is given by Equation A1–14a. This equation is the familiar *rule for the equivalent resistance of parallel resistors.* Note that the parallel resistance is always *less* than the smallest resistor in the group.

It should be clear how to generalize the series and parallel resistance formulas of Equation A1–11 and Equation A1–14a, respectively, for the case of more than three resistors: Add more R_i or $1/R_i$ terms in each respective formula. If only two resistors are involved, the parallel resistance case becomes

$$R_p = \frac{R_1 R_2}{R_1 + R_2} \tag{A1–14b}$$

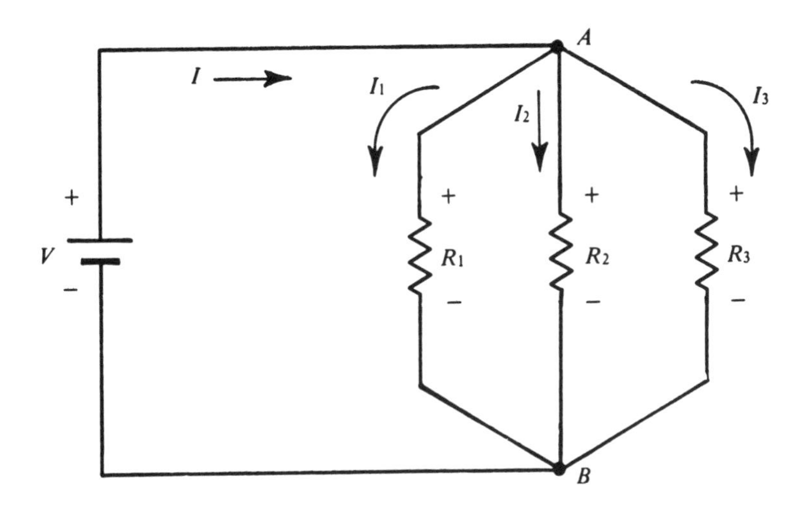

FIGURE A1–9. *Parallel Resistance Circuit*

We next review by example circuits consisting of resistor combinations that can be considered in series or parallel.

Example: Consider the circuit given in Figure A1–10a. Suppose both the *current drain* from the battery and the *current through the 5-ohm resistor* are desired. Our goal is to replace the circuit with one equivalent resistance connected to the battery.

Solution: The 1-ohm and 2-ohm resistors are clearly in parallel and, in fact, are in parallel with the 3-ohm resistance, too. We compute their effective resistance:

$$\frac{1}{R_{p1}} = \frac{1}{1\ \Omega} + \frac{1}{2\ \Omega} + \frac{1}{3\ \Omega} = \frac{11}{6\ \Omega} \qquad R_{p1} = 0.545\ \Omega$$

The parallel combination is now replaced with a single 0.545-ohm resistor to obtain the simpler circuit of Figure A1–10b. We now see the 0.545-ohm resistor in series with the 4-ohm resistor, which gives an equivalent resistance of 4.545 ohms. This in turn is in parallel with the 5-ohm resistance to give a resistance (using Equation A1–14b)

$$R_{p2} = \frac{(4.545\ \Omega)(5\ \Omega)}{4.545\ \Omega + 5\ \Omega} = 2.38\ \Omega$$

and the circuit then reduces to the one shown in Figure A1–10c. Now there are three resistors in series that have the equivalent resistance of

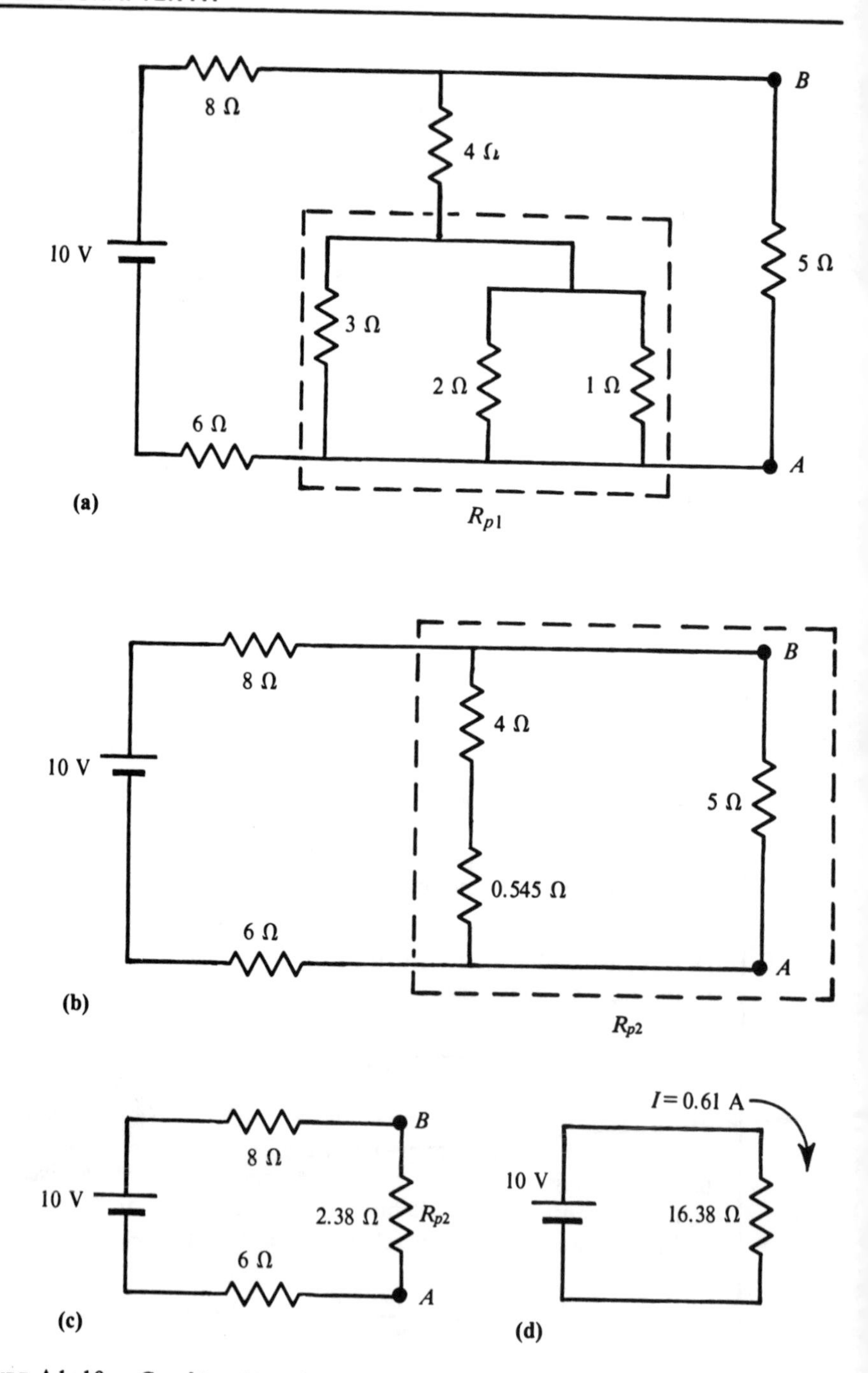

FIGURE A1–10. *Combination of Series and Parallel Resistors*

$8\ \Omega + 6\ \Omega + 2.38\ \Omega = 16.38\ \Omega$. The completely simplified circuit is shown in Figure A1–10d. The *current drain* from the battery is $I = 10\ \text{V}/16.38\ \Omega = 0.61$ A.

The second quantity desired is the current through the 5-ohm resistor. We need to figure out the voltage V_{AB} that appears across it; then the current through it is simply $V_{AB}/5\ \Omega$. We can follow points A and B down to Figure A1–10c. The 0.61-ampere current from the battery must pass through the 2.38-ohm equivalent resistor that is connected between A and B. Then the voltage between A and B must be

$$V_{AB} = IR = 0.61\ \text{A} \times 2.38\ \Omega = 1.45\ \text{V}$$

Finally

$$I_{5\ \Omega} = \frac{V_{5\ \Omega}}{5\ \Omega} = \frac{V_{AB}}{5\ \Omega} = \frac{1.45\ \text{V}}{5\ \Omega} = 0.29\ \text{A}$$

It is important that the student be thoroughly familiar with this sort of circuit analysis before studying electronic circuits.

Voltage Divider

So many situations in electronics are quickly understood in terms of *voltage divider action* that a review of this circuit and concept is important. Basically the circuit is a poor person's way to make a lower voltage source out of a battery. In Figure A1–11a we have two resistors in series that are connected to a battery of voltage V. Let us think of the circuit as placed in a box, and wires connected to the ends of one resistor, R_2, are brought out of the box. The voltage between these wires is called V_{out}. We intend to connect some *load resistance* R_L in a moment, but leave it off at first.

We wish to find V_{out}. First the current I flowing in the loop is given by

$$I = \frac{V}{R_1 + R_2}$$

The voltage across R_2 must be this current times its resistance, and the output voltage is the same:

$$V_{out} = V\left(\frac{R_2}{R_1 + R_2}\right) \qquad \textbf{(A1–15)}$$

The quantity $R_2/(R_1 + R_2)$ in Equation A1–15 is always a fraction, so that V_{out} is this fraction of the battery voltage V. In fact, the voltage V is divided between R_1 and R_2; hence the name *voltage divider*.

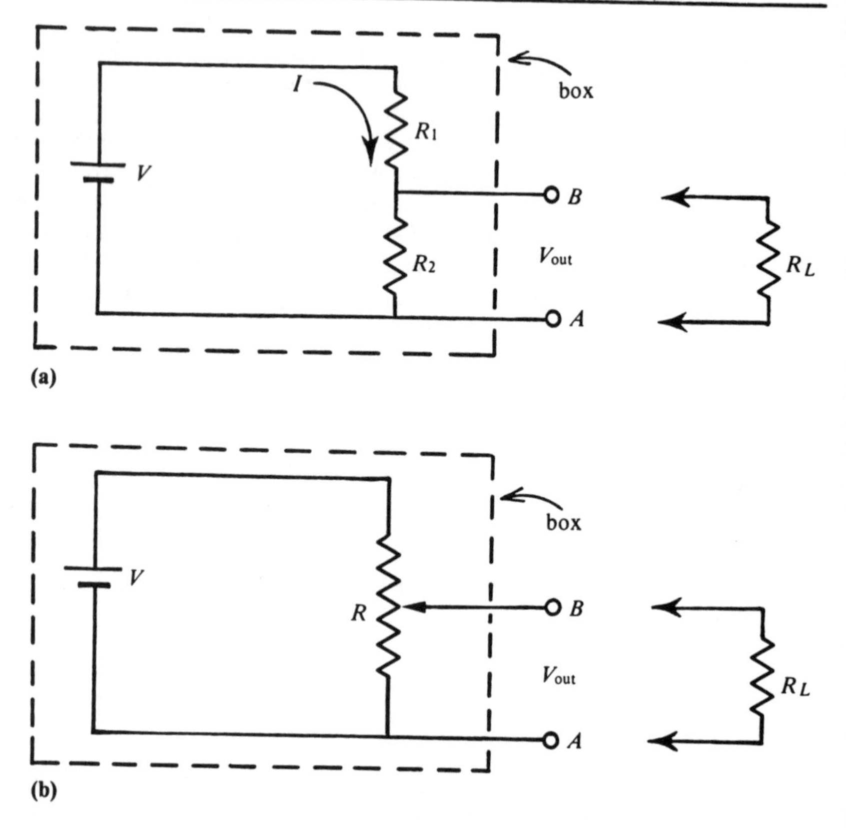

FIGURE A1–11. *Voltage Divider*

If a variable resistor is used as illustrated in Figure A1–11b, the resistance $R_1 + R_2$ is a constant R and the fraction R_2/R becomes the fractional setting of the slider along the potentiometer. This is a quick and easy way to produce a continuously adjustable voltage. However, the voltage V_{out} is *not* independent of the current drawn by R_L (as it would be for a good battery). If R_L is connected we have a parallel resistance combination between A and B. This resistance is always less than R_2 was alone, and we can easily evaluate V_{AB} in this case and find it to be *less than* the voltage V_{out} before R_L was connected (the so-called open-circuit voltage).

Review of ac Circuits

Until this point the voltage sources discussed were constant, which is the case for good batteries, and the resulting currents are then also constant, or direct currents. But many electronic applications deal with voltages and currents that change with time and may even alternate in direction or polarity. Such changing currents are called *alternating currents* or *ac*. This section of our review covers basic ac circuits and properties of important elements other than resistors used in ac circuits. This section concludes with a discussion of ac filter circuits, which may be new rather than review material.

Basic Concepts

A source of voltage that is changing in time is represented schematically by a circle as shown in Figure A1–12a. This changing-voltage source is called an ac voltage. Time graphs of some common types of ac voltage are shown in Figure A1–12b along with the common names for these waveforms. The shape of one cycle of the waveform is generally drawn within the circle that represents the ac voltage source.

The *sinusoidal waveform* is very basic; indeed, more complex forms can be thought of as the combination of many sinusoidal frequencies (called Fourier analysis). The mathematical expression for this voltage $v(t)$ as a function of time t is

$$v(t) = V_p \sin(\omega t + \phi)$$

where $\omega = 2\pi f = 2\pi / T$, V_p is the peak value of the voltage, ω is the circular frequency in radians per second, ϕ is the phase angle in radians, f is the frequency in cycles per second or hertz, and T is the period in seconds.

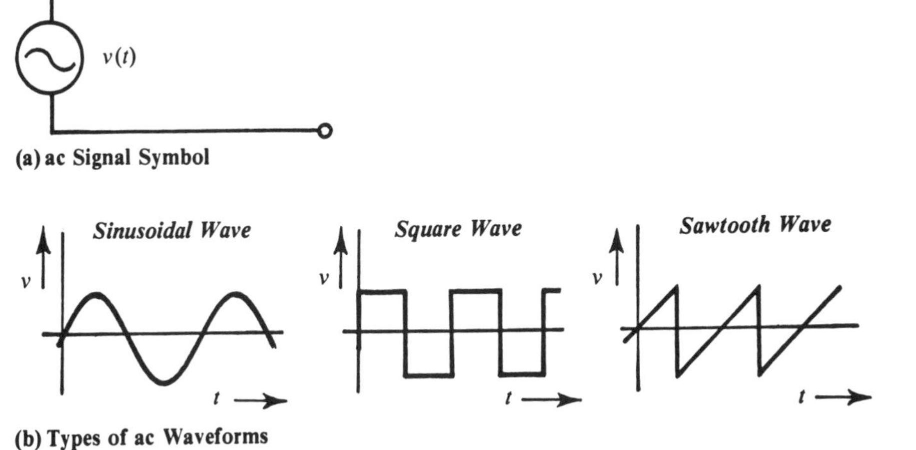

FIGURE A1–12. *ac Signal Source Symbol and Common Waveform Types*

NOTE: Lowercase letters are used for time-varying values, while capital letters are used for peak or constant values.

If $\phi = 0$ we have the simplest sine function graph, and we think of this as the *reference case.* If ϕ is a negative number, the graph shifts so that peaks occur later, and we say the voltage now *lags the reference.* Conversely, if ϕ is positive the voltage *leads the reference* (see Figure A1–13). Note that if $v(t)$ leads by $\pi/2$ radians (or 90°), the peak occurs at $t = 0$ and we have a cosine graph, whereas if it lags by 90°, a negative or inverted cosine results.

If an ac voltage is applied to a resistor as shown in Figure A1–14, the instantaneous current follows the instantaneous voltage faithfully. Ohm's Law applies at every instant:

$$v(t) = i(t)R \tag{A1–16}$$

In the sinusoidal case for $v(t)$, this law becomes

$$V_p \sin \omega t = i(t)R$$

so

$$i(t) = \left(\frac{V_p}{R}\right) \sin \omega t \tag{A1–17}$$

The peak value of current I_p is clearly the multiplier of the sine function in Equation A1–17 so that

$$i(t) = I_p \sin \omega t$$

where $$I_p = \frac{V_p}{R} \tag{A1–18}$$

The power $p(t)$ dissipated in the resistor at every instant is changing in time:

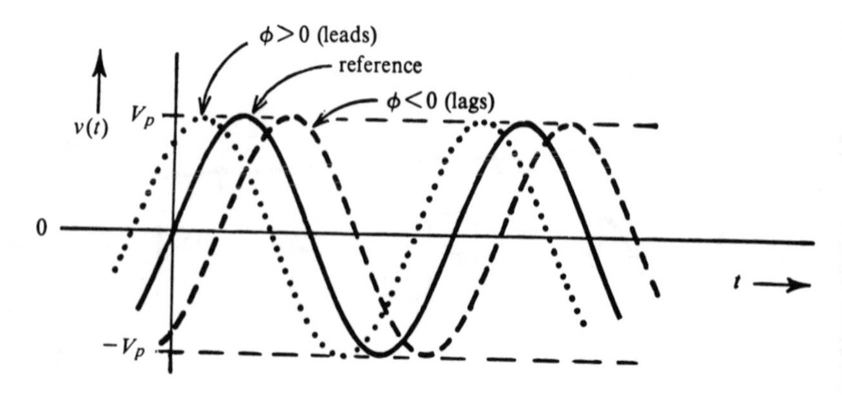

FIGURE A1–13. *Reference, Leading, and Lagging Sinusoidal Waves*

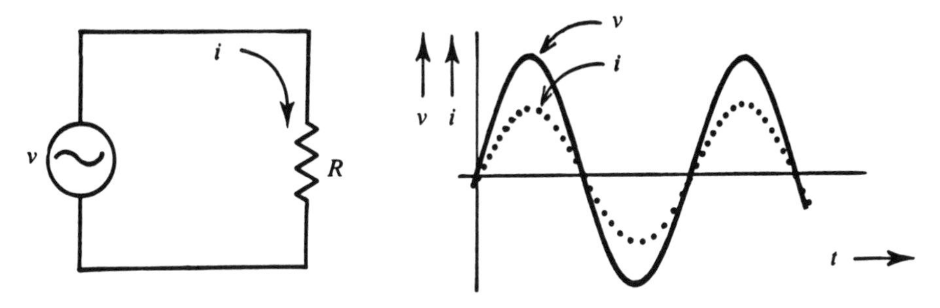

FIGURE A1–14. *ac Signal Applied to Resistor*

$$p(t) = v(t)i(t) = \left(\frac{V_p^2}{R}\right) \sin^2 \omega t \qquad \textbf{(A1–19)}$$

The time-average power $\bar{p}$ determines the heating of the resistor and is given by

$$\bar{p} = \overline{\left(\frac{V_p^2}{R}\right) \sin^2 \omega t} = \left(\frac{V_p^2}{R}\right) \overline{\sin^2 \omega t} \qquad \textbf{(A1–20)}$$

since a constant may be factored from an average. But a trigonometric expansion states that

$$\sin^2 \omega t = \tfrac{1}{2} - \tfrac{1}{2} \cos 2 \omega t \qquad \textbf{(A1–21)}$$

Since the average of the cosine in Equation A1–21 is zero, we see the average of $\sin^2 \omega t$ is ½. Then

$$\bar{p} = \frac{V_p^2}{2R} = \frac{(V_p/\sqrt{2})^2}{R} = \frac{V_{rms}^2}{R} \qquad \textbf{(A1–22)}$$

if

$$V_{rms} = \frac{V_p}{\sqrt{2}} \qquad \textbf{(A1–23)}$$

The average power is given by a familiar dc-like formula *provided* the rms (root-mean-square) voltage, V_{rms}, is used; this voltage is 70.7% of the peak voltage (Equation A1–23) for the sinusoidal waveform.

Alternatively, we can examine the power using $i^2(t)R$ and find

$$p = \left(\frac{I_p}{\sqrt{2}}\right)^2 R = I_{rms}^2 R \qquad \textbf{(A1–24)}$$

where $$I_{rms} = \frac{I_p}{\sqrt{2}} \qquad \textbf{(A1–25)}$$

Again the formula is a dc-like formula for power. In fact, most ac meters for voltage and current give the rms values rather than peak values. But a word of caution: The rms current and voltage are in general related by a factor $\sqrt{2}$ to the peak values *only* for the sinusoidal forms. For example, such meters usually do not tell the true rms current for a sawtooth waveform.

Two additional important circuit elements for ac circuits need to be reviewed: *capacitors* and *inductors.* The analysis of ac circuits with these elements is most powerfully done using the concepts of complex impedance, voltage, and current. However, to meet the objectives of this short book we content ourselves with the limited *vector approach* (such as is done in the typical college physics prerequisite course).

Capacitors and Capacitance

The *capacitor* (*condenser* is an older term) of common use consists basically of two parallel metal plates, usually with an insulator or *dielectric* in between. When a battery of voltage V is connected between the plates, equal and opposite charges $\pm Q$ very quickly flow onto the opposite plates (see Figure A1–15a). The charge Q on one plate—that is, on the capacitor—is proportional to the applied voltage through the proportionality constant C, which is called the *capacitance* of the unit (see Figure A1–15b). We write

$$Q = CV \qquad \textbf{(A1–26)}$$

where Q is in coulombs, V is in volts, and C is in farads (after Michael Faraday). If d is the distance between the plates, it can be shown that the electric field E that is in this region is given by

$$E = \frac{V}{d} \qquad \textbf{(A1–27)}$$

If the area of each plate is A, it is not difficult (with Gauss's Law) to show the capacitance is given by

$$C = \frac{\epsilon A}{d} \qquad \textbf{(A1–28)}$$

where ϵ is the *permittivity* of the dielectric material between the plates. The permittivity of vacuum is ϵ_0, which may be calculated through the expression

$$\frac{1}{4\,\pi\,\epsilon_0} = 9.0 \times 10^9 \left(\frac{\mathrm{Nm}^2}{C^2}\right) \qquad \textbf{(A1–29)}$$

Air is very close to this value, but other materials have a larger ϵ and enhance the capacitance by some factor k, which is called the *relative dielectric constant.*

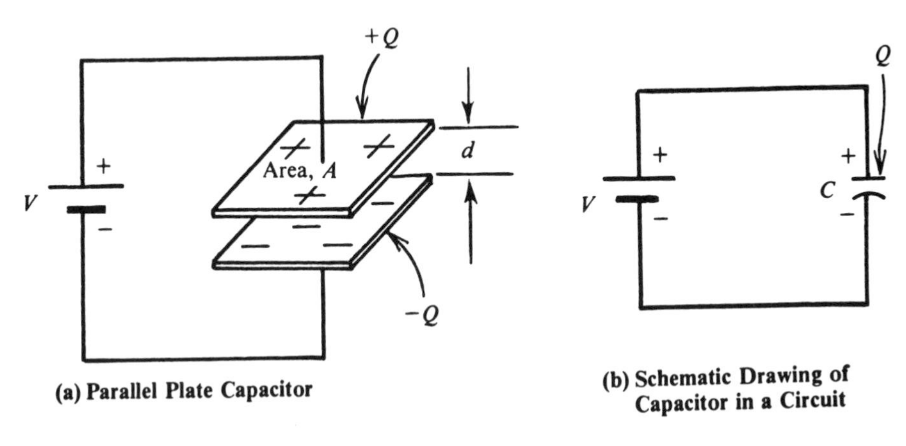

FIGURE A1–15. *Essential Parallel Plate Capacitor and Schematic Drawing*

We have

$$k = \frac{\epsilon}{\epsilon_0} \tag{A1–30}$$

so that

$$C = \frac{k\epsilon_0 A}{d} = \frac{kA}{4\pi 9 \times 10^9 d} \tag{A1–31}$$

A large area produces a large capacitance (and a physically large capacitor); a small distance d between the plates also produces a large capacitor, but increases the chance for a spark to jump through the dielectric (*voltage breakdown*). Capacitors have printed on them the maximum safe *working voltage*.

One Farad is a very large capacitor. Typical values range from one picofarad (1 picofarad = 1 pF = 10^{-12} F or 1 $\mu\mu$F) to some thousands of microfarads (1 microfarad = 1μF = 10^{-6}F).

Paper, mica, mylar plastic, and polystyrene plastic are common dielectrics. They have relative dielectric constants of the order of 2 to 3. Capacitors between 50 picofarads and about 5 microfarads often use these materials. Paper is common and cheap, mica is high quality but expensive, while the plastics have a good combination of price and quality. The term *high quality* in this instance refers to such factors as *low leakage* of charge through the dielectric, stability of capacitance against time and against temperature changes (low *temperature coefficient*), and low ac power absorption (low *power factor*). Usually sheets of aluminum foil with a dielectric sheet in between are rolled up to make a compact cylinder. Wires protrude from the ends; the wire connected to the outside sheet is often marked with a bar printed around that end (connect this end to ground for shielding purposes). The precision of common capacitors of this group is typically ±10%.

A certain ceramic, *barium titanate,* has a high relative dielectric constant of about 1000. Compact capacitors using two small discs with this dielectric in between are called *ceramic capacitors* or *disc capacitors,* and range in value from 5 microfarads at 1000 V to 0.1 microfarads at 25 V (these are working voltages for the particular sizes). Their low inductance makes them desirable radio frequency capacitors.

To obtain compact capacitors with capacitance larger than a few microfarads, *electrolytic capacitors* are usually used. The conventional electrolytic capacitor is constructed from a *single* aluminum foil rolled up with paper that is saturated with a certain chemical electrolyte. When an initial *forming voltage* with the proper polarity is applied between the electrolyte and the foil, an extremely thin oxide film forms on the foil; this becomes the very thin dielectric or insulator. The electrolyte/paper is in contact with the outer aluminum housing and becomes the one plate of the capacitor, while the aluminum foil is the other plate. Because of the very thin dielectric film, the capacity of the device is very large for its size.

However, an electrolytic capacitor can only be used with one end more positive than the other. The required positive terminal is marked (+), and the device is called a *polarized capacitor.* If the voltage should be reversed to an electrolytic capacitor, the electrolyte decomposes into gases that can cause the unit to explode! Heat and age also cause deterioration of the electrolytic film, and these capacitors are often the culprits in the radio or stereo that gives out a loud buzz after a few years of use (or even nonuse). The precision of electrolytic capacitors is typically −10% to +100%; note the large possible excess capacity. Also, the capacity decreases sharply below −20°C.

In recent years, *tantalum electrolytic capacitors* have become available that have a precision of ±10%. This type of electrolytic capacitor also has lower inductance, smaller leakage current, and better aging and temperature characteristics than the conventional electrolytic type. Since these features are provided along with large capacity in very small size, tantalum electrolytic capacitors are becoming quite widely used. However, they have been significantly more expensive than conventional electrolytic capacitors. For both electrolytic types, the maximum operating voltage is often quite low in order to obtain small size; this limitation must be carefully observed for a particular unit.

To obtain *variable capacitors,* two sets of parallel plates are often made to move by each other as a shaft is turned (thus changing the effective area A of the capacitor). Some small *trimmer* units have plates squeezed together by a screw; turning the screw causes C to be variable through the variable distance d between plates. Trimmer capacitors are used to make small adjustments to the fixed capacitance in an electronic circuit.

Figure A1–16 illustrates commonly encountered shapes for the various types of capacitors: ceramic (Figure A1–16a), mica (Figure A1–16b), paper and plastic (Figure A1–16c), and electrolytic (Figure A1–16d).

When capacitors are connected in series and to a battery, the same charge must accumulate on each capacitor (see Figure A1–17a). It is left as an exercise to show that the group behaves like a single capacitor with the effective capacitance given by Equation A1–32.

$$\frac{1}{C_{\text{eff}}} = \frac{1}{C_1} + \frac{1}{C_2} + \frac{1}{C_3} \qquad \textbf{(A1–32)}$$

The effective capacity is *smaller* than any contributing value. On the other hand, when capacitors are connected in parallel, the effective capacity is simply the sum as given in Equation A1–33 (see Figure A1–17b). This is reasonable if we think of the larger effective area that results.

$$C_{\text{eff}} = C_1 + C_2 + C_3 \qquad \textbf{(A1–33)}$$

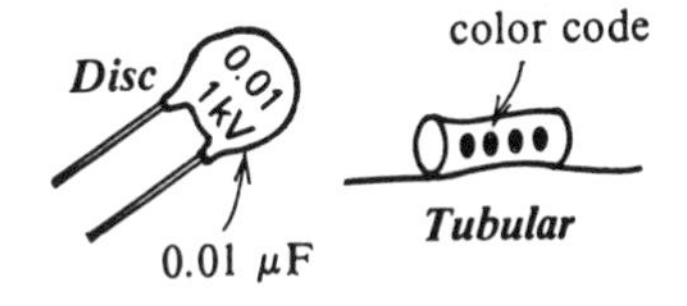

(a) Ceramic Types

(b) Mica Type

(c) Paper, Plastic Type

(d) Electrolytic Types

FIGURE A1–16. *Various Common Capacitors*

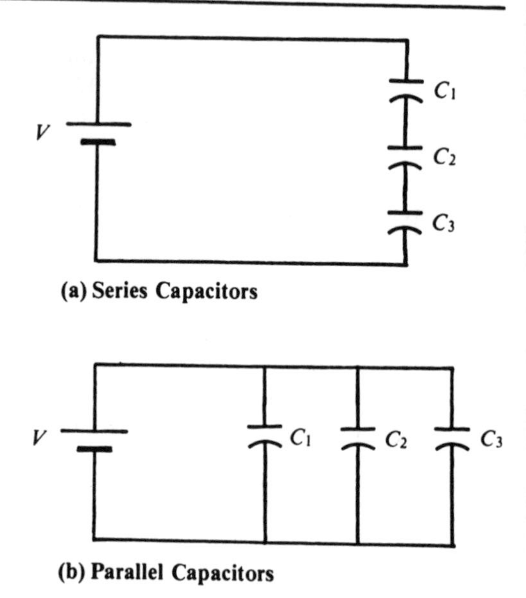

FIGURE A1–17. *Series Capacitors and Parallel Capacitors*

RC Circuit

When a battery is connected through a resistance R to a capacitance C, the charge accumulates exponentially in the capacitor and the voltage across the capacitor grows toward the battery voltage accordingly. The *characteristic time* involved is equal to the product RC; this is the time to reach 63% of the final charge or voltage. This feature is the basis for most timing circuits used in electronics.

Figure A1–18a illustrates a battery connected through a single-pole–double-throw (SPDT) switch to the series connection of a resistor R and capacitor C. The switch is alternately moved from position (1) to position (2) and back. Then the voltage v_{AB} that appears between points A and B in the circuit has a time graph as shown in Figure A1–18b. In fact, the graph shows that the circuit within the dotted box in Figure A1–18a can be thought of as a form of *square-wave generator* (also see Figure A1–12b).

The voltage waveform v_C that develops across the capacitor is shown in Figure A1–18c. Notice that while the switch is in position (2) and the capacitor is charging, the v_C voltage grows exponentially (with time) toward the battery voltage V. When the switch is in position (1), the capacitor discharges exponentially toward 0 volts. The characteristic time RC is illustrated for the charging and discharging curves.

Note that the mathematical expressions for the respective charging and discharging curves are shown in Figure A1–18c in conjunction with the curves. For their derivation, the interested reader is referred to the supplement to this chapter. The expressions apply to this case if the square-wave period

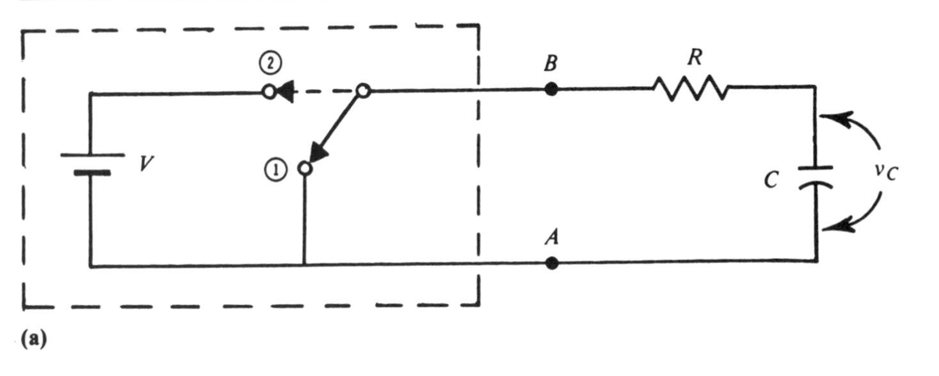

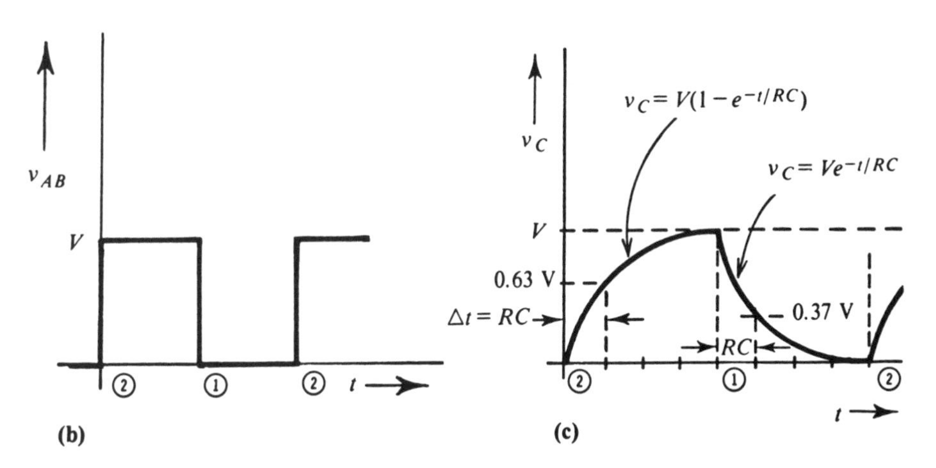

FIGURE A1–18. *Response of Capacitor-Resistor Combination to Applied Square-Wave Voltage*

is sufficiently long enough to permit complete charging or discharging in each half cycle. As a rule of thumb, the time for nearly complete charging or discharging can be taken to be $4RC$, since the capacitor voltage moves within 1.8% of its asymptotic value in this time interval.

Capacitive Reactance

When an ac voltage is connected to a capacitor as illustrated in Figure A1–19a, the capacitor alternately charges in opposite senses as the polarity of the applied voltage changes. Consequently, an ac current flows. An important feature of this analysis is that the voltage applied to the capacitor is required to be sinusoidal. We then find that the ac current flowing is also

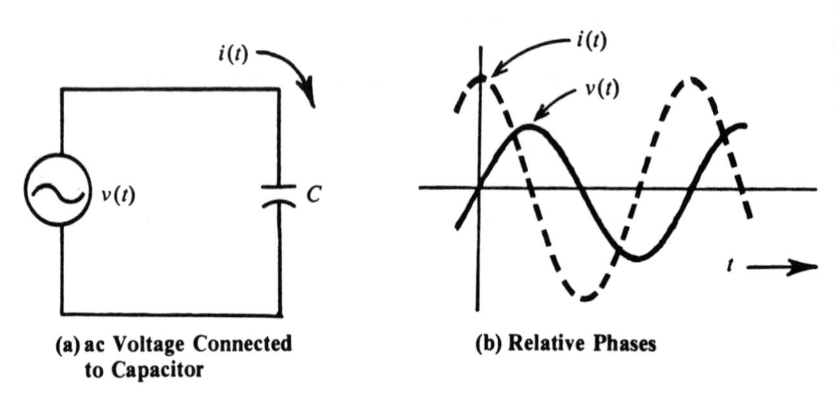

FIGURE A1–19. *ac Voltage Applied to a Capacitor and Phase Relationship Between Applied Voltage and Resulting Current in a Capacitor*

sinusoidal in time, although it is shifted in its phase relative to the applied voltage.

At every instant $v(t) = q(t)/C$. If the voltage is sinusoidal, we have $V_p \sin \omega t$ as this voltage, so that

$$V_p \sin \omega t = \frac{q(t)}{C}$$

or

$$q(t) = CV_p \sin \omega t \tag{A1–34}$$

In general, instantaneous current $i(t)$ is the rate that charge moves past a point in the circuit. That is, if an increment of charge Δq flows in a time interval Δt about the instant in question, the instantaneous current is the ratio $\Delta q/\Delta t$ in the limit that Δt approaches zero. In the circuit of Figure A1–19a, this instantaneous current (or charge movement rate) must also be the rate that the charge on the capacitor is changing. Then it follows from differential calculus that the current $i(t)$ is the derivative with respect to time of the time-dependent expression for charge on the capacitor:

$$i(t) = \lim_{\Delta t \to 0} \left(\frac{\Delta q}{\Delta t}\right) = \frac{dq}{dt} \tag{A1–35}$$

(The reader without a differential calculus background may skip to the conclusion for $i(t)$, which is obtained in Equation A1–36.) The current in the circuit is the time derivative of Equation A1–34.

$$i = \frac{d}{dt}(CV_p \sin \omega t) = \omega CV_p \cos \omega t$$

or

$$i = \left(\frac{V_p}{(1/\omega C)}\right) \sin\left(\omega t + \frac{\pi}{2}\right) \qquad \textbf{(A1–36)}$$

Note that the current is not in phase with the voltage, but leads it by 90°, as evidenced by the $+\pi/2$ radians in Equation A1–36. The relative phases are illustrated in Figure A1–19b. Letting

$$X_C = \frac{1}{\omega C} \qquad \textbf{(A1–37)}$$

we see that the peak current is the multiplier of the sine function in Equation A1–36:

$$I_p = \frac{V_p}{X_C} \qquad \textbf{(A1–38)}$$

X_C is called the *reactance of the capacitor* and has the dimensions of ohms. Equation A1–38 shows that an expression similar to Ohm's Law holds for the voltage across a capacitor in relation to the current through it; that is,

$$V_p = I_p X_C \qquad \textbf{(A1–39)}$$

We can also divide Equation A1–39 by $\sqrt{2}$ to find

$$V_{rms} = I_{rms} X_C \qquad \textbf{(A1–40)}$$

Notice that increasing either the frequency or the capacitance lowers the reactance and therefore increases the current in the circuit of Figure A1–19.

Inductors

The basic *inductor* is a coil of wire wound on a tube as shown in Figure A1–20a. When current passes through the wire, a magnetic field B is produced that threads through the interior of the coil. If the current is changing in time, the magnetic field in the coil's interior is also changing. Now the *magnetic flux*—that is, the number of magnetic field lines—in the coil is proportional to the field strength, which is in turn proportional to the coil current. A changing current thus implies a changing magnetic flux in the circuit and, by Faraday's Law of Electromagnetic Induction, an emf is induced around the circuit. Mathematically, Faraday's Law states:

$$v_{emf} = -\frac{d\phi_B}{dt} \qquad \textbf{(A1–41)}$$

where ϕ_B is the magnetic flux. Since ϕ_B is proportional to the current i flowing, we write

$$\phi_B = Li \qquad \textbf{(A1–42)}$$

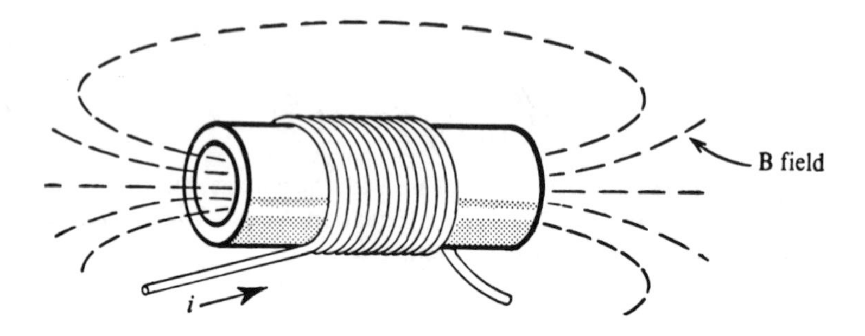

(a) Basic Inductor

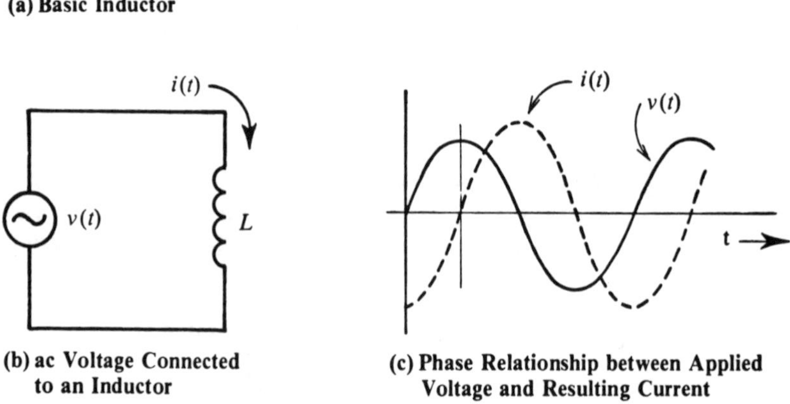

(b) ac Voltage Connected to an Inductor

(c) Phase Relationship between Applied Voltage and Resulting Current

FIGURE A1–20. *The Inductor and Its Basic Circuit Behavior*

where the proportionality constant L is called the *inductance of the coil.* In MKS units, ϕ_B is in webers, i in amperes, and L is in henries. Substituting Equation A1–42 into Equation A1–41 gives

$$v_{\text{emf}} = -L\frac{di}{dt} \tag{A1–43}$$

The minus sign indicates that the induced voltage is a *back emf*—that is, an induced voltage that opposes the current change and the voltage that caused it.

Inductive Reactance

Consider now that an ac voltage is connected to an inductor L, as shown in Figure A1–20b. (This figure also shows the conventional circuit symbol for an inductor; the symbol is suggested by the side view of a solenoid such as the one shown in Figure A1–20a.) We assume an ideal inductor where the resistance of the wire composing the solenoid is negligible; only the back emf of the inductor opposes current flow, and this only if the current is changing. Then the applied voltage equals the induced emf:

$$v = V_p \sin \omega t = L \frac{di}{dt} \tag{A1–44}$$

To solve for the current, we *separate the variables* by multiplying each side by the differential dt and take the indefinite integral of each side of the equation. We also move L to the other side; that is,

$$\int di = \int \frac{V_p}{L} \sin \omega t \; dt \tag{A1–45}$$

Then

$$i(t) = \frac{V_p}{L} \left(\frac{-\cos \omega t}{\omega} \right) + \text{CONST} \tag{A1–46}$$

where CONST is the constant of integration. We'll discard this constant on physical grounds: If it were not zero, the current would have a dc component—that is, it would be larger in one direction than the other without any cause. Using $-\cos \omega t = \sin (\omega t - \pi/2)$, Equation A1–46 can be written:

$$i(t) = \frac{V_p}{\omega L} \sin \left(\omega t - \frac{\pi}{2}\right) \tag{A1–47}$$

We see the current *lags* behind the applied voltage by 90° because of the $-\pi/2$ radians. This is diagrammed in Figure A1–20c.

The peak current is again the multiplier of the sine function in Equation A1–47; that is,

$$I_p = \frac{V_p}{\omega L} = \frac{V_p}{X_L} \tag{A1–48}$$

$$\textit{where} \quad X_L = \omega L \tag{A1–49}$$

is called the *reactance of the inductor*. Again an Ohm's Law relationship is seen in Equation A1–48; that is,

$$V_p = I_p X_L \tag{A1–50}$$

Indeed if ω is in radians/second (that is, 2π times the frequency in hertz or cycles per second) and L is in henries, the reactance X_L is in ohms. Notice that if either the frequency or the inductance is increased, the inductive reactance increases and the ac current that flows in the circuit of Figure A1–20b decreases.

Practical inductors range from radio frequency suitable units of a few microhenries up to audio frequency units of several henries. The radio frequency inductors are often wires wound on ferrite (iron oxide) cores that enhance the magnetic field and therefore the inductance. Laminated iron core is used to make large inductors, but the iron is frequency responsive only in the audio region. In either case the wire is chosen so that the resistance is small compared to the reactance at frequencies of use; resistance also determines the current handling capabilities. The resistive heating must remain within safe levels. Figure A1–21 illustrates some common inductors.

When inductors are placed in series, the inductances add in the same way as resistances; that is,

$$L_s = L_1 + L_2 + \ldots \tag{A1–51}$$

There is also a corresponding rule for parallel inductors:

$$\frac{1}{L_p} = \frac{1}{L_1} + \frac{1}{L_2} + \ldots \tag{A1–52}$$

RL Circuit

Suppose a battery V is connected through a resistor R to an inductor L as shown in Figure A1–22a. After the switch in the circuit is closed, the current in the circuit does not begin flowing immediately. The inductor's back emf proportional to di/dt (the rate of change of current) prevents this. It can be shown that the current grows exponentially to the value determined by the inhibiting resistance R. The graph of current versus time for the circuit is shown in Figure A1–22b along with a mathematical expression for the current. The characteristic time for the exponential growth is given by the ratio L/R. Although this behavior could be the basis for timing circuitry, the

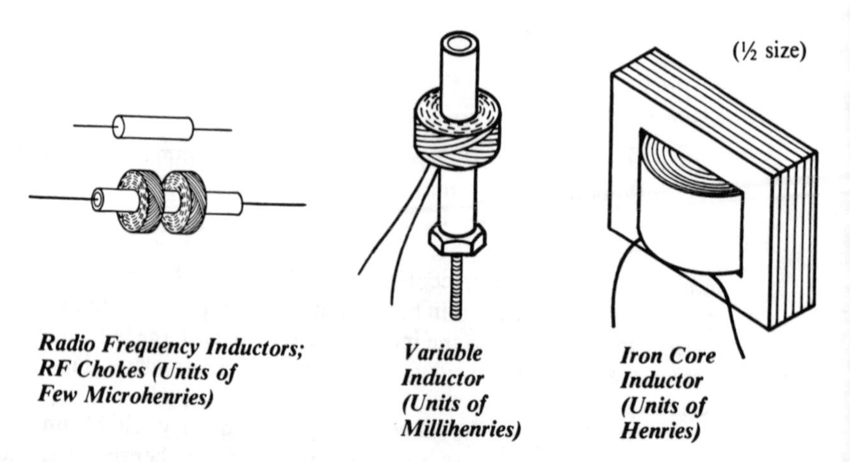

FIGURE A1–21. *Common Inductors in Electronics*

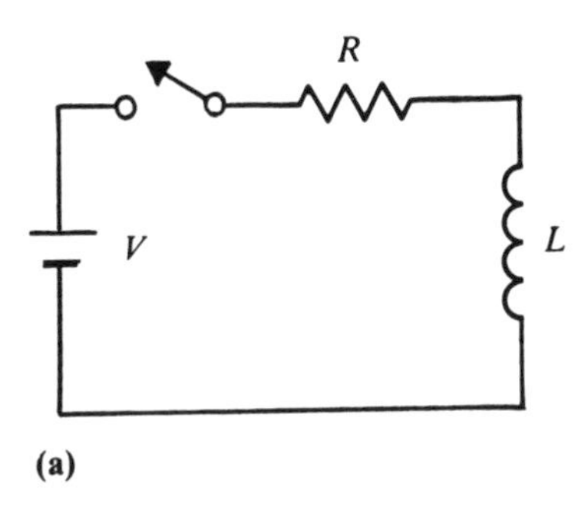

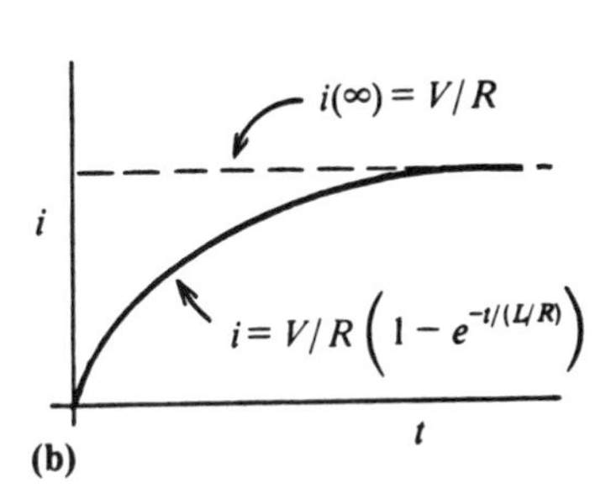

FIGURE A1–22. *Series* RL *Circuit Response to dc Voltage*

RL circuit is seldom used because of the inconvenient size and weight of the inductors generally required. In fact, the contemporary use of inductors is primarily in radio frequency circuitry (100 kilohertz and above) where the reactance of a millihenry is significant.

Transformers

If two coils of wire are placed near each other, a changing current in one coil produces a changing field that threads through the other. Faraday's law of induction then requires that an emf be induced around the circuit of which the second coil is a part. The coils are usually wound so that all the field produced by the first coil (which we call the *primary winding*) passes through the second coil (the *secondary winding*). The applied emf, V_p, across the primary is proportional to the magnetic flux through it, which in turn is proportional to the number of turns of wire in it (N_p). Similarly, the emf V_s across the secondary is proportional to the number of turns N_s of the secondary. Therefore, the simple rule for ideal transformers is that the ratio of secondary voltage to primary voltage is just the same as the ratio of the windings:

$$\frac{V_s}{V_p} = \frac{N_s}{N_p} \qquad \textbf{(A1–53)}$$

Such a coil arrangement is called a *transformer*. In this way ac voltages can easily be *stepped up* or *stepped down* to meet various requirements.

The coils of a transformer for low frequencies are wound on a laminated-iron core to increase the inductance of the windings. In this way, if no load is connected to the secondary, very little current will flow in the primary windings. Indeed for an ideal transformer, the power into the primary should equal the power passing out from the secondary:

$$V_p I_p = V_s I_s \qquad \textbf{(A1–54)}$$

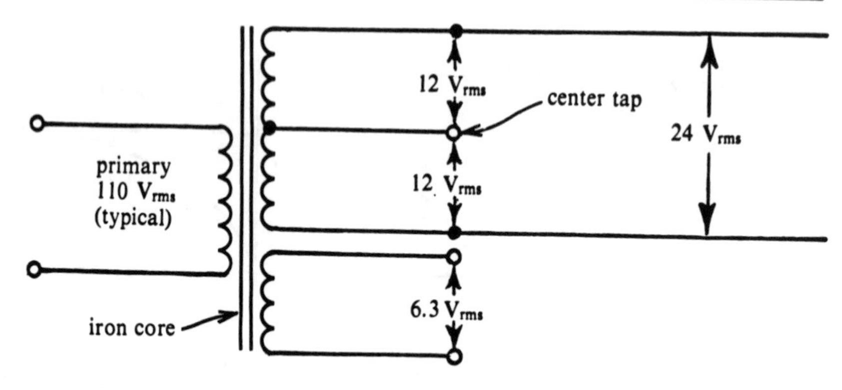

FIGURE A1–23. *Schematic Representation of a Multiwinding Transformer with One Tapped Winding*

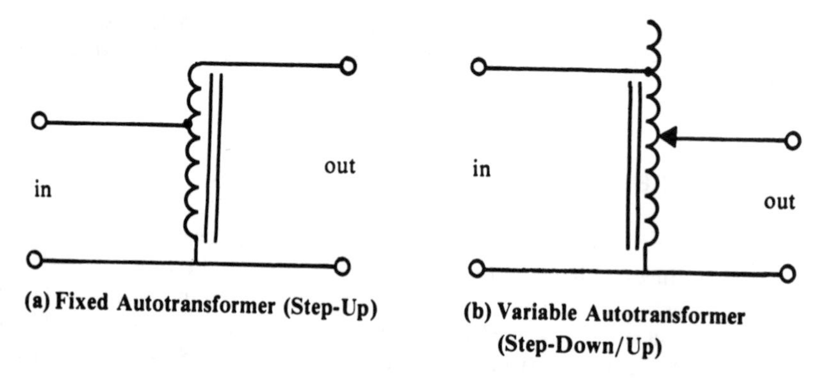

FIGURE A1–24. *Fixed and Variable Autotransformer Schematic Symbols*

The schematic representation of a transformer suggests the coils and the iron laminations (see Figure A1–23). It is also possible to have several separate windings to achieve several voltages simultaneously. If the voltages can have a common potential in a given application, the windings need not be separate, but can be portions of the same coil; that is, wire *taps* may come out from the coil. A *center tap* is illustrated in Figure A1–23.

Autotransformers use a single winding for both the primary and the secondary. These are cheaper but the secondary is *not* then electrically isolated from the primary circuit. Figure A1–24a shows the schematic symbol for a *fixed autotransformer.* Notice a fixed tap position is made to the coil winding; the ratio of output to input is therefore also fixed. A *variable autotransformer* (Variac is a common trade name) uses a sliding tap to produce a convenient variable ac voltage. See Figure A1–24b for the schematic symbol of a variable autotransformer. In fact, the unit diagrammed implies that a small amount of step-up is possible, as well as step-down. This is a common configuration.

Series Alternating-Current *RC* Circuit

We next wish to consider series connections involving *reactive circuit* elements and examine how they respond to an applied ac voltage. The most elegant and powerful approach uses the notion of complex impedance and the *imaginary number* $\sqrt{-1}$. However, for the series ac circuits that primarily concern us in this book, it is possible to use an easier approach.

The principal complexity of ac circuits over dc circuits is that the current through a capacitor or inductor is *not* in phase with the applied voltage, as already reviewed. Consider the case of a resistor R and capacitor C connected in series to a sinusoidal voltage V as shown in Figure A1–25a. The following is true about this circuit:

1. At every *instant* the voltage v_R across the resistance and the voltage v_C across the capacitance is equal to the applied voltage v_{in}; that is,

$$v_{in} = v_R + v_c \qquad \textbf{(A1–55)}$$

2. At every instant, the current i is the same in the resistor as in the capacitor; that is, they have the same frequency, phase, and amplitude.
3. If the applied voltage is sinusoidal, the current flowing is expected to be sinusoidal in time.
4. The voltage across the resistor is in phase with the current, but the voltage across the capacitor lags behind the current (or the current leads the voltage) by 90°.

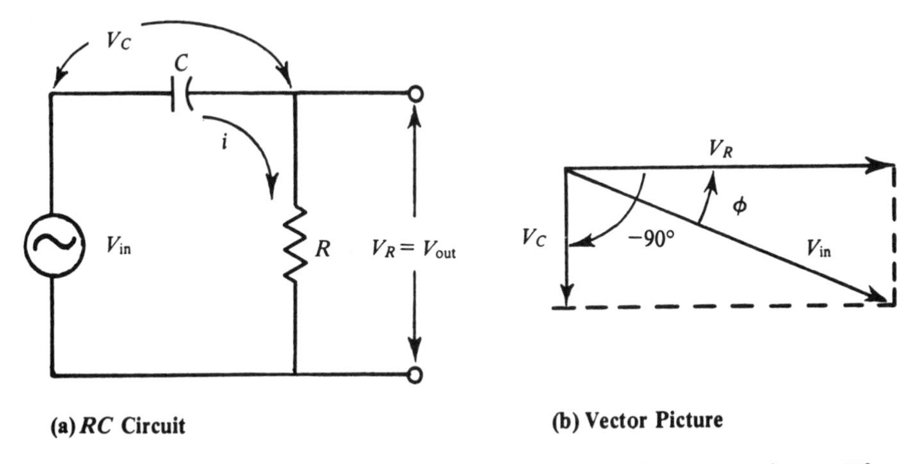

FIGURE A1–25. *Series* RC *Connection with Output Connection for High-Pass Filter*

The central point we now make is that when two sine waves of the *same* frequency (but differing in *phase*) are added together, another sine wave of the same frequency results. With a bit of trigonometry, it is possible to show that the amplitude of the resultant sine wave is given by a *sum of vectors rule*. The two component vector amplitudes should be the same as the two respective sine wave amplitudes, and the angle between the two vectors should be the same as the phase angle between the two sine waves. The resultant of these vectors then tells the amplitude and phase of the sine wave that results.

If V_R and V_C represent the peak voltage across the resistor and capacitor, respectively, then the magnitude of the total voltage across the two elements is the magnitude of the vector sum of V_R and V_C vectors. This in turn is equal to the input voltage V_{in} in Figure A1–25a. The angle between the vectors should be chosen the same as the phase angle between the time-varying voltages. The voltage across the capacitor lags the voltage across the resistor by 90° according to point 4 above. Consequently, the vector picture is as shown in Figure A1–25b.

By the Pythagorean theorem, the hypotenuse V_{in} is given by

$$V_{in} = \sqrt{V_R^2 + V_C^2} \qquad \textbf{(A1–56)}$$

However, $V_R = IR$ and $V_C = IX_C$. Substituting these into Equation A1–56 and solving for I gives the current flowing:

$$I = \frac{V_{in}}{\sqrt{R^2 + X_C^2}} \qquad \textbf{(A1–57)}$$

We let

$$Z = \sqrt{R^2 + X_C^2} \qquad \textbf{(A1–58)}$$

where Z is called the *impedance* of the circuit. Then $I = V_{in}/Z$ or $V_{in} = IZ$, and an expression like Ohm's Law results.

NOTE: Since peak and rms values are proportional, Equation A1–57 holds for rms voltage and current as well. However, the total impedance is somewhat more complicated than the simple sum of R and X_L that we might expect for a series circuit from the dc resistance analog.

The phase angle of the current in this circuit (compared to the applied voltage) is the same as the angle between V_{in} and V_R in Figure A1–25b. Since V_R is at an angle $+\phi$ relative to V_{in}, the current *leads* the applied voltage by

$$\phi = \arctan \frac{V_C}{V_R} = \arctan \frac{X_C}{R} \qquad \textbf{(A1–59)}$$

This "leading" feature is characteristic of capacitive circuits (recall $\phi = 90°$ for the case of capacitor only, mentioned earlier).

No power is dissipated in ideal capacitors or inductors—only in resistors. The power dissipated in the RC circuit of Figure A1–25 is therefore

$$P = I^2R = IIR = \left(\frac{V_{\text{in}}}{\sqrt{R^2 + X_C^2}}\right) IR$$

$$= V_{\text{in}}I\left(\frac{R}{\sqrt{R^2 + X_C^2}}\right) = V_{\text{in}}I\cos\phi \qquad \textbf{(A1–60)}$$

where the *power factor* $\cos\phi = R/\sqrt{R^2 + X_C^2}$. The current and voltage must be in rms to give the proper average power. Notice that the power expression is more complicated than the simple VI product of Equation A1–8a (the dc case) due to the phase angle and power factor. Indeed, if $R = 0$ we have $\phi = 90°$ and $\cos\phi = 0$, so the power is zero; this is correct since the circuit is then purely capacitive. If the frequency is such that $X_C = R$, we can show (assigned problem) that the power is half of what it is at very high frequencies. Recall that at high frequencies the capacitive reactance is small and the current is greatest. This frequency is called the *half-power frequency* of the circuit.

Low-Pass/High-Pass *RC* Filters

The voltage appearing across just the resistor R in the series RC circuit of Figure A1–25 can be put to good use. In fact, wires are pictured as connected to these points and leading to the right. We call the voltage between these points V_{out} or the *output voltage* from the circuit; the wires could well lead to other electronic circuits, meters, or whatever.

The voltage V_{out} is just the current flowing times the resistance:

$$V_{\text{out}} = IR = \left(\frac{V_{\text{in}}}{\sqrt{R^2 + X_C^2}}\right) R = V_{\text{in}} \left(\frac{R}{\sqrt{R^2 + X_C^2}}\right) \qquad \textbf{(A1–61)}$$

In general the output voltage is less than the input voltage because of the voltage divider type action that occurs between the resistance R and the reactance X_C.

It is useful to consider the behavior of the output voltage as the frequency of the input voltage is varied from very low to very high frequencies. Since the capacitive reactance X_C decreases with increasing frequency as shown in Figure A1–26a, the output voltage V_{out} as given by Equation A1–61 must go from nearly zero to essentially V_{in}. Then the ratio of the output to the input voltage, $V_{\text{out}}/V_{\text{in}}$, goes from zero to one as the frequency is changed from zero to very high frequencies, as shown in Figure A1–26b. At high frequencies the voltage out is the same as the voltage in, whereas at low frequencies the input voltage is virtually blocked from passing to the output. This feature is very useful in a situation where an unwanted low frequency is mixed in with a

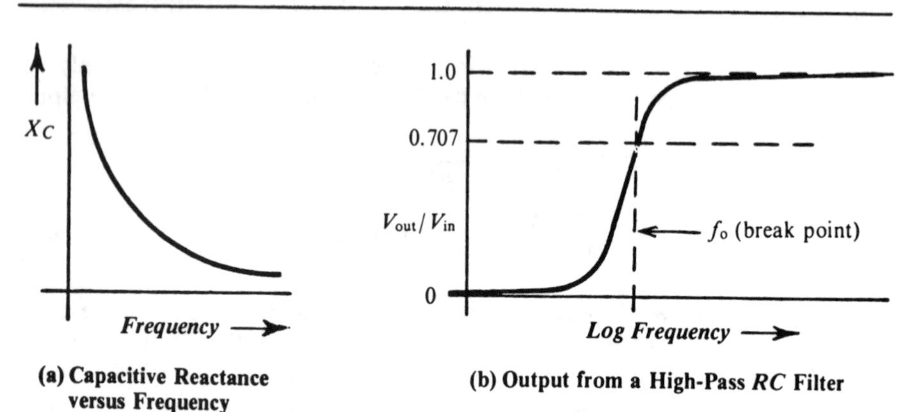

(a) Capacitive Reactance versus Frequency

(b) Output from a High-Pass *RC* Filter

FIGURE A1–26. *Behavior of the High-Pass* RC *Filter*

desired high-frequency signal. The circuit can be used to filter out the unwanted low frequencies and pass only the desired higher frequencies; hence the name *high-pass filter*. See Figure A1–27 for a diagram of this feature.

The frequency that divides the high-pass region from the low-frequency (reject) region is usually taken to be that for which $X_C = R$. This is the half-power frequency mentioned earlier and is also referred to as the *break-point frequency* f_0. From Equation A1–61 we can see that

$$\frac{V_{out}}{V_{in}} = \frac{1}{\sqrt{2}} = 0.707$$

at the breakpoint frequency, and f_o is so indicated in Figure A1–26b.

A "*low-pass filter*" may be obtained by interchanging R and C in Figure A1–25a. In other words, take the output from across the capacitor rather than the resistor. In this case, unwanted high frequencies may be removed from the signal.

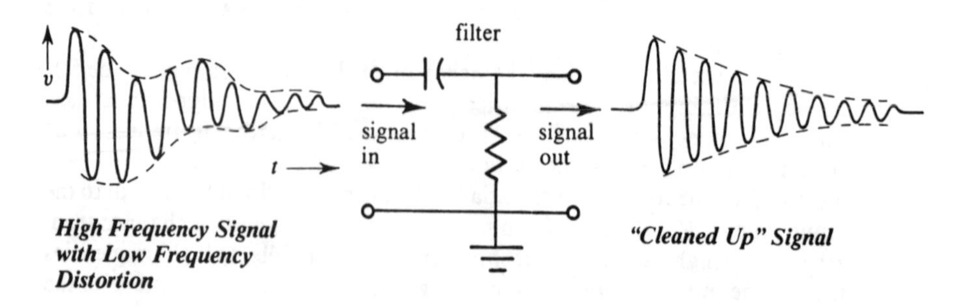

High Frequency Signal with Low Frequency Distortion

"Cleaned Up" Signal

FIGURE A1–27. *Action of a High-Pass Filter to Reject Low Frequency Distortions in a Signal*

It is also possible to construct low-pass and high-pass filters from series combinations of resistors and inductors. In all cases the frequency behavior is easily understood qualitatively by thinking of voltage divider action and the frequency behavior of the reactance. Circuits and derivations are left to the Problems at the end of the chapter.

Network Analysis

This last principal section of the chapter may not be review material. It considers some techniques for analysis of electrical networks that cannot be analyzed by elementary series/parallel arguments. However, we will find the notions developed to be of general utility in our work with electronic amplifiers and systems. The techniques that may be applied include use of Thevenin's Theorem, the Superposition Theorem, Norton's Theorem, and others. Only Thevenin's Theorem is discussed here, but the Superposition Theorem and Norton's Theorem are briefly considered in the Problems at the end of this chapter for the interested reader.

Thevenin's Theorem

There is a remarkable theorem about circuits containing *linear electrical elements*—namely, elements for which current is directly proportional to voltage—that is, used again and again in doing electronic work. The theorem is called Thevenin's Theorem and is simply this: If two connections lead away from an arbitrary connection of batteries and resistors (we may think of the batteries and resistors as being in a box, either real or imaginary), then the electrical effects of the circuit in the box on whatever is connected to the box is just the same as if in the box were merely a single battery connected in series to a single resistance (and connected to the leadout wires).

We speak of finding the Thevenin equivalent circuit for the given circuit, which consists of the Thevenin equivalent voltage, V_{Th}, and the Thevenin equivalent resistance, R_{Th}. Figure A1–28 illustrates the concept. Although dc batteries are drawn in the figure, the theorem is also valid for a collection of ac voltage sources and resistors. (If capacitors are present, the theorem applies only at sufficiently low frequencies.) This diagram should impress on us the fact that although a box may contain a complicated voltage/resistor circuit, the electrical behavior of the two terminals from the box is just the same as if the circuit were very simple: merely a voltage source in series with a resistor. Chapter A5 exploits this remarkable fact; it is basically what permits us to characterize and work rather easily with integrated circuit amplifiers (indeed any amplifier) that may in fact contain a complicated voltage/resistor circuit. The actual interior design of these devices does not greatly concern us.

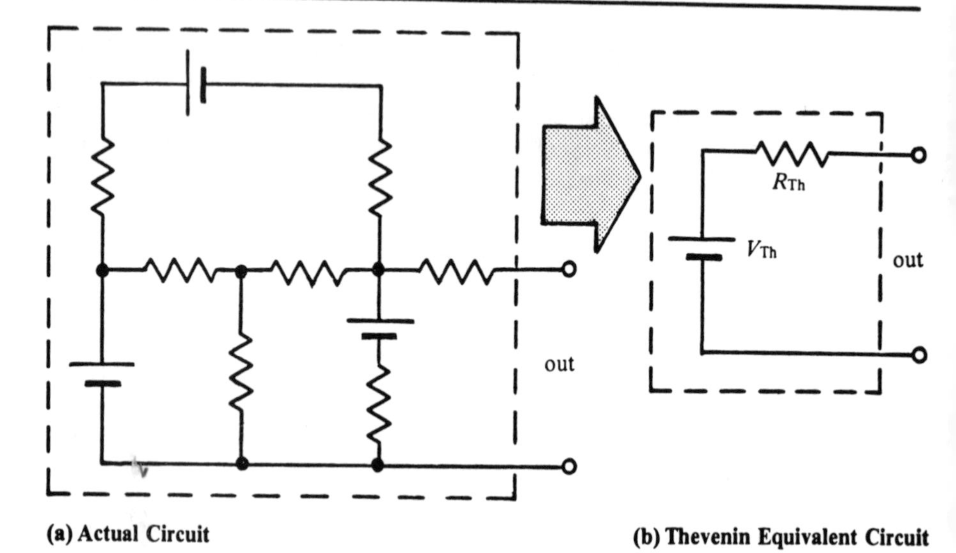

FIGURE A1–28. *Conceptual Equivalence of Thevenin's Theorem*

However, Thevenin's Theorem can also be a very effective technique for analyzing or understanding circuits. Thus, later in this section Thevenin's Theorem is demonstrated in contrast to the rather traditional circuit analysis technique based directly on Kirchhoff's Laws of Networks.

The question immediately before us is: How do we find the voltage of the single battery (the Thevenin equivalent voltage) and the resistance of the single resistor (the Thevenin equivalent resistance) that are in the Thevenin equivalent box?

We may be faced with (1) a theoretical circuit, for which the Thevenin equivalent is desired; or (2) a laboratory situation where a circuit is literally in a box and the elements of the equivalent circuit must be found by experimental measurements. The rules for the two situations will now be considered.

Theoretical case

1. Replace all batteries in the circuit with wires and compute the theoretical resistance that appears between the two output terminals. This is the Thevenin equivalent resistor R_{Th}.
2. Compute the voltage that appears between the output terminals when no load is connected; that is, the voltage as would be measured by an ideal voltmeter connected to the terminals. This is the Thevenin equivalent voltage V_{Th}.

We will not prove the theorem or even this approach to finding the equivalence elements. However, notice that if the actual circuit is as simple as a battery in series with a resistor, this approach clearly gives the correct answer; namely, the circuit itself. It should be noted also that for a circuit as complicated as the one drawn in Figure A1–28a, it may be no easy task to find the solutions. Kirchhoff's laws may need to be applied to a multiloop circuit.

Experimental case

1. Connect a very high impedance voltmeter (approaching the ideal) to the output terminals. Its reading is the Thevenin equivalent voltage V_{Th}.

2a. Connect a very low impedance (approaching the ideal) ammeter to the output and measure the *short circuit current* I_{short}. The Thevenin resistance must be given by

$$R_{Th} = \frac{V_{Th}}{I_{short}} \qquad \textbf{(A1–62)}$$

2b. *Alternate way.* Connect a variable resistance across the output, and monitor the voltage across this resistance with the quality voltmeter. Adjust the resistance until the voltmeter reading is half the open circuit reading of step (1). The value of the attached resistance now equals the Thevenin equivalent resistance. This method is the technique most commonly used.

The generality of these experimental techniques can be proved by assuming Thevenin's Theorem is true. Then the box behaves like the circuit of Figure A1–28b—only a battery in series with a resistance need be imagined. A very good voltmeter will indeed measure the battery voltage since no voltage drop occurs across R_{Th} (no current flows through it). Further, when a load is connected to the equivalent circuit, it should be clear that voltage divider action occurs. The output voltage is reduced to just half the battery voltage V_{Th} when the load resistance equals R_{Th}.

Networks Requiring Kirchhoff's Laws or Thevenin's Theorem for Solution

Consider the circuit shown in Figure A1–29a. We wish to find the current flowing from the 4-volt battery. This is *not* a problem that can be solved by elementary techniques. That is, it is not possible to replace certain resistors with a parallel equivalent or a series equivalent so that a simple circuit is obtained. The circuit *can* be solved by applying Kirchhoff's laws; that is, by obtaining several simultaneous equations from application of Kirchhoff's laws. It can also be solved by use of Thevenin's Theorem, and both approaches are illustrated.

1. *Example:* (Solution by application of Kirchhoff's laws) A current must be chosen for each circuit leg and a direction assumed (see Figure A1–29a). (If a direction is in fact wrong it does not matter; the solution for that current will come out negative.) This is done in Figure A1–29b. Next, Kirchhoff's Current Law is applied to junction J in Figure A1–29b.

$$I_1 + I_3 = I_4 \qquad \textbf{(A1–63)}$$

To solve for the three unknown currents (even if only the current in the 3-ohm resistor I_3 is desired) we must find two additional simultaneous equations. These are obtained by applying Kirchhoff's Voltage Law for voltages around two different loops in the circuit. These loops are shown as dashed lines in Figure A1–29b. First consider the left-hand loop that goes counterclockwise around the circuit. Beginning at point B and traveling entirely around and back to B, we have potential rises and falls as follows:

$$+I_4 4\ \Omega + I_1 1\ \Omega - 2\ \text{V} = 0 \qquad \textbf{(A1–64)}$$

For the right-hand loop, beginning at point A and moving counterclockwise again, we have

$$+4\ \text{V} - I_3 3\ \Omega - I_4 4\ \Omega = 0 \qquad \textbf{(A1–65)}$$

Notice that the trip through the 4 Ω resistance is from (−) to (+) for the path chosen from B, while it is the opposite for the path chosen from A. Thus the signs of the $I_4 4\ \Omega$ term differ in the two equations. Eliminating I_1 and I_4 from the three equations A1–63, A1–64 and A1–65 now gives the solution for I_3. More generally, the three equations can be solved by the method of determinants. The final numerical solution from this method is left to the student to solve in the exercises for this chapter.

2. *Example:* (Solution by application of Thevenin's Theorem) This problem can be solved rather quickly by application of Thevenin's Theorem and serves to illustrate the method and utility of this approach. Imagine that all of the circuit of Figure A1–29a is placed in an imaginary box, *except* for the 4 V battery. This is drawn in Figure A1–30a.

First we work on the Thevenin equivalent resistance. The battery in the box is replaced by a short, and the resistance between points A and B should be calculated, as diagrammed in Figure A1–30b. We see 1 ohm in parallel with 4 ohms (equivalent to 0.8 ohms), which is in turn in series with 3 ohms to give a Thevenin equivalent resistance of 3.8 ohms. Next the Thevenin equivalent voltage is obtained by calculating the voltage appearing between points A and B in Figure A1–30c. This calculation is trivial. The voltage at B relative to A must be the *same* as the voltage at J relative to A (no voltage drop occurs across the

3-ohm resistance since no current flows through it in *this* circuit). But the voltage at J is clearly just 4/5 of 2 V because of voltage divider action between the 1-ohm and 4-ohm resistors. Hence the Thevenin equivalent voltage $V_{Th} = 1.6$ V.

Finally, the current that flows from the 4 V battery in the original circuit is found by connecting the 4 V battery to the simple Thevenin equivalent box as shown in Figure A1–30d. There we see two opposing batteries. Applying Kirchhoff's Voltage Law around this single loop gives

$$4\text{ V} - I_3 3.8\ \Omega - 1.6\text{ V} = 0$$

or

$$I_3 = \frac{4\text{ V} - 1.6\text{ V}}{3.8\ \Omega} = 0.63\ A$$

The solution here certainly comes more quickly than by Kirchhoff's laws/simultaneous equations approach. We can almost work out I_3 in our heads!

NOTE: The solution of a network problem by use of Thevenin's Theorem is particularly time-saving when only one network current is required, as in the case of the example above. The use of Kirchhoff's laws and solution of the associated simultaneous equations may be equally effective if *all* network currents are desired.

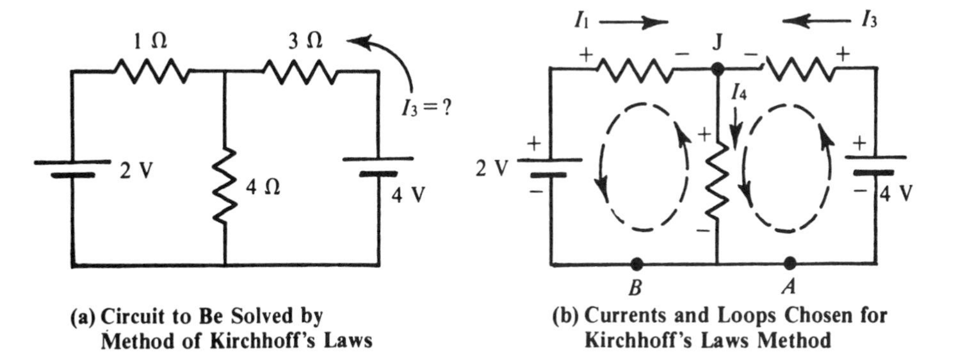

FIGURE A1–29. *Circuit with Application of Kirchhoff's Laws*

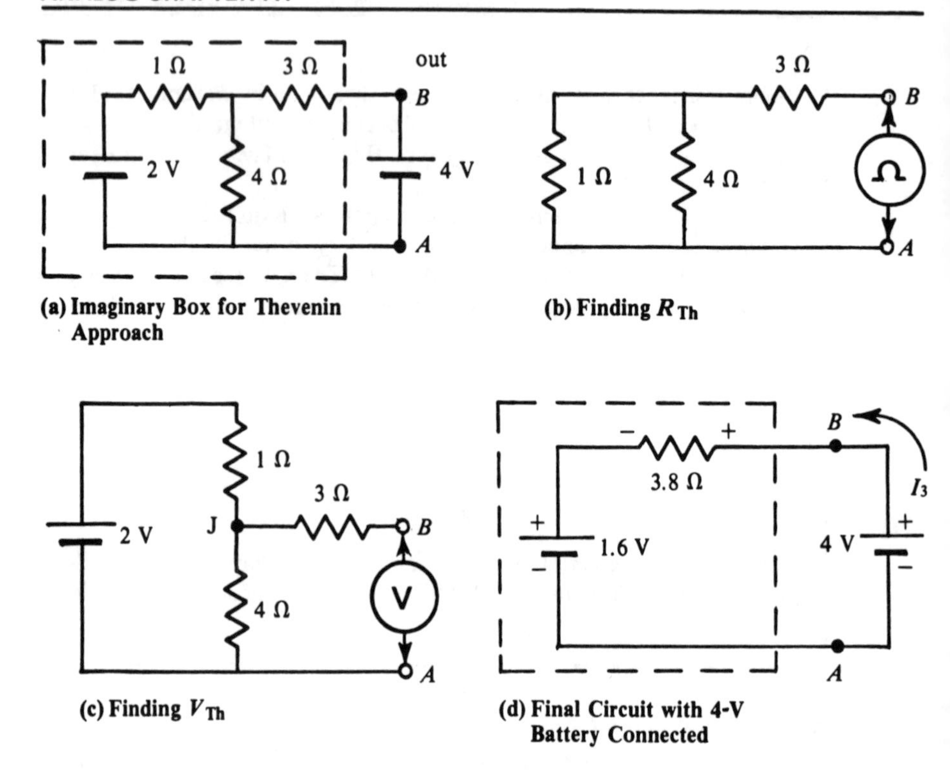

(a) Imaginary Box for Thevenin Approach

(b) Finding R_{Th}

(c) Finding V_{Th}

(d) Final Circuit with 4-V Battery Connected

FIGURE A1–30. *Application of Thevenin's Theorem to a Circuit Problem*

Supplement to Chapter: Derivation of Exponential *RC* Charging Law

We now derive the exponential charging character of an *RC* circuit such as that shown in Figure A1–18a. Suppose the switch is moved from position (1) to (2) in that diagram at time $t = 0$. A time-varying current $i(t)$ then begins to flow through R, and a charge q consequently begins to accumulate on the capacitor. We can write Kirchhoff's Voltage Law around the circuit at any time:

$$V = v_R + v_C \quad \textbf{(A1–66)}$$

$$= iR + \frac{q}{C} \quad \textbf{(A1–67)}$$

But the current i is the rate that charge arrives on the capacitor, or $i = dq/dt$. Taking the derivative of Equation A1–67 and substituting for i gives

$$0 = R\frac{di}{dt} + \frac{i}{C} \tag{A1–68}$$

Now we separate the variables i and t and perform the definite integral on both sides:

$$\int_0^t \frac{di(t)}{i(t)} = -\int_0^t \frac{dt}{RC}$$

which gives

$$\ln\left(\frac{i(t)}{i(0)}\right) = \frac{-t}{RC}$$

Letting each side be the exponent of e now yields

$$\frac{i(t)}{i(0)} = e^{\frac{-t}{RC}} \tag{A1–69}$$

The initial current $i(0)$ is determined by Equation A1–67 in conjunction with the initial charge on the capacitor. If the capacitor is initially discharged, we have $i(0) = V/R$. Then $i(t)$ is determined and, working from Equation A1–66, we have

$$v_C = V - v_R = V - iR = V - Ve^{\frac{-t}{RC}}$$

or

$$v_C = V\left(1 - e^{\frac{-t}{RC}}\right) \tag{A1–70}$$

This is the voltage across a charging capacitor as indicated in Figure A1–18c. The student should check that it has the correct behavior by mentally substituting $t = 0$ and $t =$ infinity into the equation (to note v_C goes from 0 to the steady battery voltage V).

To find the solution for the case that the switch is moved from (2) to (1) in Figure A1–18a, we notice that we can regard the problem as the same as before but with $V = 0$. But the differential equation A1–67 would still be the same, so that the $i(t)$ solution must again be that of Equation A1–69; only the initial condition $i(0)$ will be different. If the capacitor had become fully charged so that $v_C = V$, then from Equation A1–66 we have at this *new* initial time that

$$0 = v_R + v_C = i(0)R + V \quad \text{or} \quad i(0) = -\frac{V}{R}$$

Then at later times

$$v_C = -iR = Ve^{\frac{-t}{RC}} \quad \text{(A1–71)}$$

The expression of Equation A1–71 just represents an exponential decay beginning at V and is also shown in Figure A1–18c. When $t = RC$ (the characteristic time), we have $v_C = Ve^{-1} = 0.367$ V and the capacitor voltage has fallen to 37% of its initial value.

REVIEW EXERCISES

1. (a) What is the volt a measure of?
   (b) What are its fundamental units?
2. Criticize the following statement: "The common automobile battery has an emf of 12 V, therefore the lead-acid cell emf is 12 V."
3. Consider the battery types: A. alkaline-manganese, B. carbon-zinc, C. lead-acid, D. mercury, E. nickel-cadmium.
   (a) Which are readily rechargeable?
   (b) Which is noted for a stable terminal voltage during discharge?
   (c) Which is the inexpensive "flashlight" battery?
   (d) Which use paste electrolytes?
   (e) Which has a 2.0 V emf?
   (f) Which has a 1.35 V emf?
   (g) Which types might be used in applications requiring currents of 10 amperes?
   (h) Which type exhibits a quick initial fall in emf while discharging, but remains quite constant for most of the remaining discharge?
4. (a) What is a voltage drop?
   (b) How is it always related to current direction?
5. Which of the following is not expressed in volts? A. potential drop, B. emf, C. potential difference, D. power, E. standard cell voltage.
6. A certain resistor is marked with 3 orange stripes, followed by a gold stripe. Its resistance can be expected to be between what values?
7. Consider the resistor types: A. carbon, B. wirewound, C. metal film, D. oxide film.
   (a) Which two are commonly color coded?
   (b) Which two are used for high precision and stability?
   (c) Which one is not appropriate for radio frequencies?
   (d) Which one is readily available with power ratings over 5 watts?
   (e) Which two have lowest electrical noise?
   (f) Which one is often 10% precision?
8. Select the false statement:
   (a) The resistance of 3 resistors in parallel is less than that of any of the resistors.
   (b) The resistance of 3 resistors in series is less than that of any of the resistors.

(c) The resistance of 4 identical resistors in parallel is less than that of 3 of said resistors in parallel.
(d) The resistance of 4 identical resistors in series is more than that of 3 of said resistors in series.

9. (a) State Kirchhoff's Current Law.
   (b) State Kirchhoff's Voltage Law.
10. (a) What is the difference between an emf and a voltage drop?
    (b) Can a voltmeter detect which is which?
11. Which term does not belong? A. potentiometer, B. transformer, C. rheostat, D. variable resistor.
12. What happens to the capacitance of a capacitor if:
    (a) The area is doubled.
    (b) The distance between the plates is doubled.
    (c) The dielectric constant of its dielectric is doubled.
13. Consider the following capacitor types: A. ceramic, B. electrolytic, C. mica, D. mylar, E. paper, F. polystyrene, G. tantalum.
    (a) What does the term for each type refer to?
    (b) Which types can be found with a capacitance of 500 microfarads?
    (c) Which types are polarized (proper voltage polarity must be used)?
    (d) Which type is useful because of its low inductance (and is not polarized)?
    (e) Which types are useful for low leakage and low price?
    (f) Which often has a disc shape?
14. What is the most common electrical circuit that is the basis for timing (causing certain time delays, and so forth) in electronic circuits?
15. Consider the circuit shown in Diagram A1–1. Each resistor is 10 Ω, the capacitor is 1 μF, and the battery is 2 V. The switch is closed and we wait a few seconds before reading the voltmeters. What do voltmeters V_1 and V_2 read?

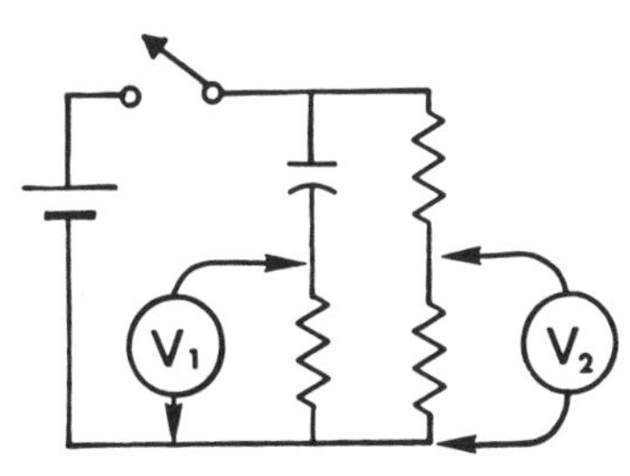

DIAGRAM A1–1.

16. Name the electrical quantity corresponding to each of the following units:
    (a) coulomb
    (b) ampere
    (c) volt
    (d) ohm
    (e) watt
    (f) farad
    (g) henry
17. Why does increasing the number of wire turns in a coil (with length and diameter held constant) increase the inductance of the coil? (In fact, the inductance is approximately proportional to the square of the number of turns.)
18. (a) If the frequency of the voltage that is applied to a capacitor is doubled, what happens to the current through the capacitor? (Assume other factors remain constant.)

(b) Substitute inductor for capacitor in part (a).

19. (a) What is the effect of an ideal inductor in a dc circuit?
    (b) What is the effect of an ideal capacitor in a dc circuit?
20. (a) What is the difference between a standard transformer and an autotransformer?
    (b) Why might you risk touching *one* output lead from a 200-V secondary of a standard transformer, but not a lead from even a 2-V autotransformer secondary?
21. A transformer (alone) cannot "transform" a dc voltage to another dc voltage. Why not?
22. (a) What is the phase relationship between the ac current passing through a capacitor and the voltage across it?
    (b) Substitute inductor for capacitor in part (a).
23. The concept of capacitor reactance may be applied directly for the voltage-current relationship of a capacitor in which situation? A. sinusoidal applied voltage, B. square-wave applied voltage, C. sawtooth applied voltage, D. dc applied voltage.
24. Consider a resistor and capacitor connected in series, with 3 ac voltmeters connected as shown in Diagram A1–2. A certain ac current is flowing.
    (a) It is *not* true that $V = V_R + V_C$, where the Vs here are voltmeter readings. Discuss briefly why it is not.
    (b) Write a correct relationship between the three voltmeter readings.

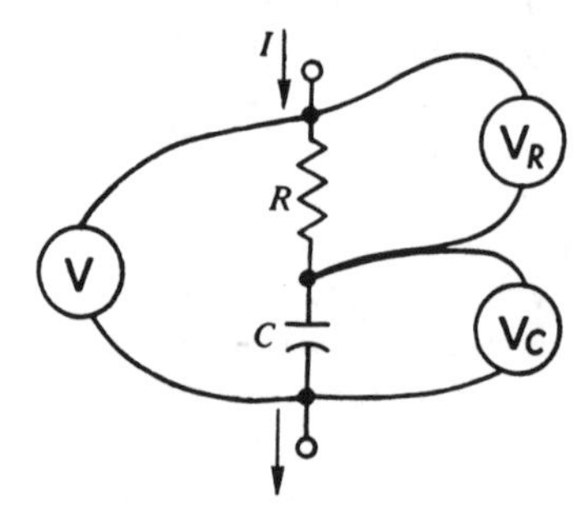

DIAGRAM A1–2.

25. Where is the term *half-power frequency* used? What does it refer to?
26. What type of filter is represented by the circuit drawn in Diagram A1–3? Make a verbal (qualitative) argument to justify your answer.

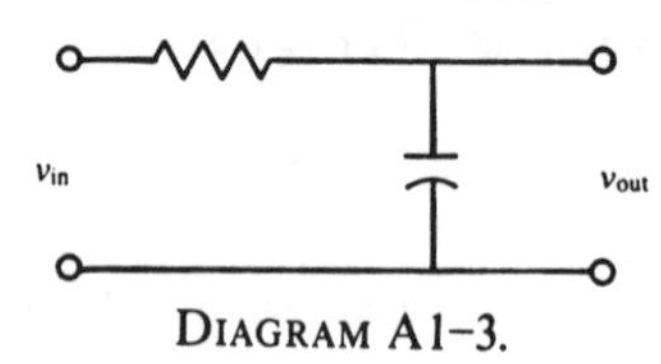

DIAGRAM A1–3.

27. Sketch the output/input voltage versus frequency for a low-pass filter. Identify the half-power frequency f_o quantitatively in the sketch.

PROBLEMS

1. What is the resistance of a rod of carbon (to be used in a carbon composition resistor) that has a diameter of 2 mm and a length of 1 cm?
2. At times it is useful to wind your own resistor from nichrome wire. If a 0.100-ohm

resistor is to be made from number 22 wire-gauge nichrome wire, what length should it have? Look up the diameter of number 22 wire in a reference such as *The Handbook of Chemistry and Physics* or *The Radio Amateur's Handbook.* It is useful to note the various standards of the past; assume AWG (American Wire Gauge), which is the same as B&S (Brown and Sharp).

3. This exercise suggests the use of an OEM (original equipment manufacturer) electronic catalog. (Indeed, thumb through it to note the range of items available.) Examples in the United States are the Newark, Allied, Cramer, and Hamilton-Avnet electronics catalogs. Look up the prices for the AA-size ("penlight") battery for each of the types listed below. You may need to find AA dimensions and use them for some types. Compute the price ratio of each of the others to carbon-zinc.
   (a) carbon-zinc
   (b) nickel-cadmium
   (c) alkaline
   (d) mercury
4. What is the (warm) resistance of a 100-watt light bulb that operates from 110 V?
5. An inexpensive and widely used resistance box contains 1-watt 10% carbon resistors. A switch selects the desired resistor.
   (a) What is the maximum current that may pass respectively through the 47-ohm and 47-kilohm resistors within it?
   (b) What is the maximum voltage that may be applied respectively to the 47-ohm and 47-kilohm resistors?
6. Suppose 3 resistors of 1 kΩ, 2 kΩ, and 3 kΩ are connected in series to a 6-V battery.
   (a) Diagram the problem (schematic diagram).
   (b) What current flows from the battery, and what power is delivered by it? (Express in mA and mW, respectively.)
   (c) What is the voltage across each resistor?
   (d) Verify that they add up to the applied voltage.

   NOTE: Volts = milliamps × kilohms; it is usually convenient in electronics to work in mA and kΩ units for current and resistance, respectively.
7. Suppose 3 resistors of 1 kΩ, 2 kΩ, and 3 kΩ are connected in parallel to a 6-V battery.
   (a) Draw the schematic circuit for the problem.
   (b) What is the effective resistance connected to the battery?
   (c) What is the battery current drain?
   (d) Find the current in each resistor and verify that they sum up to the answer in part (c).
8. (a) Consider the circuit pictured in Diagram A1–4. Find the current in the 3-Ω, 6-Ω, and 5-Ω resistors. Also find the voltage across each.
   (b) Multiply each resistance in the figure by 1000; that is, change from Ω to kΩ. Again find all currents and voltages.

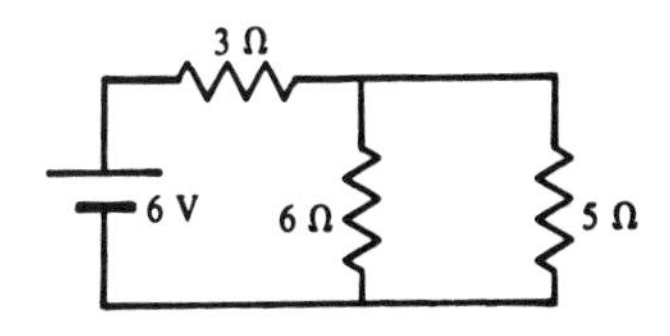

DIAGRAM A1–4.

9. Consider the circuit pictured in Diagram A1–5. Find the current flowing in each resistor. Note all resistances are in *kilohms.*

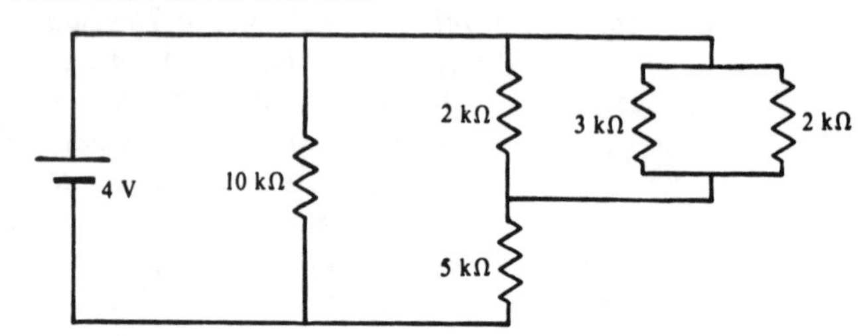

DIAGRAM A1–5.

10. The 3 resistors pictured in Diagram A1–6 are in some op-amp circuit, but the details of the rest of the circuit are not drawn and do not matter here.
    (a) What are the readings of the 3 voltmeters?
    (b) Diagram the circuit and show the more positive test prod of each voltmeter (this is conventionally the red test prod).

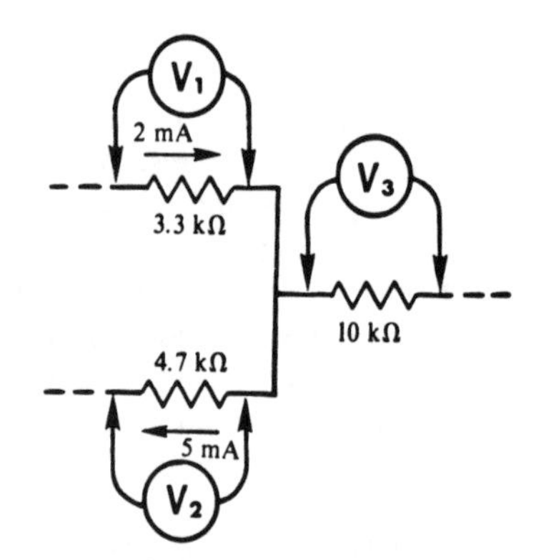

DIAGRAM A1–6.

11. An electronic device of some sort is connected in the circuit shown in Diagram A1–7. The milliammeter *A* reads 12 mA. What is the reading of the voltmeter *V*?

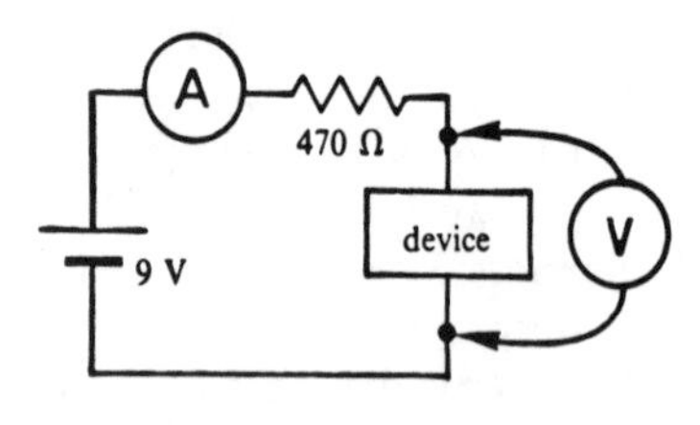

DIAGRAM A1–7.

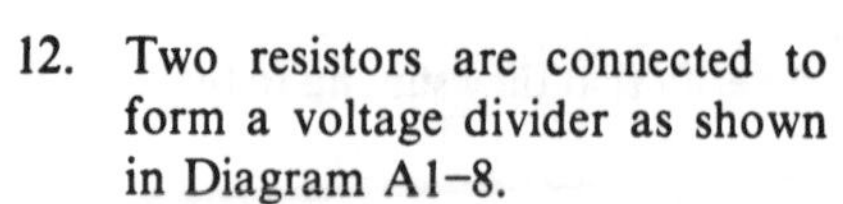

12. Two resistors are connected to form a voltage divider as shown in Diagram A1–8.
    (a) What is the voltage out when nothing is connected to it?
    (b) Assume a 6-kΩ resistance is connected to the output. What voltage appears across the output?
    (c) When an unknown resistance is connected to the output, its

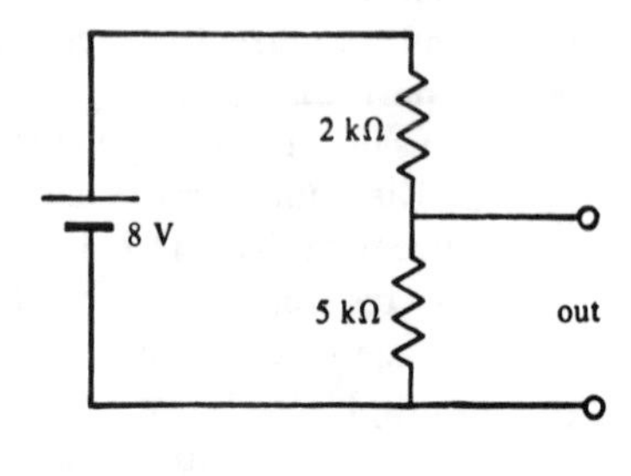

DIAGRAM A1–8.

voltage is observed to decrease to 80% of the value in part (a). What is the value of the resistance?

13. Suppose a paper capacitor is to be constructed with a capacitance of 0.01 μF. If the thickness of the paper is 0.1 mm and its (relative) dielectric constant is 2, what area is required for each plate?

14. Suppose a disc-ceramic capacitor is to be constructed with a capacitance of 0.01 μF. The thickness of the ceramic is 0.1 mm and its relative dielectric constant is 1000. What area is required for each plate?

15. Suppose a 10-kΩ resistor, a 2-μF capacitor, and a switch are connected in series to a 20-V battery, as shown in Figure A1–18a. Also suppose the switch has been in position (1) for a long time and is switched to position (2) at $t = 0$. Calculate and plot the voltage across the capacitor v_C for the time interval from 0 up to 3 characteristic times. Perhaps a dozen points will do. You should obtain the rising part of the charge/discharge plot in Figure A1–18c.

16. An important integrated circuit discussed in Chapter A7 (the 555 timer) makes use of an *RC* charging circuit. In one connection, it performs a test on a charging capacitor, asking "When does the capacitor's voltage reach 2/3 of the charging voltage?" Thus, in the circuit pictured in Diagram A1–9, what time is required for the capacitor voltage to reach 6.00 V after switch *S* is opened?

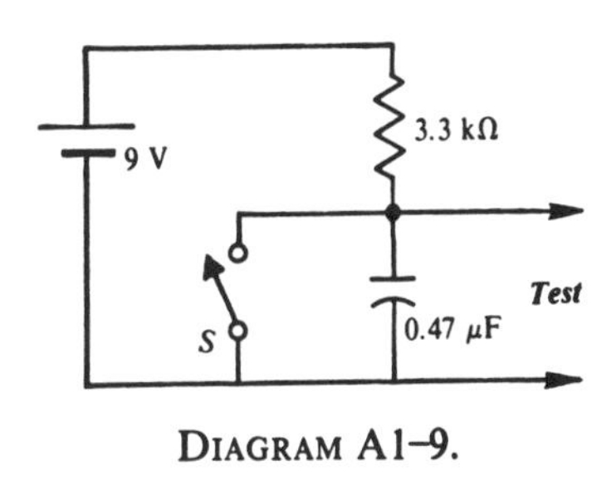

DIAGRAM A1–9.

17 A certain voltage is characterized by the expression

$$v = 5\ \text{V} \sin (0.0628\ \text{sec}^{-1}\ t)$$

The argument of the sine function is in radians. Give the following characteristics for it:

(a) average voltage
(b) peak voltage
(c) peak-to-peak voltage
(d) rms voltage
(e) frequency
(f) period

18. Consider the circuit pictured in Diagram A1–10. Make a moderately accurate sketch of the voltage between the terminals versus time.

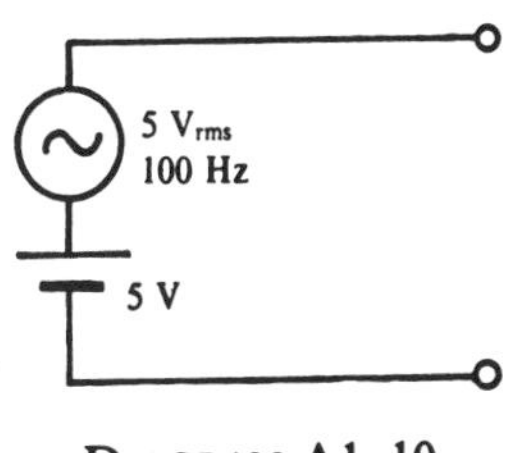

DIAGRAM A1–10.

19. Write out a derivation for the effective capacitance of 3 capacitors connected in series; that is, derive Equation A1–32 for Figure A1–17.
 NOTE: The same charge flows onto each capacitor.
20. (a) Assume that a 1-μF capacitor is connected to a 6.3-V_{rms} voltage. Find the rms current flowing through the capacitor if the frequency is 60 Hz, and also if it is 10 kHz.
 (b) Assume a 1-mH (millihenry) choke is connected instead to the 6.3-V_{rms} source. Find the currents for the two frequencies.
21. Consider the transformer with the primary and secondary windings pictured in Diagram A1–11. Tap B is actually connected to the earth and is a true ground.
 (a) If there are 5000 turns on the primary, what is the total number of turns in the secondary?
 (b) How many windings exist between tap B and lead C?
 (c) Make a sketch of the voltage (relative to ground) versus time for tap A and also for tap C. Draw the voltage from tap A as a solid line and from C as dashed, and on the same graph.
 NOTE: The voltage at C is 180° out of phase with the voltage at A.

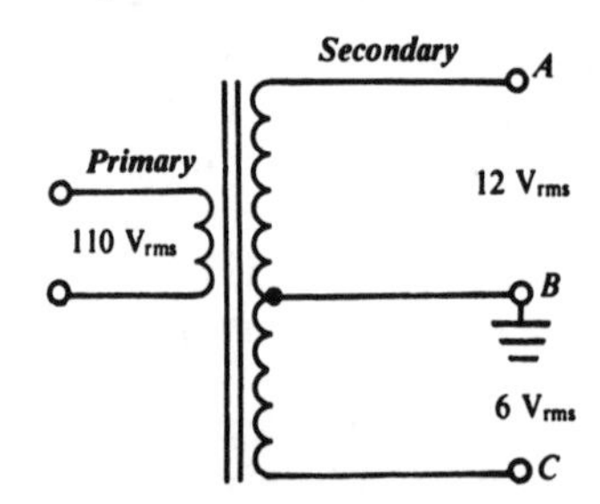

DIAGRAM A1–11.

22. Assume that a resistance R_s is connected to the secondary of a transformer, that Equation A1–53 is valid for the transformer, and that the transformer is ideal; that is, all power presented to the primary of the transformer is passed to the output of the secondary. Show that the impedance R_p, presented by the primary to any ac voltage source connected to it, is given by

$$R_p = \left(\frac{N_p}{N_s}\right)^2 R_s$$

 where R_s is the resistance connected to the secondary
 NOTE: Evaluate the ratio of the primary voltage to primary current. This ratio is the *input impedance* of the transformer; we see that a transformer may be used to "transform impedance" in ac circuits (but not for dc circuits). R_p may be smaller or larger than R_s.
23. Consider an RC high-pass filter such as shown in Figure A1–25. Assume that $V_{in} = 10\ V_{rms}$, $C = 0.01\ \mu$F, and $R = 10$ kΩ.
 (a) If the voltage frequency is 1 kHz, find:
  (1) current flowing
  (2) voltage across the resistor
  (3) voltage across the capacitor
  (4) ratio V_{out}/V_{in}
  (5) power factor angle

(6) average power delivered by the ac voltage source

(b) What is the half-power frequency for this circuit?

24. Suppose we construct an RC high-pass filter from a 0.01-μF capacitor and a 4.7-kΩ resistor.
    (a) Diagram the circuit.
    (b) What is the half-power frequency for this filter?
    (c) At what frequency is V_{out}/V_{in} equal to 0.1?
    (d) At what frequency is V_{out}/V_{in} equal to 0.9?
    (e) Could this filter be used to reduce 60-Hz line frequency interference in a signal?
25. (a) Consider an RC high-pass filter such as shown in Figure A1–25. Give an argument to support the fact that the maximum power delivered by the constant voltage source must occur at high frequencies. Deduce a formula for this maximum power in terms of just V and R.
    (b) Show that when the frequency is such that $X_C = R$, the power delivered by the constant voltage source is half of the value given by the formula in part (a).
26. (a) Design an RC high-pass filter circuit that has $V_{out}/V_{in} = 0.5$ at 5 kHz. At very high frequency, the filter is to present an impedance of 1000 Ω at its input.
    (b) What is the phase of V_{out} relative to V_{in} at this frequency (5 kHz)?
27. (a) Design an RC low-pass filter that has $V_{out}/V_{in} = 0.707$ at 1 kHz. At very high frequencies, the impedance at the filter input is to be 500 Ω.
    (b) What is the phase of V_{out} relative to V_{in} at this frequency?
28. Consider an inductor connected to a resistor as shown in Diagram A1–12.
    (a) Make a qualitative argument for the type of filter this is.
    (b) Derive an expression for V_{out} in terms of V_{in}, R, and X_L. Include a diagram similar to Figure A1–25b in your derivation.

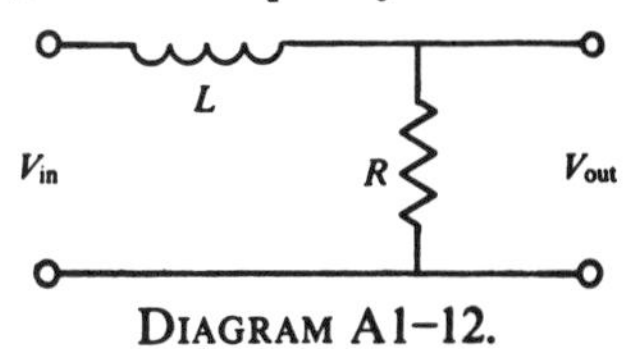

DIAGRAM A1–12.

NOTE: Although it does not influence the expression deduced here, V_L should point upward, rather than down as V_C does. Why?

29. Solve the simultaneous equations A1-63, A1-64, and A1-65 to find the unknown currents in Figure A1–29. The current I_3 should agree with that found in the text by application of Thevenin's Theorem.
30. Find the Thevenin equivalent circuit for the circuit pictured in Diagram A1–13.

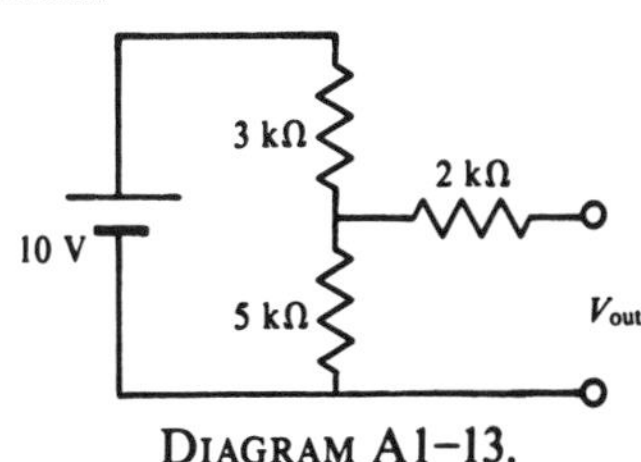

DIAGRAM A1–13.

31. Consider the circuit of Figure A1–29.
    (a) Calculate the current through the 2-V battery by applying Thevenin's Theorem. Assume the 2-V battery to be outside an imaginary box that contains all the other circuit elements (properly connected).
    (b) Finally, assume the 4-Ω resistor is outside the Thevenin box. Calculate the current through it. Now all the circuit currents have been obtained.
32. Consider the circuit drawn in Diagram A1–14.
    (a) Find the Thevenin equivalent circuit for the circuit in the box in Diagram A1–14. Make a drawing of the equivalent circuit, with the values of the components specified.
    (b) Find the current through the 1-V battery. Is the battery charging or discharging?

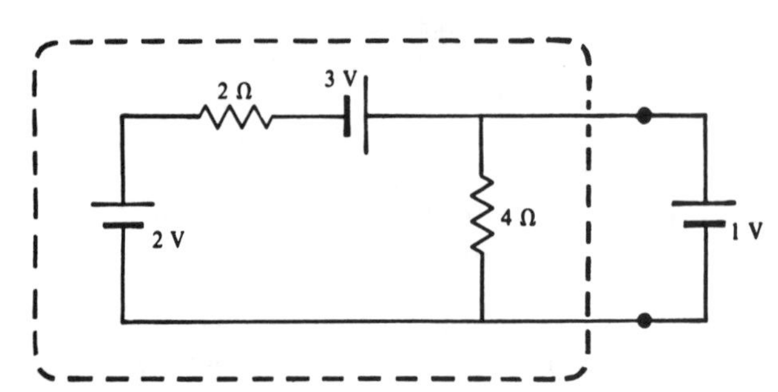

DIAGRAM A1–14.

33. The *superposition theorem* for electrical circuits is rather simple and makes a problem like that in Figure A1–29 rather easy to solve. The theorem is this: "The current in each leg of a circuit can be obtained by calculating the current in each leg, assuming only one battery in the entire circuit is active at one time; all the others should be shorted. Then the actual current in any particular leg is the algebraic sum of the contributing currents in that leg, caused by all the batteries simultaneously." For example, in Figure A1–29 the current in the 1-Ω resistor is the sum of the current that would flow in it if only the 2-V battery were present (and the 4-V battery replaced by a short) plus the current if only the 4-V battery were present. Solve for the 3 currents in Figure A1–29 by use of the superposition theorem. These solutions may be checked with earlier solutions.
34. (a) An ideal battery is a constant voltage source. Show that a large voltage source in series with a very large resistor (as shown in Diagram A1–15) performs nearly like a constant current source. (For example, show that the current that passes from the source is nearly constant no matter what resistance (with-

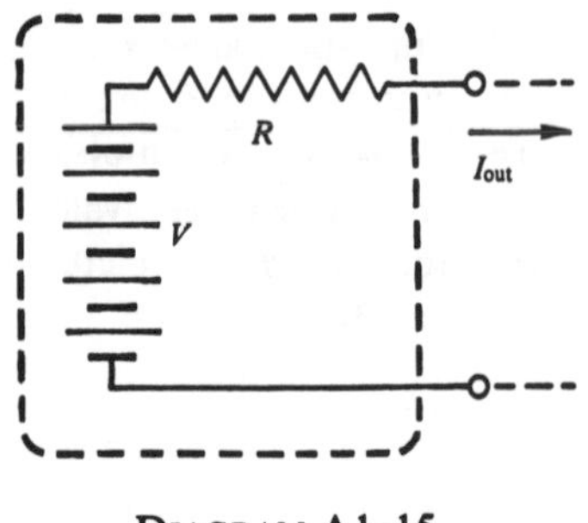

DIAGRAM A1–15.

in limits) is connected as load to the source.)

(b) What determines the value of the constant current?

(c) What limits the resistance of the load for proper operation?

(d) Are there limits on any voltages that may be present in a circuit connected to this constant-current source?

35. *Norton's Theorem* is a close relative of Thevenin's Theorem. It states that a circuit containing an arbitrary connection of batteries and resistors (and also constant-current sources) between two terminals will behave as though just a resistance in parallel with a constant-current source is connected between the two terminals (see Diagram A1–16). The circle with enclosed arrow represents the constant-current source, and I_N is the Norton equivalent current. R_N is the Norton equivalent resistance. Find the Norton equivalent circuit of a battery V in series with a resistance R (that is, for a Thevenin equivalent circuit). Justify your answers.

NOTE: Consider short-circuit output current and open-circuit output voltage of both the actual circuit and the Norton equivalent circuit.

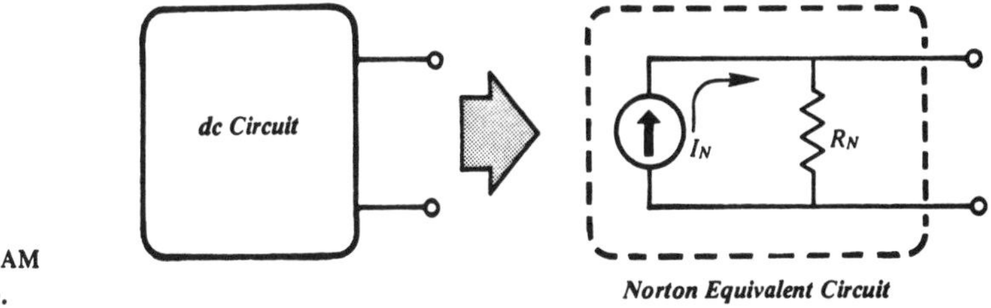

DIAGRAM A1–16.

36. (a) Consider an RC filter (either high-pass or low-pass) that is well beyond the breakpoint frequency and into the attenuation region in terms of frequency. Show that V_{out}/V_{in} changes by a factor of 10 for every factor-of-10 change in frequency.

(b) Suppose we make a plot of $\log(V_{out}/V_{in})$ versus $\log f$. Base-10 logarithms are used. This log-log plot is often called a *Bode plot.* Show that well into the attenuation region, the plot is a straight line with slope -1 for a low-pass filter and slope $+1$ for a high-pass filter.

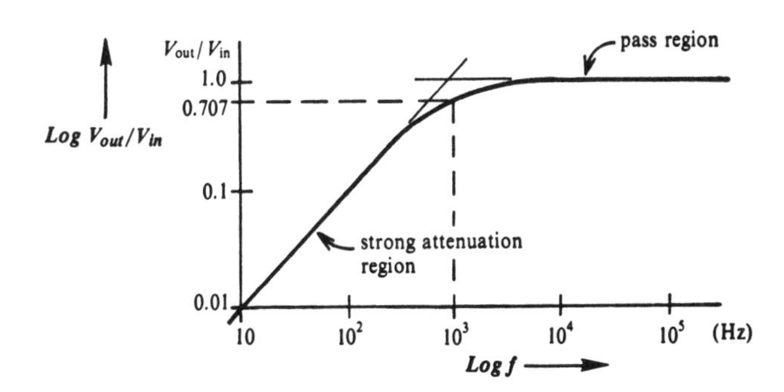

DIAGRAM A1–17.

NOTE: Then the Bode plot is simple to draw over most of the "curve." The drawing (Diagram A1–17) shows the Bode plot for a high-pass filter with a breakpoint frequency of 1 kHz. Note that the straight-line asymptotes cross at this frequency.

37. Suppose two *RC* filters are cascaded as pictured in Diagram A1–18. The buffer represents a factor-of-1 amplifier that simply passes the voltage from the output of the first filter to the input of the second. (We learn in Chapter A6 how simple it is to design one of these with an op amp.) The buffer has no effect on the output voltage of the first filter. Show that in the low-frequency attenuation region, the attenuation of the overall filter changes by a factor of 10^2 for every factor-of-10 change in frequency. This is an example of a *second-order filter*. What is its advantage?

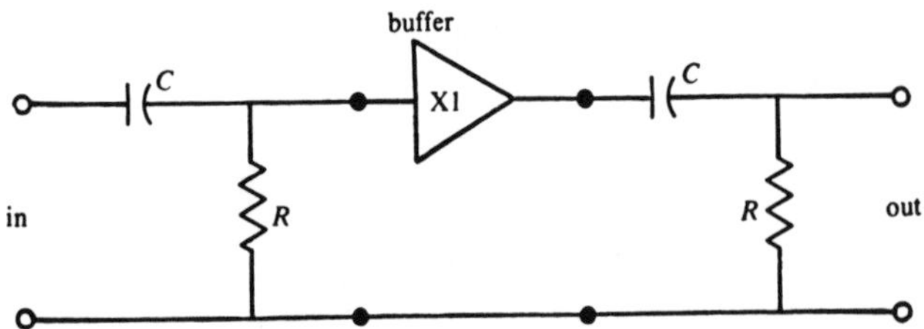

DIAGRAM A1–18.

CHAPTER

A2

Important Electronic Instruments

There are a number of electrical and electronic instruments that are routinely used in electronics practice and in electronics-based laboratory experimentation. Much of this equipment is used to measure the voltage, current, or resistance presented by an electronic circuit or by the actual laboratory phenomenon being studied. Thus, various electronic measuring devices are commonly used for testing electronic circuits either in the design phase or after they are being used in the laboratory. Moreover, an experimental phenomenon may also be measured or monitored with this equipment. If the phenomenon is not inherently electrical, it is often made so through the use of *transducers*—the topic of Chapter A3.

Consequently, it is very desirable to have a basic understanding of the principles of operation, operating techniques, and capabilities and limitations of the various common electronic measuring instruments. These aspects are discussed in this chapter for the various types of meters available, such as the VOM, FTVM, and DVM. The Wheatstone bridge is discussed in some detail, since this circuit is widely used to measure phenomena that are manifested through a changing resistance. Particular attention is paid to the modern triggered-sweep oscilloscope, since this is a very versatile laboratory instrument with a number of operating modes. The chapter also considers electronic equipment that is used to generate various forms of ac voltage. Instruments to measure the frequency of ac waveforms are then discussed. The chapter concludes with a discussion of the practical techniques used to electrically connect measuring instruments to electronic circuits; however, these methods are also important because they are used to interconnect electronic equipment generally.

Basic Meters

D'Arsonval Meter

The student should already be familiar with the essentials of the moving coil or *d'Arsonval meter*. Though now being replaced by digital (all electronic) meters, it still has a place as an inexpensive and handy device. The movement responds to current. When current flows through the windings of the armature or coil, the coil experiences a force due to nearby permanent magnets. Actually, a torque is produced on the coil that turns it against the counter-torque of a coil spring, or a *taut band* suspension. A twist of the suspension proportional to the current results, and an attached needle moves over a scale. The meter is characterized by its resistance and by the current required to move the needle to a full-scale reading or the *full-scale sensitivity;* for example, a 50-microamperes movement, or a 1-milliamperes movement. If the meter is used only in its basic form (often with zero at the center of scale of travel for the needle), it is called a *galvanometer.*

Ammeter

Frequently the "bare" d'Arsonval meter is too sensitive for common current measurements. A bypass or *shunt* resistance R_{sh} is then arranged around the meter as shown in Figure A2–1. In essence, a galvanometer is shunted to increase its range (see Figure A2–1a). The desired full-scale reading I is related to the shunt current I_{sh} and movement current I_m by

$$I = I_{sh} + I_m \tag{A2–1}$$

The full-scale sensitivity I_m and resistance of the movement coil R_m must be known to proceed in a design. The voltage across the movement V_m for full-scale deflection is calculated to be

$$V_m = I_m R_m \tag{A2–2}$$

This voltage also is responsible for the shunt current I_{sh}. We have

$$V_m = I_{sh} R_{sh} = (I - I_m) R_{sh} = I_m R_m \tag{A2–3}$$

Then

$$R_{sh} = \frac{R_m}{(I/I_m) - 1} \tag{A2–4}$$

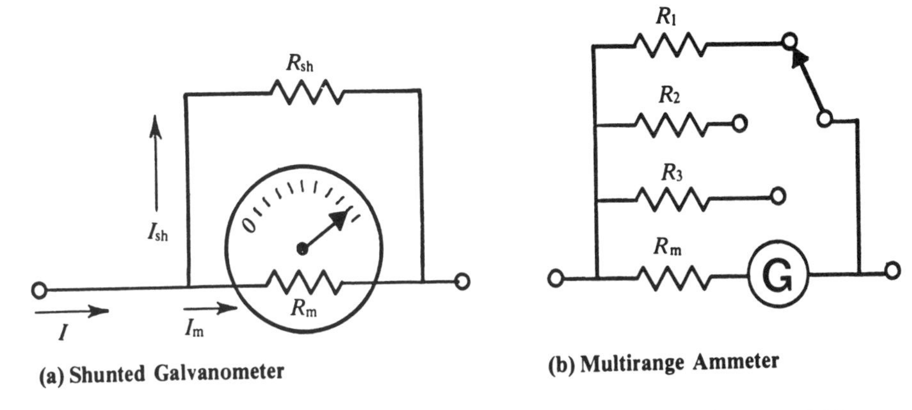

(a) Shunted Galvanometer

(b) Multirange Ammeter

FIGURE A2–1. *Construction of an Ammeter from a Sensitive D'Arsonval Meter*

A switch can be used to make a more versatile multirange instrument as shown in Figure A2–1b. There will typically then be several rows of numbers along the scale; an appropriate row is used (with the markings) depending on the switch setting.

Voltmeter

The d'Arsonval meter is basically an ammeter, but it may be used to measure voltages if the movement is quite sensitive. A series resistance R_s is connected to the movement such that the desired voltage must appear across the meter when the full-scale current I_m passes through the meter (see Figure A2–2a). Let the desired full-scale voltage be V. We have $V = I_m(R_s + R_m)$ from which the required series resistance in terms of the known I_m, R_m, and V is obtained; that is,

$$R_s = \frac{V}{I_m} - R_m \qquad \text{(A2–5a)}$$

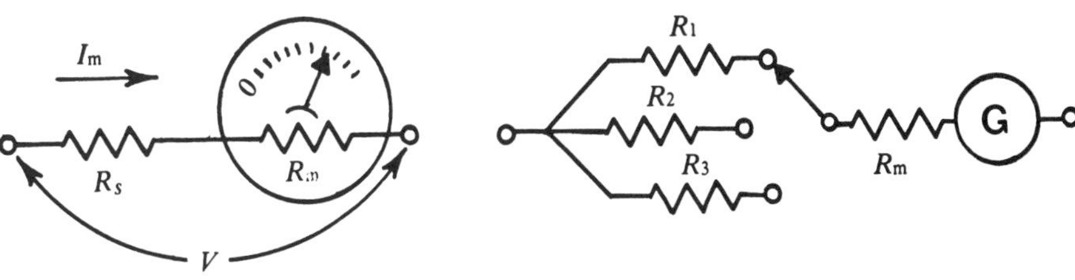

(a) Series "Multiplier" Resistor Added to Movement

(b) Multirange Voltmeter

FIGURE A2–2. *Construction of a Voltmeter from a D'Arsonval Meter*

But it is useful to examine the total resistance of the voltmeter when used with the proper multiplier resistor to get a certain full-scale voltage capability. It is obtained from Equation A2–5a as

$$R_s + R_m = V\left(\frac{1}{I_m}\right) \qquad \textbf{(A2–5b)}$$

Notice that the total resistance is obtained by multiplying the full-scale voltage by the reciprocal of the full-scale current sensitivity. In fact, this reciprocal has the dimensions of ohms/volt. Thus if $I_m = 50$ microamperes, the meter is a 20,000 ohms/volt voltmeter. If such a movement is used to make a 10 V full-scale voltmeter, the resistance between the *testing prods* of the voltmeter will be 10 V × 20,000 ohms/volt = 200 kilohms. Notice that the higher the voltage range is, the larger is the meter resistance.

To construct a versatile meter with several ranges, a switch that connects to one of several multiplier resistors is commonly used, as shown in Figure A2–2b.

Ohmmeter

If a known voltage is placed across an unknown resistance and the resulting current through the resistor is measured, the resistance may be calculated from Ohm's Law: $R = V/I$. Typically, ohmmeters use a carbon-zinc battery to supply the voltage and the movement responds to the current. A scale is then marked off to read resistance directly; however, the scale is not linear, since the resistance is proportional to the reciprocal of the current.

Carbon-zinc dry cells show a marked decrease in terminal voltage over their useful life. Therefore a potentiometer is typically placed in the circuit to *zero adjust* the meter (see Figure A2–3a and b). When the leads of the ohm-

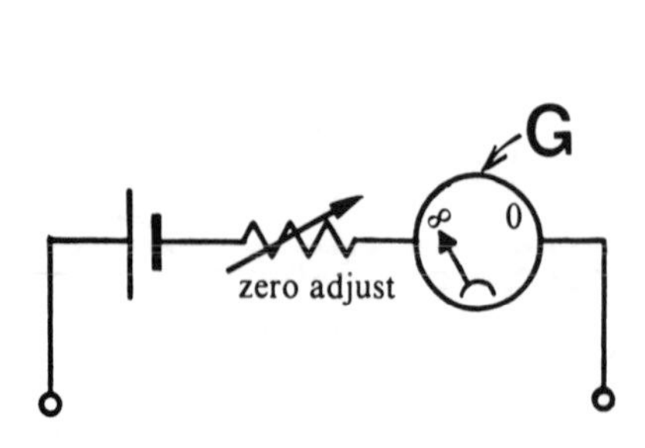

(a) Basic Circuit with Galvanometer G

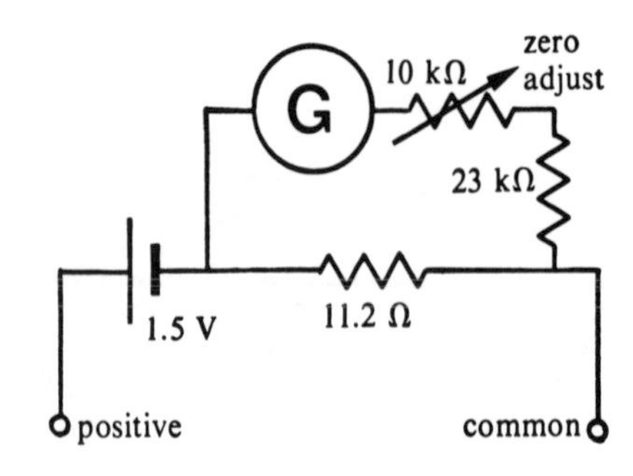

(b) Simpson VOM Ohmmeter for R × 1 Scale

FIGURE A2–3. *Ohmmeter*

meter are connected together, the zero adjust knob should be used to cause the needle to point to zero ohms (full-scale current). Note that leaving the leads shorted for a length of time will drain the ohmmeter batteries. If the zero adjustment cannot be made, the batteries are probably old and need replacement.

Measuring Instruments

VOM, FTVM, and VTVM

The volt-ohm meter, commonly called a VOM or a multimeter, is a handy combination ac/dc voltmeter, ohmmeter, and ammeter with switch selectibility. A sensitivity of 20,000 ohms/volt is quite common and implies a 50-microamperes movement so that we can expect the most sensitive current range to be 50 microamperes as well. Resistors of 1% accuracy are typical, and a 2% overall dc accuracy is therefore common. D'Arsonval-type meters with an accuracy of 1/2% exist, but are rare.

To obtain a much higher resistance—and thus approach the ideal more closely—electronic amplifying circuits are used to obtain the *vacuum tube voltmeter* (VTVM) or the more modern *field-effect transistor voltmeter* (FTVM). A maximum *grid-leak resistor* or *gate-leak resistor* of about 10 megohms is typically found in the input amplifier of these units; it is literally connected between the two test prod inputs and determines the input impedance of these meters. Therefore it is constant on most ranges. These voltmeters typically can measure a wide range of dc or ac voltage, but not current.

The typical scheme for ac voltage measurement in both VOMs and VTVMs is to rectify and filter the ac voltage to produce a dc voltage equal to the peak-to-peak ac voltage. Indeed, a *peak-peak ac scale* typically exists on VTVMs. However, the rms scale is simply set to be 0.707/2 times the peak-peak scale; this is correct for the sinusoidal waveform, but will be wrong for other waveforms. The peak-peak scale, however, is correct for any waveform.

VTVMs require some minutes for the tubes to warm up after turn-on. Furthermore, a zero adjust occurs for use on all scales, not just the ohmmeter scale. There is also an *infinity adjust* for the ohms setting on VTVMs. Even though VTVMs must be plugged into a wall socket for tube power, they typically have a battery for the ohmmeter circuit. FTVMs are battery-operated and therefore free of the bothersome line cord; they also have little warm-up drift.

Wheatstone Bridge

For more precise measurement of resistance than the few percent afforded by an ohmmeter, the *Wheatstone bridge* has been commonly used since the turn of the century. A square of four resistors is used, with one resistor being the unknown R_x. An *excitation voltage* V is connected across one diagonal, and galvanometer or sensitive voltmeter is connected across the other diagonal as shown in Figure A2–4. Measurement of the unknown R_x uses a null comparison approach and assumes that the other three resistors are known and that one is adjustable.

The bridge is balanced by adjustment of R_3 until no current flows through the galvanometer G. When balanced or nulled, no current flows through G because the voltage at point A (relative to ground gnd, for example) is the same as the voltage at B. But we have two voltage dividers and thus have

$$V_A = \frac{R_2 V}{R_1 + R_2} = V_B = \frac{R_x V}{R_3 + R_x}$$

so that

$$R_x = R_3\left(\frac{R_2}{R_1}\right) \qquad \textbf{(A2–6)}$$

For precise work the resistors can be 0.1% or even 0.01% with the adjustable resistance being an accurate decade resistance box. Notice that in the *balanced condition,* a change in the excitation voltage V will have no effect.

NOTE: A *decade resistance box* uses a series of switches in order to provide an accurate variable resistance. Each switch S selects one of ten precision resistances for its decade of range. These ranges are connected in series so that a resistance such as 347.6 ohms may be readily selected by adjusting the four corresponding switches.

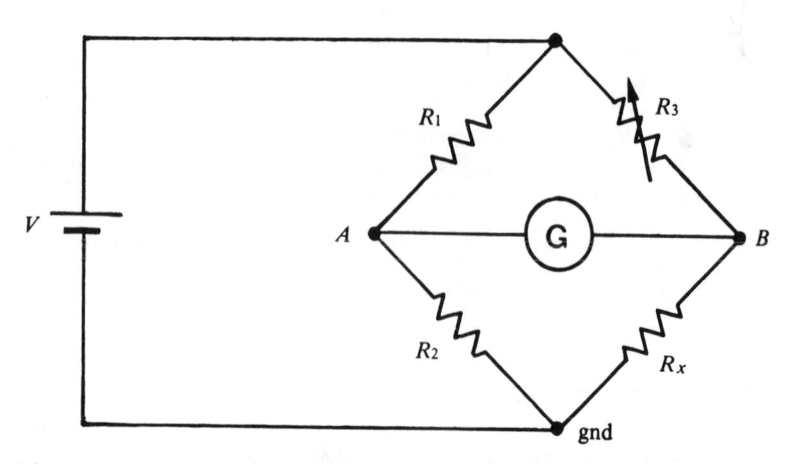

FIGURE A2–4. *Schematic Diagram of the Basic Wheatstone Bridge*

However, many contemporary applications of the Wheatstone bridge use it in an *out-of-balance* mode. (Accurate measurements of resistance are now made more quickly with a good digital voltmeter.) Many experimental parameters can be monitored through a resistive effect, as discussed in Chapter A3. That is, a change in the parameter causes a change in the resistance of a *resistive transducer*. For changes in R_x of up to 10%, the bridge voltage V_{AB} is quite linearly proportional to ΔR_x. Furthermore, suppose the transducer is susceptible to a second parameter such as temperature. The effect of such variations can be minimized by making R_2 an identical device that experiences the same extraneous effect—for example, temperature—but not the actual parameter to be measured. We can see that even though R_2 and R_x may vary due to this same parameter, they will do so to the same extent, and no out-of-balance condition will result.

Suppose that the resistances R_1, R_2, and R_3 in these three arms of a Wheatstone bridge are all equal to R. Then it is readily seen that at balance the resistance R_x is also equal to R. This is a simple and convenient bridge configuration for use and for analysis of the out-of-balance mode of operation. Suppose the transducer resistance changes by an amount ΔR, and the excitation voltage is V. The chapter Problems show that the out-of-balance voltage (or signal) V_{out} from the bridge output is given by

$$V_{out} = V_B - V_A = \left(\frac{V}{4}\right)\left(\frac{\Delta R}{R}\right) \qquad \textbf{(A2-7)}$$

provided $\Delta R/R$ is much less than one. Thus the output signal is proportional to the fractional change in the transducer resistance. Also notice that the signal depends on the excitation voltage; therefore, this voltage must be stable for accurate out-of-balance operation.

Potentiometer

The *potentiometer* is a null comparison device that measures voltage and has been in use where high precision is needed since the turn of the century. It has now been largely replaced by the *digital voltmeter,* which is just as accurate and a good deal faster to use. However, voltage standardization still often uses a potentiometer. Also, the potentiometer has the interesting characteristic that at null, no current is drawn from the source being measured. In this sense it becomes an ideal voltmeter. (It does draw or deliver current when off balance, however.)

The circuit for the basic potentiometer is shown in Figure A2–5. The R_A *working current* adjustment potentiometer is adjusted until the voltage at point T is, for example, 1.500 V. Then the unknown voltage V_x is connected through a galvanometer to the movable contact of the main potentiometer R (from which the name *potentiometer* derives). The contact is then moved

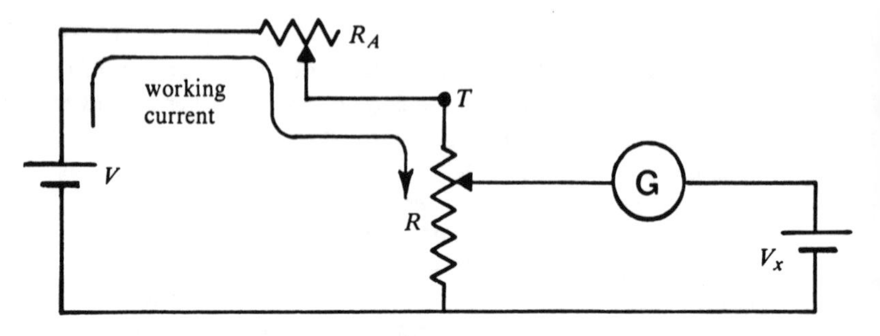

FIGURE A2–5. *Basic Potentiometer Voltage-Measuring Circuit*

until G reads zero. At this balance the unknown voltage V_x must equal the voltage at the movable contact; this is the reason no current flows through the detecting meter.

If the potentiometer R has high linearity, the fractional position of the slider can be used to calculate the voltage at its position by taking this fraction of the known voltage at T. There is one problem remaining, and that is to adjust the voltage at T accurately by adjustment of R_A. This may be done by connecting a *standard cell*—that is, a standard single cell battery for which the voltage is accurately known—to the *unknown* V *terminals* V_x. However, in this case V_x *is* accurately known. Then the potentiometer R is set to the fraction corresponding to this accepted voltage, and R_A is adjusted until G reads zero. The potentiometer is now calibrated. An unknown voltage may be determined by connecting it to the V_x terminal in place of the standard cell and balancing the potentiometer as discussed above.

Digital Voltmeter (DVM)

If for no other reason, the d'Arsonval meter is limited in accuracy by the degree to which the scale can be read by the eye (of course, bearing friction, nonlinearity of the magnetic field or of the coil spring are other likely problems). *Digital voltmeters* (DVM) use electronic circuits to convert a voltage or current to a number that is then read from a numeric display. Any number of digits can in principle be read by the eye in this case. The trick is to design electronic circuits of sufficient accuracy and linearity.

Details of the way a voltage or current can be digitized are part of a digital electronics course; however, certain features should be mentioned here. Since the voltage to be measured is presented to the input of an amplifier, the resistance of the voltmeter is generally independent of the range used. Ten megohms is a typical value. Some DVMs have *autoranging,* meaning the measuring range—or decimal point—is selected automatically to keep the number in the range of the display. *Autopolarity* is quite common: A plus or minus sign lights up to indicate the polarity of the voltage. The DVM accuracy is typically on the order of the least significant digit. However, the accuracy of the ac range is almost always less than that for dc. When using the ohmmeter capability, a definite current (for example, 1 milliampere) is generally sent through the unknown resistance. The meter then reads the voltage across the resistance; for example, if 1-milliampere current is used, the voltage is numerically equal to its resistance in kilohms.

Oscilloscope

Perhaps the most versatile single electrical instrument is the *oscilloscope.* With it one can measure dc voltage and ac voltage quite readily with an accuracy of a few percent. However, its greatest service is drawing a time graph of an arbitrary voltage waveform from which we can see relative voltage levels as they occur in time. Inherent in this application is the ability to measure the time interval between voltage changes that can occur as quickly as a fraction of a microsecond.

Figure A2–6 illustrates the essential parts of the *cathode-ray tube* (CRT) that is the heart of an oscilloscope. Beginning at the left, a heater or filament glows red-hot and boils electrons out of a cathode structure. These electrons are accelerated toward two accelerating anodes that are at rather high positive potentials. However, a negative voltage applied to the grid close to the cathode repels some electrons back toward the cathode and in this way controls the number of electrons that strike the screen at the far right; this in turn controls the spot brightness or intensity. The relative voltages of the two accelerating anodes determine the sharpness of *focus* of the spot formed.

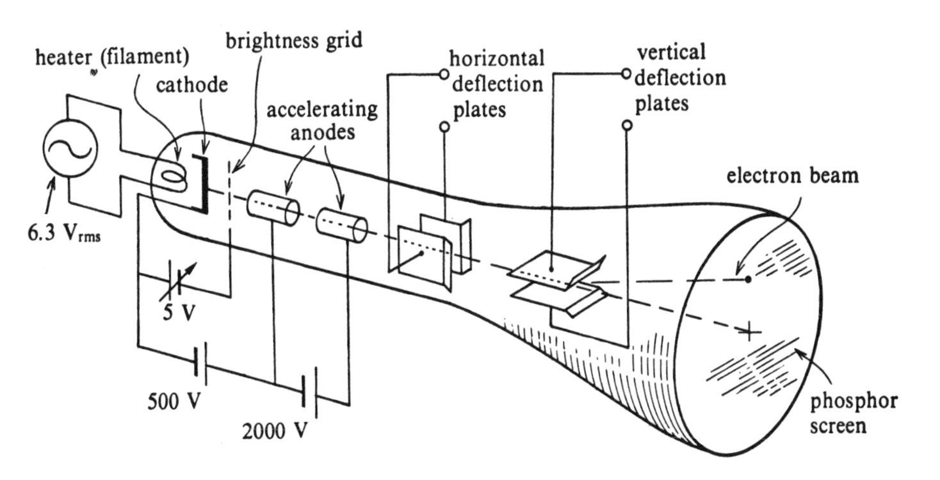

FIGURE A2–6. *Basic Components of a Cathode-Ray Tube*

The anodes are cylinders and consequently the electrons can zip on through toward the deflection plates. If a voltage is connected between a given set of plates, the electrons are attracted toward the more positive plate and the beam is deflected from the straight-through course. When the electrons strike the end of the tube, a phosphor coating on the end emits visible light at the spot of electron bombardment. Figure A2–7 is representative of the controls commonly found on a laboratory-quality oscilloscope. The assortment can be bewildering at first, but they are actually quite understandable.

Most contemporary oscilloscopes are of the *triggered-sweep* variety, with the vertical axis calibrated in volts and the horizontal axis calibrated in time. A *vertical* control selects the desired volts/centimeter for the vertical scale, and a *horizontal* control selects the microseconds/centimeter (or perhaps milliseconds/centimeter or seconds/centimeter) for the horizontal scale. The consequence is that the spot of the electron beam is deflected upward—or downward if the polarity is reversed—by say 1.5 centimeters if a 1.5-V battery is connected to the VERT IN jacks. Meanwhile, a sawtooth voltage generated inside the scope is applied to the horizontal deflection plates, causing the spot to move from left to right at a definite rate of so many centimeters/second.

NOTE: The rate that the spot moves across the screen is the reciprocal of the scale setting selected by the horizontal control knob.

There are usually several knobs that control the *triggering mode* of the scope, and these are a bit more elaborate to understand. The central point to appreciate is that a triggered-sweep scope causes the horizontal sweep of the spot to occur at the same point of the voltage waveform time after time. Only in this way is a stable picture drawn on the scope face (no double exposure or ghost-like traces). When the TRIGGERING MODE is set to one of the *internal modes*—for example, ac, dc, or auto—a monitoring circuit within the scope measures the incoming voltage. Only when the signal reaches a certain *triggering level* selected by the LEVEL knob does it instruct the sweep circuit to begin a sweep. The graph of the waveform then begins at the left of the scope display at this *trigger point*. These features are illustrated in the scope picture of Figure A2–8a. During the *retrace* of the spot from right to left, a *blanking* circuit in the scope extinguishes the unwanted trace. Figure A2–8b shows the sawtooth-like HORIZONTAL SWEEP voltage V_H generated within the scope; this produces the desired horizontal motion of the spot to trace out the graph of the input voltage waveform.

If the LEVEL knob is examined in Figure A2–7, a (+) and a (−) will be noted to either side of a zero. When the LEVEL knob is set to zero, triggering will take place near the point of zero voltage on the input waveform. Turning the knob toward (+) moves the triggering point to a higher (more +) voltage on the waveform, while turning toward (−) moves it to a lower point.

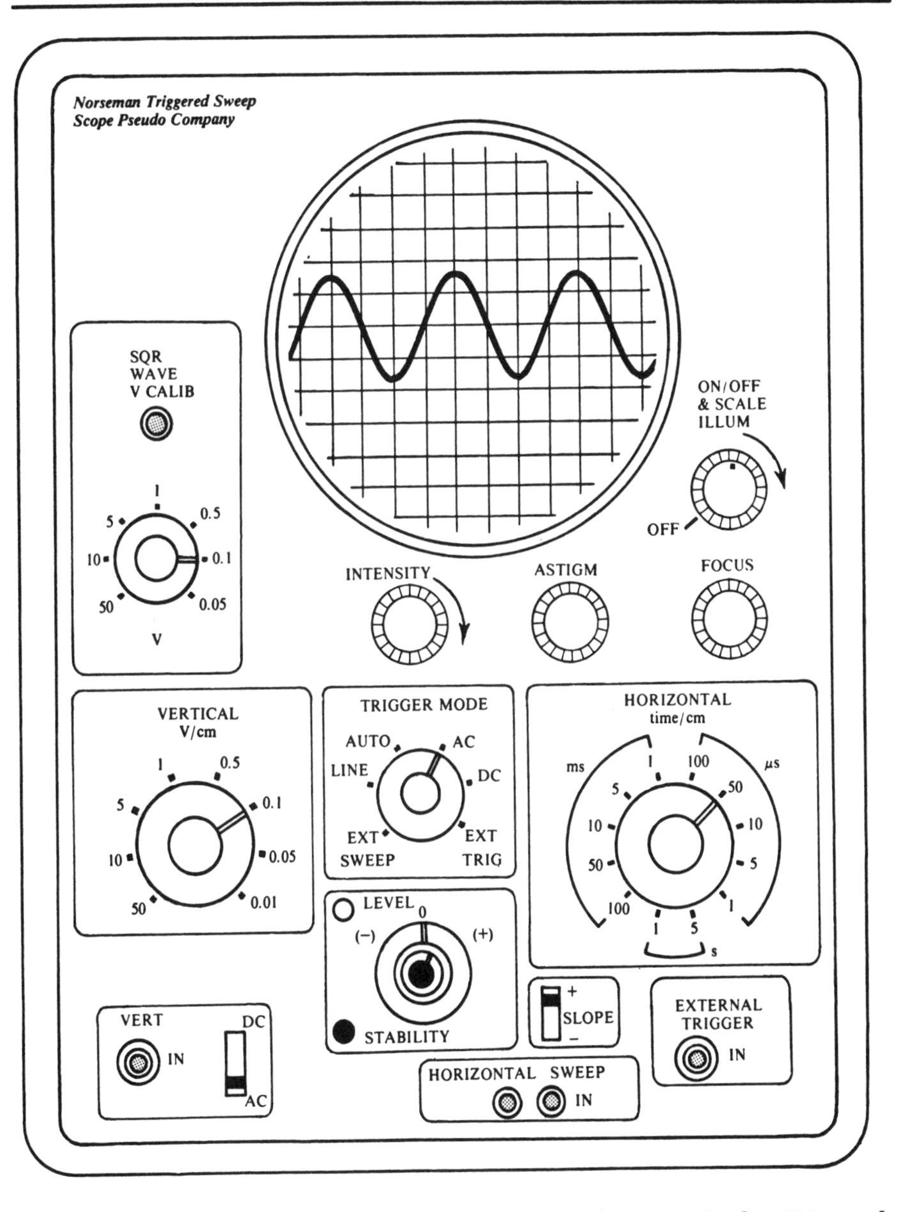

FIGURE A2–7. *Controls Commonly Found on a Laboratory-Quality Triggered-Sweep Oscilloscope*

In addition, most scopes have a (+), (−) SLOPE switch. With this switch, we can choose to begin the trace on the upward (+) slope of the waveform, or the (−) downward slope. Note the (+), (−) indicated points in Figure A2–8a. The scope picture shown assumes that the SLOPE is set to (+).

We now consider the meaning of the various TRIGGER MODE settings.

1. If the mode is in ac or dc, no sweep occurs—that is, the scope is dark—until a voltage with an amplitude above the triggering level is at the scope vertical input.

2. The AUTO mode is convenient in that a trace—that is, a line—is present even with no signal in. It can be reassuring to see that the scope is active and just where the trace is; the absence of vertical deflection shows that there is no substantial signal in the circuit to which the scope is connected. However, we usually lose control over the triggering level in this mode because the scope automatically triggers at somewhere near the middle of the waveform. For some waveforms the trace may be unstable.

3. When the mode is placed to LINE, the sweep occurs in synchronism with the 60-hertz line voltage coming to the scope through the power cord. This is only useful if the signal being observed is itself somehow derived from the domestic power lines, so that it will be stabilized on the scope.

4. In the EXTERNAL TRIGGER mode, the sweep sawtooth is triggered *not* by the signal into the vertical input jack, but rather by whatever waveform is presented to the EXTERNAL TRIGGER IN jack. The triggering level knob now operates relative to this signal. For example, a voltage pulse into a *pulsed nuclear magnetic resonance* (PNMR) spectrometer trig-

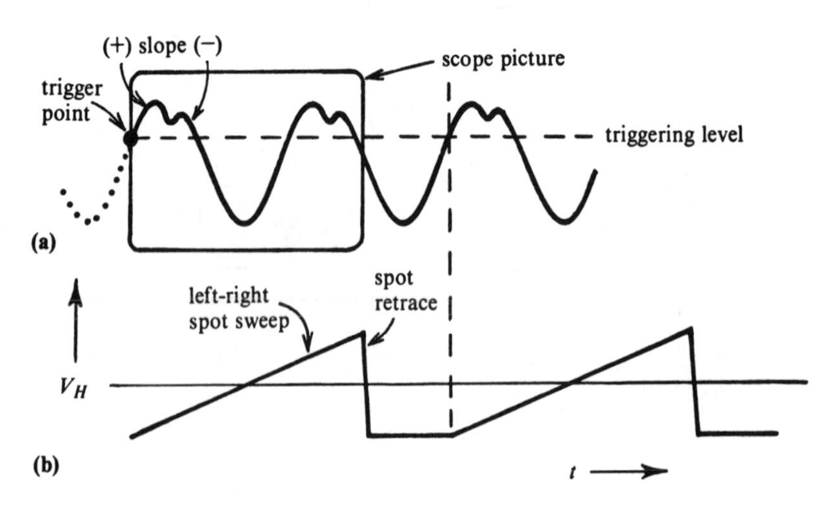

FIGURE A2–8. *Sweep Timing of a Triggered-Sweep Scope*

gers a chain of events that ultimately causes a *transient signal*—that is, a decaying voltage—to be produced by the sample being studied. The signal can be reliably displayed on a scope by connecting the voltage pulse also to the external trigger input and using external trigger mode. The sample signal is input as usual to the vertical input and the sweep will display it nicely. The horizontal time/centimeter knob is still operative.

5. When the TRIGGER MODE is set to EXT SWEEP (external sweep), the horizontal deflection system is disconnected from the sawtooth oscillator and its time/centimeter knob control. Rather it is connected to the HORIZONTAL SWEEP IN jack, and the spot will now move horizontally in the X direction in response to this input signal (in the same manner as the vertical Y motion is controlled by the voltage to the VERTICAL IN jack). The scope is then being used in the *X-Y mode.*

In fact, a true X-Y mode scope has a volts/centimeter knob for the horizontal direction in addition to the one for the vertical direction. It is fun to connect two sine wave oscillators to the vertical input and horizontal sweep inputs and watch the so-called *Lissajous figures* on the scope. See pp. 31–34 of Malmstadt, Enke, and Toren (1963) for a careful pictorial justification of the various figures.

Dual-trace scopes can display time graphs of two different signals simultaneously. Then two vertical input jacks are present, each controlled by a vertical volts/centimeter knob. For example, a dual-trace scope is desirable for displaying the signal into a system and its signal output simultaneously for comparison. It is rather like two scopes for the price of one. However, the two traces are actually displayed alternately one above the other—*alternate sweep*—which works well for signals occurring at rather high frequency so that no flicker is evident. For low-frequency signals the *chopped mode* can be used, where the trace rapidly moves up and down and shows a sample of each input as it moves across the scope face. An additional nice capability of most dual-trace scopes is that the two volts/centimeter controls and jacks are used in the X-Y mode to give true X-Y scope performance (again, the time-base sawtooth is then internally disconnected from the Y axis). The scope pictured in Figure A2–7 is *not* a dual-trace scope.

Triggered-sweep scopes sometimes also have a *stability* control that sets the sensitivity of the sawtooth oscillator. If set too high, the oscillator produces a continuous sawtooth without waiting for a synchronizing trigger point. The sweep then shows a confusing multiple exposure of the waveform connected to its vertical input. In fact, this is the way less-expensive *recurrent-sweep* scopes operate; a stable trace results only if the frequency of the incoming signal is a multiple of the sawtooth frequency, or more correctly, if the sawtooth frequency is so adjusted. Unfortunately, the horizontal scale is not accurately calibrated on this type of scope.

Finally, comment should be made about the *ac/dc mode* switch for the vertical input. For example, it is *not true* that the switch must be set to ac to make measurements on an ac voltage. When set to dc, the spot (trace) will move through a proper vertical deflection even when a dc voltage such as a battery is connected to the vertical input. The trace will also respond to an ac signal presented to the vertical input. However, when set to ac, a capacitor is inserted in series with the input line within the scope. The capacitor serves to block dc voltage but pass ac voltage on through to the amplifiers and scope face. Therefore, in the ac input mode, the scope trace will *not* deflect in response to a battery or other dc voltage that might be connected to the vertical input.

The purpose of the ac mode setting is to permit the vertical scale to be set to a sensitive level so that a small ac voltage "riding atop" a large dc voltage remains on the scope display to be observed. Figure A2–9a illustrates the scope display for a scope in the dc input mode; the scope is receiving an input voltage with both an ac part and a dc part as shown. The dashed line at the bottom of the display represents the trace for zero input voltage. Suppose we want to magnify the display in order to examine the small ac signal more carefully. Unfortunately, if the scope is left in dc input mode and the vertical setting is changed from 5 V/centimeter to 0.5 V/centimeter to accomplish the magnification, the trace will disappear off the top of the visible display. But switching the input mode to ac blocks the offending large dc voltage from reaching the oscilloscope amplifiers, and the enlarged ac voltage is nicely displayed as shown in Figure A2–9b. To summarize: When set to dc, the scope can be used to measure both ac and dc waveforms, but when set to ac, the scope can be used to measure only ac waveforms.

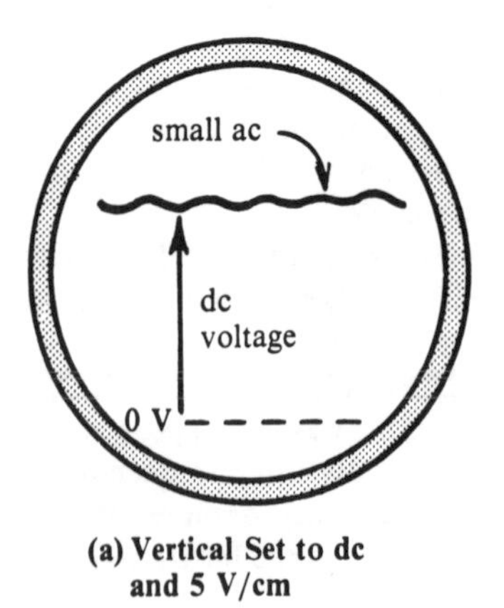

(a) Vertical Set to dc and 5 V/cm

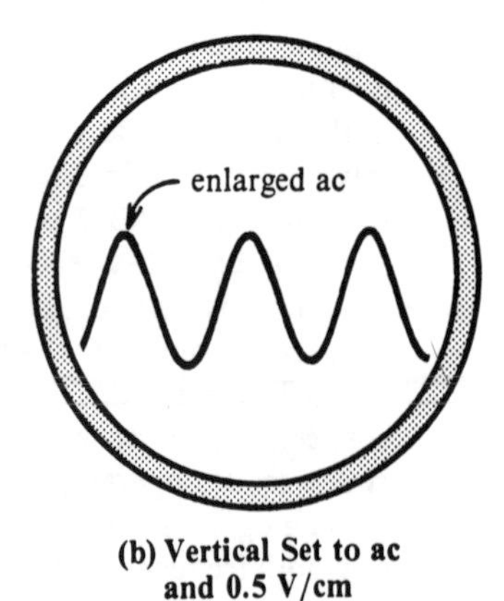

(b) Vertical Set to ac and 0.5 V/cm

FIGURE A2–9. *Use of the ac/dc Switch on a Scope*

Recorders

There are many circumstances in both science and industry that require a continuous recording over a period of time of some parameter; example parameters are voltage, pressure, temperature, and speed. The recorded data may be analyzed later for an understanding of the phenomenon, or it may be reviewed to be confident that the parameter was maintained within certain tolerances. Frequently, it is helpful to have a time graph of the phenomenon in order to literally see what is going on.

The oscilloscope provides such a time graph, but it is useful primarily for recording rapid phenomena, that is, processes that occur in a time interval of less than a few seconds. For a permanent record, a photograph may be taken of the scope display with a camera designed to be attached to the oscilloscope display bezel. However, for processes that occur over longer time periods, other techniques are generally used.

A recording measuring instrument that has been commonplace for many years in the laboratory is the *strip-chart recorder.* This instrument draws a time graph of the input voltage on a moving strip of paper. Some units use several pens to draw several time graphs simultaneously to the respective input voltages.

A related instrument is the *X-Y recorder.* In this case, the pen moves independently in two dimensions over a fixed piece of paper. The pen moves in the perpendicular *X* and *Y* directions in response to two voltages that are input to the respective *X* and *Y* inputs. This recorder is often used to present a graph of how one parameter varies in an experiment as a function of some other parameter.

A *magnetic tape recorder* permits the scientist to record data on magnetic tape. In contrast to the recording techniques just described, magnetic tape recording does not produce a graphic picture. On the other hand, it has the advantage that the information is in a *machine-readable form*—that is, the parameter behavior can be later read by an electronic instrument like a computer for further analysis. In contrast, the techniques that directly graph the data on paper generally require a human to read the data in preparation for analysis.

Strip-Chart Recorder

Most strip-chart recorders use a pen to draw a graph of the input voltage on a strip of graph paper as it is dispensed from a roll. The pen is moved laterally across the paper as the paper moves past the pen at a constant rate (see Figure A2–10).

The paper movement rate is often determined by a *synchronous ac motor.* The same type of ac motor is used in the common electric wall clock, and derives its rate from the quite stable frequency of the 60-hertz house power. The user may select gear ratios in the chart-drive mechanism in order to obtain a chart speed that may range from as slow as 1 centimeter/hour to as fast as 10 centimeters/second. However, some more recent recorders use a *stepper motor* to advance a fixed gear train. The stepper motor is an interesting recent motor-type that rotates its shaft through a fixed angular increment in response to each input power pulse. The pulse rate can be produced by accurate digital circuitry, and thus the paper advancement can be accurately selected and controlled.

The lateral pen position must be automatically adjusted by the recorder in proportion to the input voltage. This is generally accomplished with some type of *servomechanism.* A servomechanism is a control system that monitors the position or state of a system in the process of moving the system into the desired position or state. The most common way this is accomplished in a strip-chart recorder is with the aid of a potentiometer; such recorders are generally termed *potentiometric recorders.* The pen of the recorder is

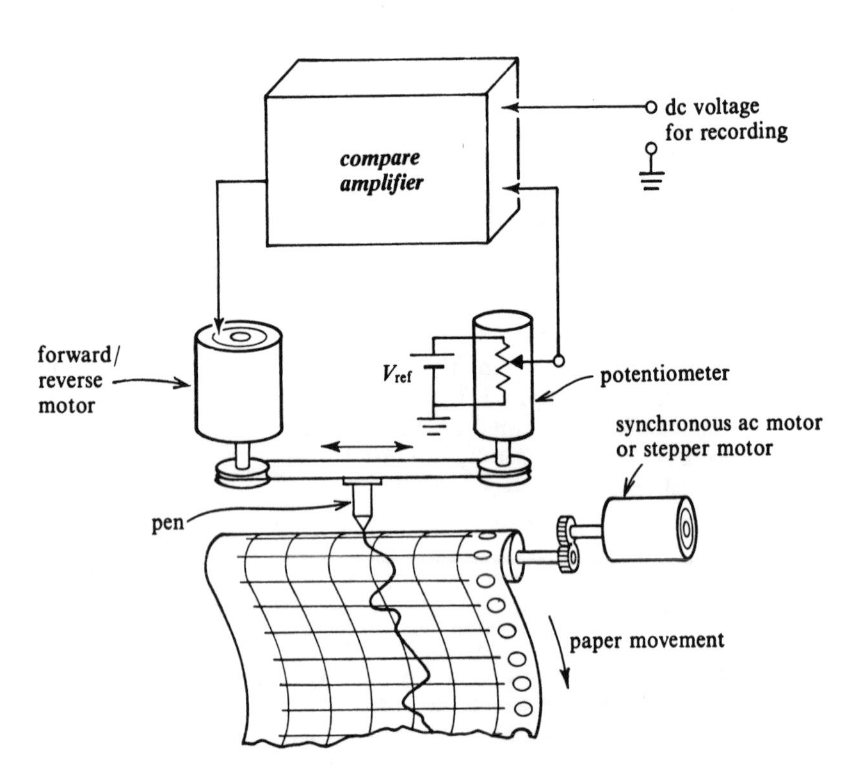

FIGURE A2–10. *System Diagram of a Strip-Chart Recorder*

attached through a cord and pulley system to an electric motor that may turn in either forward or reverse directions. The cord system also turns a multiturn potentiometer. This system is diagrammed in Figure A2–10. The motor receives power to move the pen as long as the pen is not at the proper position corresponding to the input voltage. This is done by having an amplifier compare the voltage from the potentiometer with the dc input voltage. If these two voltages are not the same, the amplifier produces an output with a polarity appropriate to turn the motor and move the pen toward the desired null position. Notice that the servosystem functions as a self-balancing potentiometer voltage-measuring system.

The typical full-scale sensitivity of the basic strip-chart recorder is 10 millivolts dc. If the input parameter is not a dc voltage, it must be converted to one by appropriate electronic circuitry. Some strip-chart recorders include this circuitry internally in order to measure current, resistance, and ac voltage. A control knob very similar to that on a VOM or multimeter is then present in order to permit selection of the function and the full-scale response. The accuracy of the pen position is typically a few tenths of 1 percent, and the response time for the pen to move full scale is generally a few tenths of a second for the motor-driven type.

More rapid response is available in recorders that used special techniques to draw the trace. One approach is to use the needle of a specially designed d'Arsonval meter as the drawing mechanism. Recorders with fast response are generally called *oscillographic recorders.* These recorders can follow a sinusoidal input signal with a frequency of up to several hundred hertz with good fidelity.

X-Y Recorder

The *X-Y* recorder, or plotter, uses two independent servomechanisms that are similar to that just described for the strip-chart recorder. These move the pen in perpendicular directions over a fixed sheet of graph paper or other paper (see Figure A2–11). The full-scale response for each direction is selected by respective control knobs. The response time of the servomechanism and accuracy of the pen position is similar to that just described for the strip-chart recorder. The *X-Y* recorder can also be used to draw a time graph (rather like a strip-chart recorder) if a voltage that increases linearly with time is presented to the *X*-axis input. Some instruments incorporate within them a *ramp voltage* source of this type in order to conveniently permit what is called the *Y-t* mode of operation.

Magnetic Tape Recorder

Nearly everyone is familiar with the common *cassette-type magnetic tape recorder* that was introduced about 1970. Many are also familiar with the high

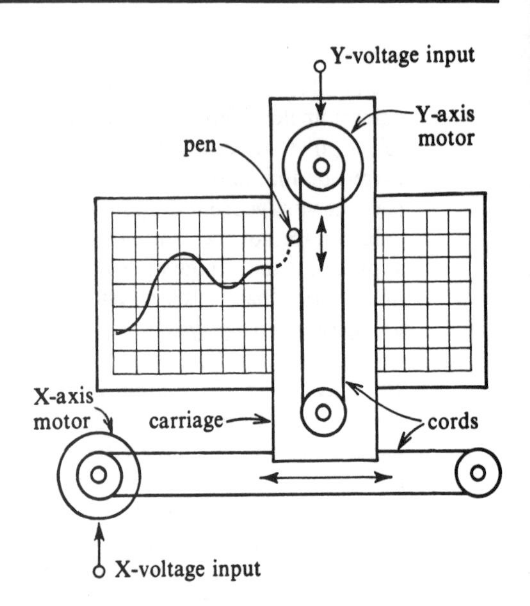

FIGURE A2–11. *Simplified Diagram of an X-Y Recorder*

quality *reel-to-reel magnetic tape recorder;* this type of recorder has been available since about 1960 for high-fidelity music recording. These devices can conveniently record ac voltages that range in frequency from about 50 hertz to 15 kilohertz (namely, the audio frequency range) and can sometimes be used in scientific applications.

NOTE: The recorder should *not* incorporate *automatic gain control* or AGC if the ac playback signal is to faithfully follow the amplitude of the original input signal. AGC automatically adjusts the amplitude of the recorded signal in order to maintain an approximately constant audio level.

Very high quality magnetic tape recorders called *instrumentation recorders* are often used in scientific work. These may be used in *direct recording mode* to record ac voltage signals with frequencies from 50 hertz to as high as 200 kilohertz (the latter frequency is *ultrasonic*). However, they may also be used to record dc voltage or low frequency ac voltage by use of an *FM recording mode.* (The FM mode is discussed later in this chapter.) The FM signal is the signal recorded on the magnetic tape. When the tape is played back, internal electronic circuitry converts back, or *demodulates,* the recorded signal to a voltage signal that corresponds to the original input voltage. Instrumentation recorders can often record several channels simultaneously; four-channel or eight-channel instruments are commonly available.

Signals that have been recorded on magnetic tape may be input to a computer through the use of one of several techniques. In one technique the voltage output from the instrumentation recorder may be input to an *analog-to-digital converter* or *ADC*. This device produces as output a number that is proportional to the input voltage, and this output number is in the form of digital voltages with which a computer works internally. (The ADC is discussed in *Digital and Microprocessor Electronics for Scientific Application.*) Another technique uses FM recording of the signal essentially as described above. At the time of playback, the frequency of the signal recorded on tape may be measured with a frequency counter. As mentioned, the frequency is proportional to the original input voltage. The frequency counter should output a number in the form of digital voltages appropriate for input to a computer. Frequency counters are briefly discussed later in this chapter, and also in detail in the *Digital and Microprocessor Electronics for Scientific Application* by Barnaal.

Other Instruments and Instrument Aspects

Signal Generator

Although not used of themselves to measure electrical quantities, instruments that generate voltage waveforms of various frequencies and amplitudes are often useful for testing equipment, as well as for incorporation into various experiment systems. It was mentioned earlier that the sinusoidal waveform is fundamental, and units that produce this form as output have long been available. Some sort of oscillating circuit is involved, and the sine-wave generator is therefore frequently referred to as an *oscillator*. A given circuit is usually limited in the range of frequencies over which it will operate; therefore, a given oscillator is usually an *audio oscillator* (frequency range about 1 hertz to 100 kilohertz), or a *radio frequency oscillator* (frequency range about 100 kilohertz to 100 megahertz).

Range and dial switches choose the frequency, and it is desirable that the voltage output be constant as the frequency is changed. Stability of amplitude for generators is generally stated in decibels (dB), and the dB will be considered later in this chapter. It is also desirable that the selected frequency be accurate (1% is typical for dial units) and stable in time (0.1% fluctuations are typical). For high-frequency stability a *crystal-based oscillator* is desirable, although this type is fixed in frequency and only available in the radio frequency range. Stability of 1 in 10^6 is then quite possible.

Oscillators may be obtained that permit internal or external manipulation of the output signal. Sine-wave oscillators with these additional capabilities are generally called *signal generators*. For example, the amplitude may vary, or be *modulated*, in proportion to some waveform of lower frequency; this is called *amplitude modulation* or *AM*. Or the frequency of the waveform may be modulated in proportion to some waveform of lower frequency; this is called *frequency modulation* or *FM.*

In recent years *waveform generators* have become available that can produce *square-wave, sawtooth* or *ramp,* and *triangle* waveforms in addition to the sine wave. A nice feature in some of these units is the *sweep frequency* capability. Often it is desirable to study the behavior of an instrument or experiment as a function of frequency. These units can continuously sweep the frequency of the output waveform from some low frequency to a high value; for example, across the audio spectrum. Then the outcome of the experiment should typically be recorded on a strip-chart recorder or other record-producing means, such as a computer. For large frequency ranges, even a logarithmic sweep is available on some units.

If the internal sweep ability is not present, a *voltage-controlled frequency input* often is. This means that the frequency out is proportional to a voltage in at this control input. Connecting a *second* waveform generator in the sawtooth mode to the control input of the first will then sweep frequency.

Finally, a relative of the waveform generator is the *pulse generator.* These units can produce a rectangular pulse of adjustable amplitude and duration, as shown in Figure A2–12. The pulses may be continuously generated at an adjustable rate or frequency, or each pulse may be produced only in response to a timing pulse supplied externally to a *trigger input.* The pulse waveform or pulse train is characterized by the length in time of each pulse, or *pulse duration,* and by the frequency of occurrence of the pulses, or *pulse frequency.* These times are illustrated in Figure A2–12. The term *pulse repetition rate* is often used on the control for the pulse frequency on a pulse generator instrument.

An important feature of any signal generator is its *output impedance.* A signal generator behaves like the Thevenin equivalent circuit shown in Figure A2–13. Consequently, when a load resistance R_L is connected to the output, only a fraction of the generator voltage V_{gen} appears across R_L. For example, if the output impedance of an audio oscillator is 600 ohms and an 8-ohm impedance loudspeaker is connected to it, the output voltage drops to only 8/608 times the signal output when a very large load resistance is connected; that is, 8/608 times the *open circuit output voltage.*

Frequency Counter

Integrated circuitry has made the *frequency counter* as inexpensive as the voltmeter, and they can now be readily found in the lab. Better types can be used in two modes: frequency and period. Consider a waveform input similar to that in Figure A2–12. In the *frequency mode,* the counter will count the number of pulses that come to its input in a fixed time duration—for example, 1 second. The number of counts is then clearly the frequency. For low frequencies, more significant digits can be obtained in a short time by measuring

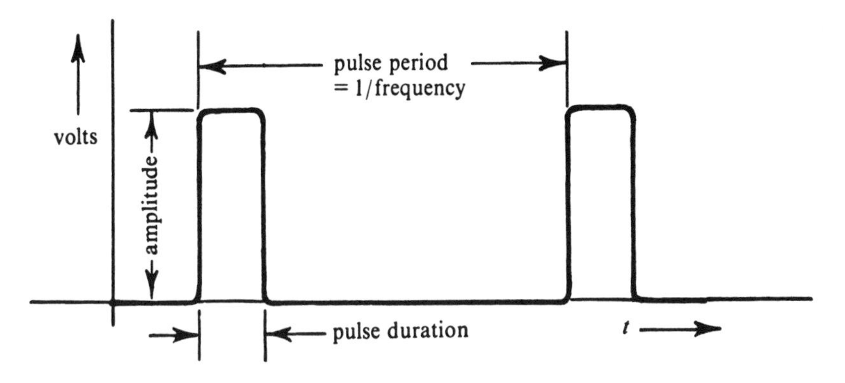

FIGURE A2–12. *Output of a Pulse Generator*

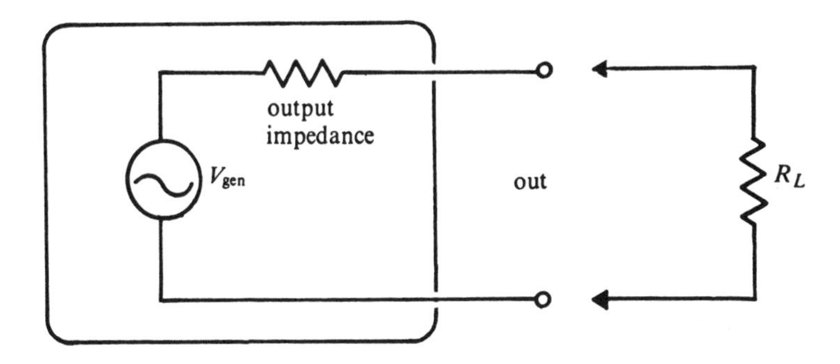

FIGURE A2–13. *Equivalent Circuit of a Signal Generator*

the time interval between two pulses. This "stop watch" approach is the *period mode* because the period of the waveform is then being measured. If we actually want the frequency, we simply calculate the inverse of this period. To count an arbitrary waveform accurately, triggering controls very similar to those on a scope are found on better frequency counters.

Attenuator

Sometimes we are fortunate to have a voltage larger than desired; then an *attenuator,* not an amplifier, is needed. A voltage divider such as the one pictured in Figure A1–11 is the simplest attenuator. Accurate attenuators use fixed resistors and a multiposition switch; for continuous adjustability a potentiometer could be used. The output voltage is clearly reduced or attenuated from what is put to the input.

However, the output impedance of the simple voltage divider is *not* constant and this may be undesirable in some applications. This fact can be seen by simply applying Thevenin's Theorem to Figure A1–11a. The output impedance is R_2 in parallel with R_1 (simply think battery-shorted to deduce this), and this value changes as the potentiometer *tap point* changes. Further, the resistance presented to the voltage source (the battery in the figure) changes when different potentiometer settings are used.

There are several clever resistance networks and switching schemes that produce an attenuator with a constant input impedance and a constant output impedance, even though the voltage reduction factor is changed. It is quite common for these units to be put in a box with a row of switches. If all switches are *down,* the voltage passes from input to output without reduction. Each switch flipped *up* causes the voltage to be reduced by a certain factor, and the factors multiply together.

It is quite common for attenuators to have the reduction factors labeled in decibels, or dB. The following section discusses this concept.

Decibel

The *Bel* is a logarithmic scale that was originally introduced in connection with sound levels because the ear perceives relative sound power levels logarithmically. One Bel contains 10 *decibels* (dB).

The decibel (and Bel) always involves a *ratio* of two power levels that we will call P_1 and P_2. Then the level of P_2 compared to P_1 in decibels is given by

$$\text{Relative Power in decibels} = 10 \log_{10}\left(\frac{P_2}{P_1}\right) \qquad \textbf{(A2–8)}$$

Example: Suppose P_2 is 100 watts while P_1 is 0.1 watts. What is the level of P_2 relative to P_1, expressed in dB?

Solution:

$$\text{Relative Power} = 10 \log_{10}\left(\frac{100\text{ W}}{0.1\text{ W}}\right) = 10 \log_{10} 10^3 = 10 \times 3 = 30\text{ db}$$

We say P_2 is 30 dB *above* P_1.

Example: Suppose P_2 is 0.1 watts while P_1 is 100 watts. What is the level of P_2 relative to P_1 expressed in dB?

Solution:

$$\text{Relative Power} = 10 \log_{10}\left(\frac{0.1\ \text{W}}{100\ \text{W}}\right)$$

$$= 10 \log_{10} 10^{-3} = 10 \times (-3) = -30\ \text{dB}$$

Here we say P_2 is 30 dB *below* P_1.

If the two power levels occur across a certain resistance R, we may substitute $P_1 = V_1^2/R$ and $P_2 = V_2^2/R$ into Equation (A2–8). Then the dB ratio in terms of the two *voltages* becomes

$$\text{Relative Power or dB ratio} = 10 \log_{10}\left(\frac{V_2^2/R}{V_1^2/R}\right) = 20 \log_{10}\left(\frac{V_2}{V_1}\right) \quad \textbf{(A2–9)}$$

Example: Suppose V_2 is 10 V while V_1 is 1 V (same resistance assumed). What is the ratio of power levels for V_2 relative to V_1?

Solution:

$$\text{dB ratio} = 20 \log_{10}\left(\frac{10}{1}\right)$$

$$= 20 \log_{10} 10^1 = 20\ \text{dB}$$

From the preceding example, we can see that *voltage* ratio and *dB ratio* are related by the rule: 20 db for every factor of 10 in voltage. This relationship is presented in Table A2–1.

Strictly speaking, the correspondence between a voltage ratio and this ratio expressed in dB, as presented in Table A2–1, is only valid when the two voltages involved appear across the same value of resistance. However, this restriction is frequently ignored in electronics practice. When large voltage ratios are encountered, it is common to express them in dB even if the two voltages appear across different resistances. This practice then simply represents a compact way to express large ratios.

TABLE A2–1. *Voltage Ratios Expressed in dB*

Voltage Ratio	0.01	0.1	1	10	100	1000
Ratio in dB	−40 dB	−20 dB	0 dB	+20 dB	40 dB	60 dB

Interconnecting Cables

Virtually all electronic and electrical equipment must be interconnected by wires or cables of some sort. Simple insulated wires or *leads* constructed with banana plugs at both ends are convenient and effective for signals with rather low frequency (dc to audio frequencies) and substantial voltage and/or current levels (tenths of a volt or more and milliamperes or more). The name *banana plug* obviously derives from the appearance of the flexible metal parts that constitute the plug (see Figure A2–14a). The plug mates with a jack that is normally insulated from the panel in which it is installed. (The 4-way jack can also be used with simple stripped-end wire or with leads using *spade lugs* or *crocodile clips* at the ends.)

At lower voltage and current than listed above, *electrical interference* can cause problems. External oscillating electric and magnetic fields (electromagnetic fields) couple to the electrical leads and introduce spurious voltages and currents. The origins of these fields can be the 110 V_{ac} house wiring and transformers, fluorescent lights, nearby computers and electrical machinery (e.g., electric motors), and radio and television stations.

The induced emf from electromagnetic induction (changing magnetic fields) can be reduced by avoiding large area circuit loops. For example, the two signal leads normally connecting two instruments can be twisted together. However, the electric fields that terminate on the connecting wires still introduce interference. The conducting wires and components in an electrical circuit literally act like aerials, receiving the radiated signals.

Electric field pick-up is eliminated (or greatly reduced) by placing the components of an electronic circuit within a metal *chassis box,* and connecting the box to earth potential (grounding the box). The electric field lines then terminate on the box, and we say the circuit is *electrostatically shielded.* Often, one point in the electronic circuit within the box is also connected to the box ground and becomes the *ground* of the circuit. Electrostatic shielding for the signal connecting wires between electronic boxes is also important and can be achieved through the use of shielded cables. A commonly used type is *coaxial cable,* which is simply a wire passing down the center and insulated from a flexible metal cylinder. The metal cylinder is often a metal braid, which is in turn enclosed in insulation (see Figure A2–14b). As with the electronic chassis box, the coaxial shield should be grounded in order to provide electrostatic shielding. Since the electronic circuit within the chasis is also usually grounded, the "coax" shield generally serves as the return path for the signal. (A complete round trip path for the electrical signal must be provided, of course.) If the signal return path must be different from ground, a shielded cable with two inner signal wires may be used, or two separate coaxial cables may be used.

A popular quick connect/disconnect connector for coaxial cables is the *BNC-series connector.* The cable plug and panel jack for this type are shown in Figure A2–14c. The outer metal shell of the plug is connected to the

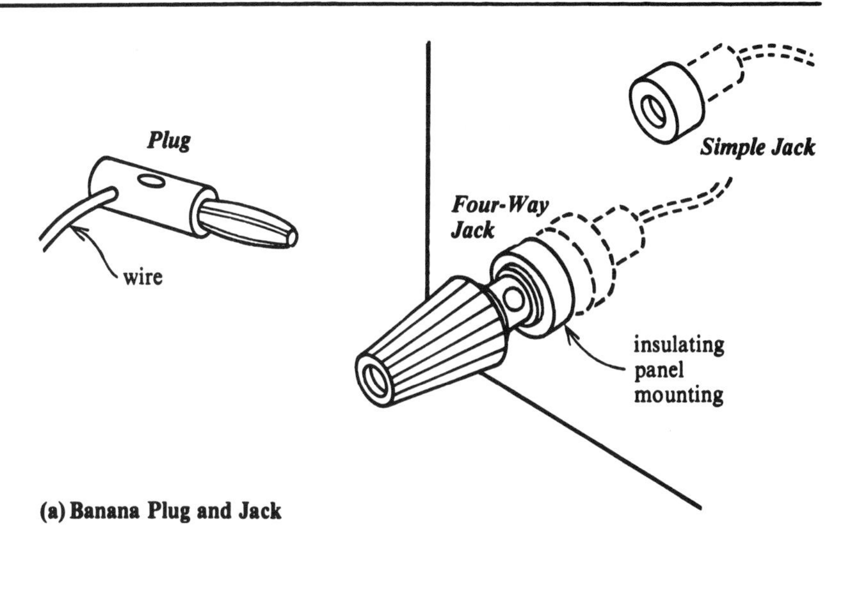

(a) Banana Plug and Jack

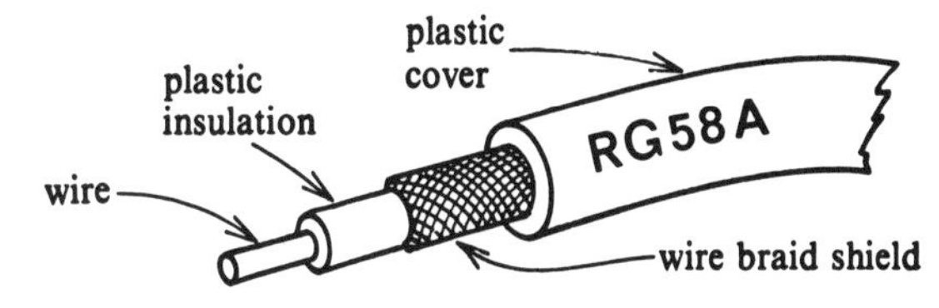

(b) Coaxial Cable

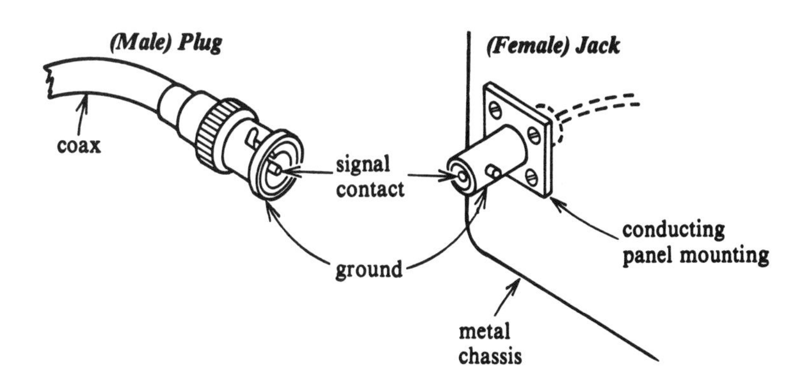

(c) BNC Plug and Jack

FIGURE A2–14. *Two Popular Quick-Connect Cables*

coax shield and mates to the metal shell of the BNC panel jack. Since the braid is to be grounded, the BNC jack is normally mounted on the metal chassis surface. In turn, the chassis is connected to earth or ground by electrical connection to the ground pin of the 3-pin ac house plug for the instrument. Thus the chassis and coax shield are both automatically grounded and provide electrostatic shielding.

This important practical fact means that one of the two signal leads for the many electronic instruments using BNC connectors (e.g., oscilloscopes, frequency counters, radio frequency generators) is automatically at ground. For example, it is *not* possible to directly connect the typical modern oscilloscope to measure the voltage that appears between two points in a circuit, if *both* points are at a voltage different from ground (we often say "are off ground"). Thus, the ground lead of the scope coax must be connected to the ground of the electronic circuit to be tested, and the signal wire from the scope coax connected to a desired off-ground point (electrically *hot*) of the circuit. If the reverse is done, the electronic circuit would be shorted to the ground, and the possible large current flow can damage components. (The author remembers well the smoke that resulted when a fellow *graduate* student in physics damaged a high quality VTVM through this error!)

At radio frequencies (100 kilohertz and above), the use of coaxial interconnecting cables become quite necessary for distances of more than a few centimeters. This is done because of electrical standing waves that can be set up on the conecting wires due to electrical wave reflections. These reflections can be eliminated by assuring that each end of the coaxial *electrical transmission line* is connected to an effective resistance equal to the *characteristic impedance* Z_0 of the cable. The characteristic impedance of a coaxial cable is determined by its radial dimensions and the dielectric constant of the interior insulation. For example standard RG58/A coax has a characteristic impedance of 50 ohms, while RG59/A coax has a 75-ohms characteristic impedance. Thus, a radio frequency oscillator connected to RG58/A coax should have an output impedance of 50 ohms.

This book will not generally be concerned with signals above 1 megahertz, and therefore these features of characteristic impedance are not of great concern. Coaxial cables do not need to be properly terminated with characteristic impedances for dc or audio frequency signals. However, it is true that the capacitance of the cable itself can cause loading problems if the cable is rather long. For example, the capacitance between the central wire and the shield is about 90 picofarad per meter for RG58/A coax. A 6-meter cable of RG58/A coax thus has a capacitance of about 600 picofarad between the central wire and shield. At 100 kilohertz, this presents an impedance of only 2600 ohms, which is a shunt between the signal wire and ground. The signal reduction due to loading when connected to a 100 kilohertz signal generator with 600 ohms output impedance, for example, becomes very significant.

To reduce the problems of capacitive loading by the connecting cable and by the scope itself, *attenuating input probes* are often used to connect oscilloscopes (also frequency counters) to high frequency circuits. Thus a typical *10-to-1 probe* uses an internal voltage divider to reduce the voltage by a factor of 10 before presenting the signal to the actual vertical input jack of the scope. The volts/centimeter setting of the vertical deflection must be multiplied by 10 for proper interpretation of the display. Sometimes this voltage reduction is necessary in order to accommodate high voltage signals to the scope. However, the real purpose for these probes is that they present a low capacitive shunt load—only about 13 picofarad—to the circuit to which the probe is connected. However, to faithfully transmit square waves (and also radio frequencies) to the scope input, a small *compensation capacitor* within the voltage divider of the probe itself must be properly adjusted. This is usually done by adjustment of the probe capacitor while observing the scope presentation as the probe is connected to a square wave source known to be "very square."

REVIEW EXERCISES

1. (a) An ammeter A and voltmeter V are connected in series to a battery as shown in Diagram A2–1a. Are either likely to be damaged? Why?
   (b) The ammeter and voltmeter are shown connected in parallel to a battery in Diagram A2–1b. Are either likely to be damaged? Why?

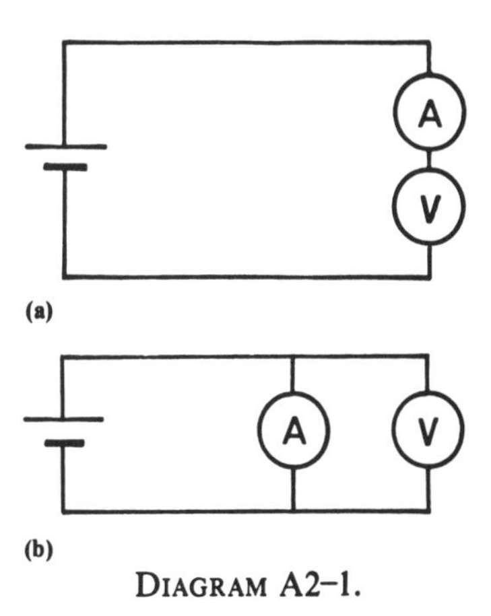

DIAGRAM A2–1.

2. (a) What do VOM, VTVM, FTVM, and DVM stand for?
   (b) Which is likely to have the lowest resistance when used as a voltmeter?
   (c) Which type can usually measure current?
   (d) What is the typical input resistance of a VTVM or DVM?
   (e) Although many VTVMs can be operated from "the mains"—that is, a wall socket—they also often contain a battery. What is the purpose of the battery?

3. You have just constructed a light-duty extension cord by connecting a 2-contact plug and receptacle to the opposite ends of a 2-conductor lamp cord.
   (a) Diagram three connections that you could make with a VOM to test the integrity of your product *before* connecting the cord to the mains.
   (b) What function of the VOM would you use?
   (c) Describe what possible problem is being eliminated in each case.
4. (a) Suppose you are considering the purchase of a VOM. You find a 1000 Ω/V meter and a 50,000 Ω/V meter with similar features, accuracy, and price. Which would you probably choose and why?
   (b) Suppose the voltmeter were to be used primarily to test batteries. Why might you consider the other choice from the one made in part (a)?
5. It is quite common to make an in-circuit test of resistance between two points in an electronic circuit, using an ohmmeter for the measurement.
   (a) Why must all voltage sources be turned off or removed from the circuit before making the test?
   (b) Suppose the ohmmeter is connected across some resistor to test it. Why is it likely that the ohmmeter's reading is less than the value of the resistor?
6. What is meant by a balanced bridge or a balanced potentiometer circuit?
7. (a) What electrical standard is required for the adjustment of a precision potentiometer?
   (b) Why is this standard not required for a precision Wheatstone bridge?
8. (a) Diagram A2–2 illustrates the ac/dc switch on the vertical input of an oscilloscope. What is the switch position for the ac setting?
   (b) Why is the ac setting sometimes desirable?
   (c) Why is the dc setting sometimes desirable?

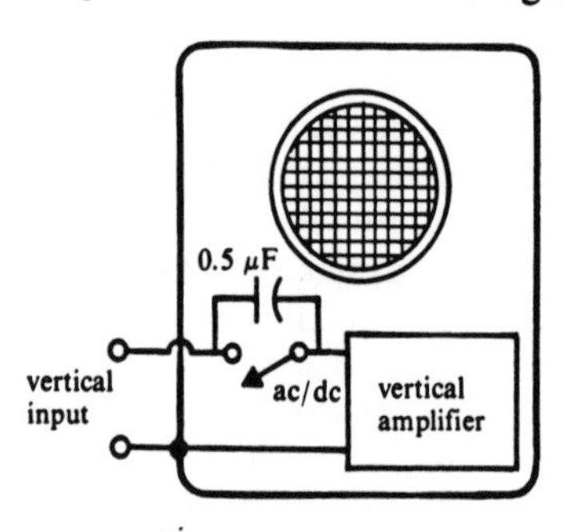

DIAGRAM A2–2.

9. What is the utility of the auto setting of the TRIGGER MODE switch of an oscilloscope? Contrast with the behavior of the scope when the triggering mode is ac or dc. What is a disadvantage of using auto mode?
10. Which one of the following is *not* true about a triggered-sweep laboratory-quality oscilloscope?
    (a) The peak-to-peak voltage of an ac waveform may be measured to within a few percent.
    (b) The voltage of a battery may be measured to within a few percent.
    (c) When the vertical input is set to dc mode, the voltage of an ac waveform cannot be measured.
    (d) When the vertical input is set to ac mode, the voltage of a battery cannot be measured.

11. Explain the difference between the external-trigger input and the horizontal-sweep input of an oscilloscope.
12. The level-control knob has a (+) and (−) associated with it. The slope switch also has (+) and (−) positions. What do (+) and (−) refer to in each case? A diagram of some waveform should be used in connection with the explanation.
13. When an oscilloscope is operated in *X-Y* mode, the scope's triggering circuitry is not required. Explain why.
14. (a) What is meant by a dual-trace oscilloscope?
    (b) Why is it evident from the front panel of Figure A2–7 that the oscilloscope pictured is not a dual-trace type?
15. (a) What is a servomechanism?
    (b) What is a stepper motor?
    (c) Most strip-chart recorders are effectively self-balancing potentiometers. Briefly explain this statement.
16. (a) The term *X-Y* occurs in both the *X-Y* oscilloscope and *X-Y* recorder. What do these two devices have in common?
    (b) What is the general nature of application of the *X-Y* feature?
    (c) Under what circumstances should one device be selected over the other in part (a)?
17. Give a relative advantage and disadvantage for each in regard to the use of the strip-chart recorder and the magnetic recorder in recording applications.
18. (a) Compare the upper frequency limits of an entertainment cassette recorder and an instrumentation recorder.
    (b) Describe the technique used by an instrumentation recorder to record slowly varying dc signals.
19. What is a principal difference between a waveform generator and the traditional audio oscillator?
20. A typical pulse generator has at least 3 control knobs. Give a possible control panel name for each of the 3.
21. What is meant by the output impedance of a signal generator?
22. Suppose you have at hand a resistance box (a selectable resistance with many known values) and a voltmeter. Describe a simple method for determining the output impedance of an audio generator that is presented to you.

PROBLEMS

1. A d'Arsonval movement with a resistance of 500 Ω and a full-scale sensitivity of 200 μA is used to construct an ammeter with 0.1-ampere full-scale capability.
    (a) Diagram the construction of the completed meter, including resistance values.
    (b) What is the voltage drop across the meter when placed in a circuit and reading full scale?

2. The d'Arsonval meter of the preceding problem is used to construct a 15-V full-scale voltmeter. Diagram the construction of the completed meter, including resistance values.
3. Suppose the needle position of a d'Arsonval meter is to be read to an accuracy of ±0.5% of full scale. If the full-scale deflection is 110°, what is the angular accuracy required (±Δ)? Draw two lines 4 cm long with the angle 2 Δ between them.
4. What will be the reading of the voltmeter in Diagram A2–3? It is a 100 Ω/V meter set to the 15-V scale.

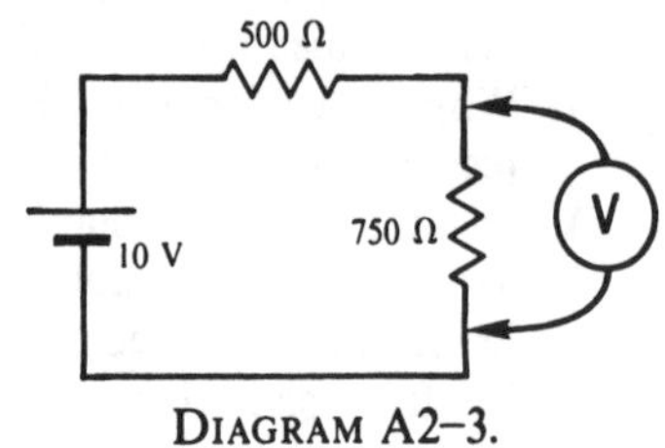

DIAGRAM A2–3.

5. Assume that the circuit shown in Diagram A2–4 is constructed to produce an ohmmeter. The galvanometer resistance is negligible.
   (a) Zero resistance is printed (as usual) at the full-scale position for the needle of the meter. What resistance *R* is required?
   (b) What value of resistance should be printed at the middle of the scale?

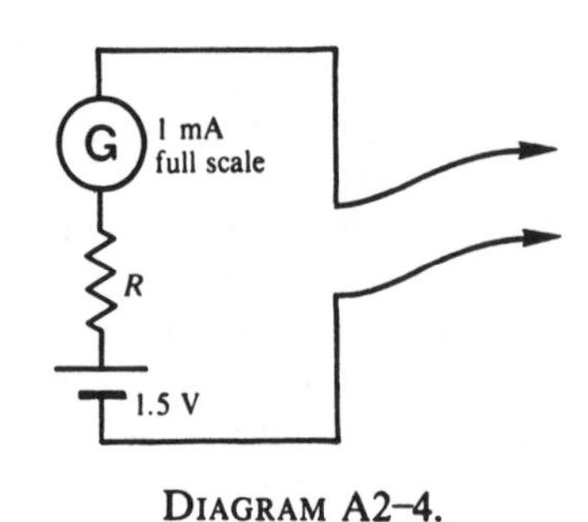

DIAGRAM A2–4.

6. (a) Suppose an ohmmeter is connected to a good capacitor (initially discharged) with a capacitance of some microfarads. What should be the behavior of the ohmmeter reading?
   (b) It is not unusual to test an electrolytic capacitor for a short by using an ohmmeter. However, what precautions should be considered? (Certainly one precaution, and possibly a second.)
7. Justify the statement made in the chapter: If 1-milliampere current is established through an unknown resistance, a voltmeter connected across the resistance shows a reading (in V) equal to the resistance in kilohms.
8. Consider the Wheatstone bridge pictured in Diagram A2–5. What is the reading of the DVM connected? Does the DVM indicate (+) or (−) voltage?

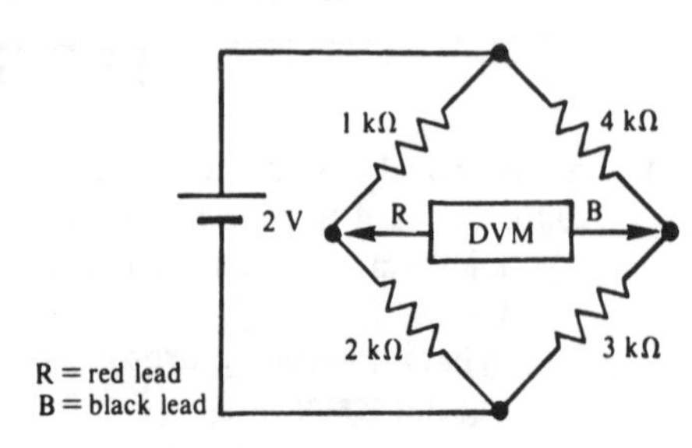

DIAGRAM A2–5.

9. Suppose the known resistances in a balanced Wheatstone bridge are each accurate to ±1.0%. What is the precision of determination of the unknown resistance?

10. (a) Show that the bridge pictured in Diagram A2–6 is in balance. K is an arbitrary constant.
    (b) Find the Thevenin equivalent circuit that presents itself to the detector terminals. (This is important in considering the impedance of the detector to use.)

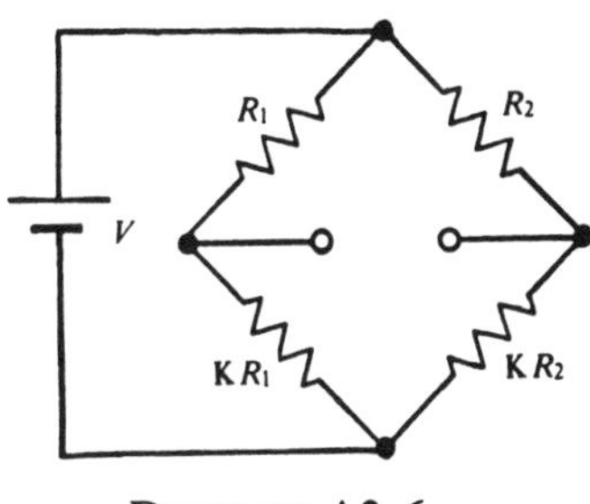

DIAGRAM A2–6.

11. Justify that if a Wheatstone bridge is in balance and R_2 and R_x (in Figure A2–4) both change with temperature in the same way, the bridge will remain in balance. You may assume that $R_x = R_x^\circ(1 + \alpha T)$ and $R_2 = R_2^\circ(1 + \alpha T)$. Here T is the temperature in Celsius, α is the temperature coefficient of resistance at 0°C, and R_x° and R_2° are the respective resistances at 0°C.

12. (a) Consider a Wheatstone bridge as diagrammed in Diagram A2–7a that is used in the out-of-balance mode to monitor the change in resistance R_x. Derive an expression for V_{out} in terms of V, R, and ΔR. For small ΔR, show V_{out} is equal to $(V/4)(\Delta R/R)$.
    (b) Suppose it can be arranged that while one transducer increases its resistance by ΔR a second transducer decreases its resistance by ΔR (in response to the given external stimulus). These may be placed in a bridge as shown in Diagram A2–7b; the configuration that results is called a *half-active bridge*. Show that V_{out} (the output signal) is twice that obtained in part (a).

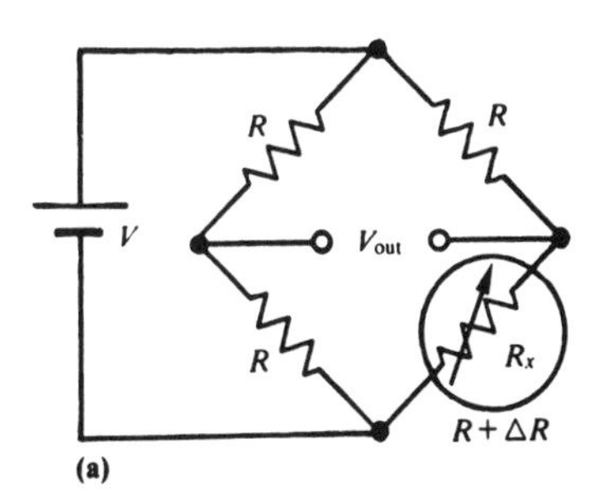

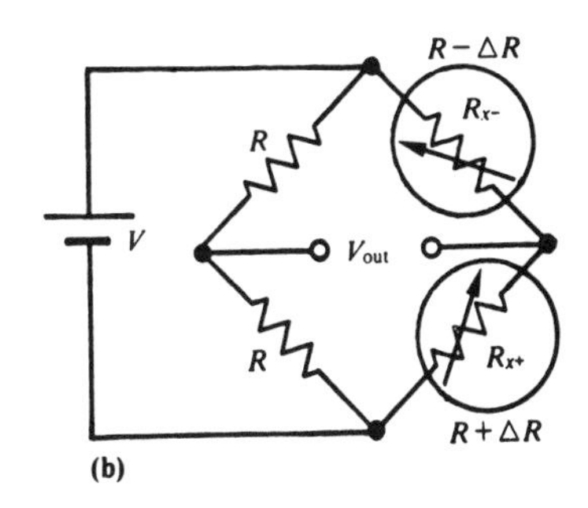

DIAGRAM A2–7.

13. Many Wheatstone bridge applications are used in the out-of-balance mode. But if a balance is not required, why not save the expense of two resistors and simply put the resistive transducer R_x in a voltage-divider connection as shown in Diagram A2–8? Criticize this "money-saving" approach.

    NOTE: Typically, the change in transducer resistance is a small fraction of its resistance.

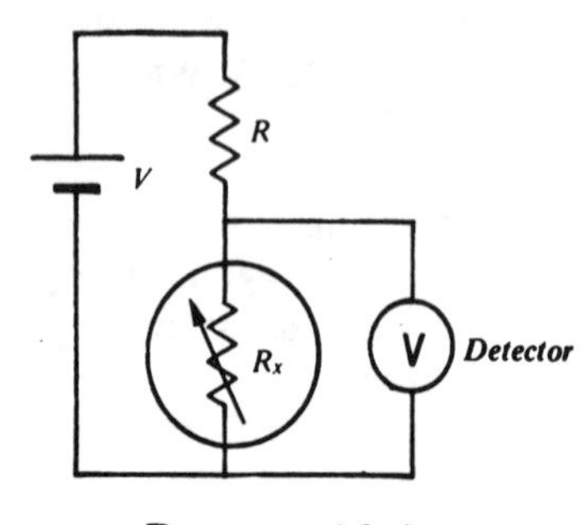

DIAGRAM A2–8.

14. A typical oscilloscope display is 6 cm high by 10 cm wide. The vertical control knob is set to 0.5 V/cm, and the horizontal control knob is set to 0.5 msec/cm. A 1-V_{rms} sinusoidal signal at 1-kHz frequency is input to the scope. Sketch with some accuracy the appearance of the scope display.

15. Suppose the scope triggering level is set for the voltage waveform as shown in Diagram A2–9. Justify that a double exposure will appear on the scope screen. Sketch what the double exposure looks like on the scope.

    NOTE: The slope switch is set to (+).

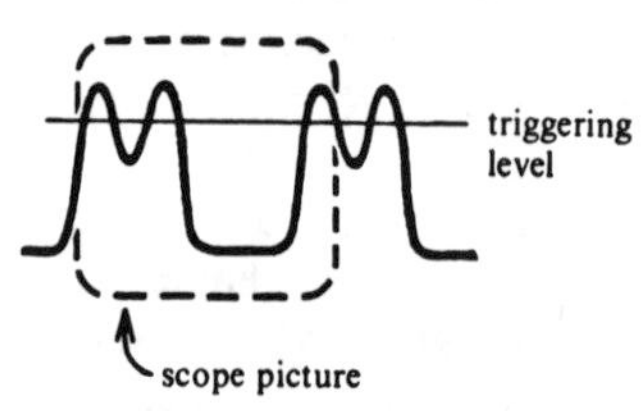

DIAGRAM A2–9.

16. A widely used audio oscillator has an output impedance of 600 Ω. Suppose the oscillator is adjusted to 1 kHz, and a DVM connected to its output reads 5 V_{rms}. Then a 0.33-μF capacitor is connected to the oscillator output. What voltage does the DVM now read?

17. A simple attenuator is constructed from 1% 1000-Ω and 100-Ω resistors. These are placed in an aluminum minibox, as shown in Diagram A2–10, for convenience in connection to cables in the lab.

    (a) Assuming an open-circuit output, the output voltage will be what fraction of the input voltage (to within 2%)?

    (b) Suppose the attenuator is connected to the output of an audio oscillator that has an output impedance of

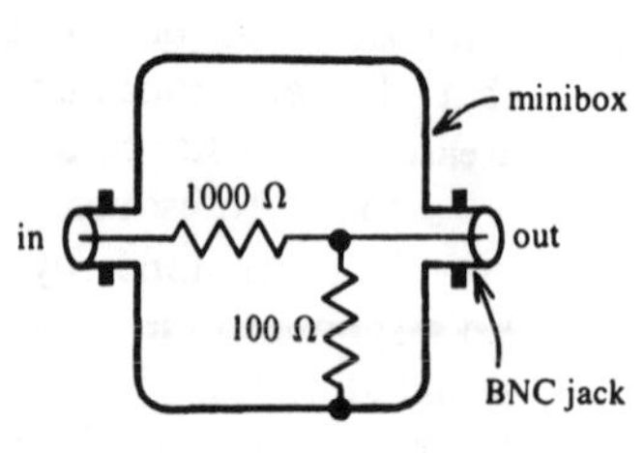

DIAGRAM A2–10.

600 Ω. The open-circuit voltage of the oscillator was 4.0 V p-p. What is the open-circuit voltage out of the attenuator?

(c) What will be the output impedance of the overall oscillator plus attenuator? Diagram the problem to help justify your work.

18. Suppose the voltage across a certain resistance changes from 4.5 V to 7 V, what is the power change in dB?
19. Suppose the power into a loudspeaker changes by −12 dB.
    (a) What is the (fractional) ratio of the power after to the power before?
    (b) What is the ratio of the voltage after to voltage before?
20. VOMs and VTVMs often have a *dBm scale* that may be used in certain situations; they actually monitor the voltage, however. A reference level of 1 mW (10^{-3} W) developed in 600 Ω is assumed, and this situation is called *0 dBm* on the meter scale.
    (a) What is the voltage across a 600-Ω resistor if a VTVM connected to it reads 0 dBm? (This voltage in fact appears across from 0 dBm on the associated voltage scale.)
    (b) Suppose the voltage across the 600-Ω resistor is 3.0 V. What is the dBm reading?
21. Consider a filter (either high-pass or low-pass). Show that at the breakpoint frequency (half-power frequency), the output voltage is down 3 dB from the unattenuated value. (Thus this frequency is also often referred to as the *−3 dB point.*)
22. The attenuation of the common commercial attenuation box is set by switches. Let 3 switches be marked F_1, F_2, and F_3, respectively. If each attenuation factor F_i is expressed as a simple ratio, the total attenuation is the product $F_1 F_2 F_3$ when all 3 switches are set. (It is just $F_i F_j$ if two switches i and j are set.) However, if the attenuation factor for each switch is expressed in dB, show that the total attenuation is $F_1 + F_2 + F_3$. Thus, the simple rule is to add together the dB values for all set switches. This is an advantage of the dB scheme.
23. (a) Consider an RC filter (or an LR filter) that is well beyond the breakpoint frequency and into the attenuation region in terms of frequency. Show that V_{out}/V_{in} changes by a factor of 10 for every factor-of-10 change in frequency. Account for the statement that the filter rolloff is −20 dB/decade.
    (b) An octave (for example, in music) is a factor-of-2 change in frequency. Account for the statement that a simple RC filter may also be called a −6 dB/octave attenuator when well beyond the breakpoint frequency.
24. Consider the second-order filter of Problem 37 in Chapter A1. Show that the attenuation region may be characterized as −40 dB/decade.
25. Consider the T attenuator shown in Diagram A2–11. Assume a voltage source characterized by V and R is connected to one set of attenuator terminals, and a load resistance also with resistance R is connected to the other set of attenuator terminals as shown in the diagram. In terms of this resistance R, the attenuator has been designed with R_1 and R_2 as defined in the diagram and assuming some fraction B.

(a) Show that $V_{out}/V_{in} = \beta$. β is thus the attenuation factor for the *terminal* voltages.

(b) Show that $V_{out}/V = \beta/2$, however. (Due to loading, V_{in} is just half of the open-circuit voltage V of the voltage source.)

(c) Show that the impedance seen between the input terminals of the attenuator is just R. A load R should be assumed connected at the attenuator output.

(d) Show that the output impedance of the network (voltage source plus attenuator) to which the load is attached is also R.

(e) Illustrate how an attenuator that reduces its input terminal voltage by β^n (n is any integer) could readily be constructed and that it has the input and output properties of parts (c) and (d).

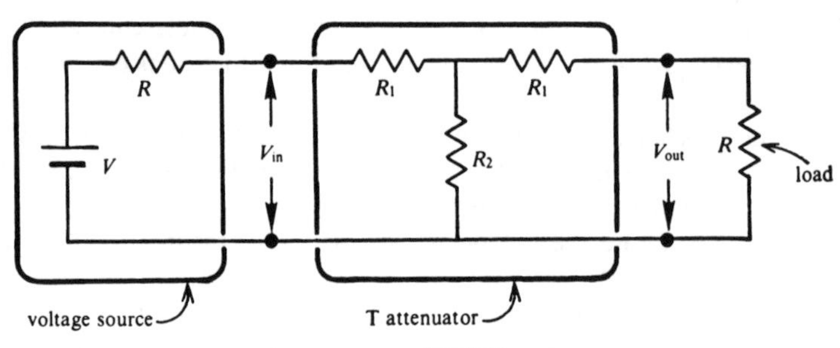

$$0 \leq \beta \leq 1$$

$$R_1 = \frac{R(1-\beta)}{(1+\beta)}$$

$$R_2 = \frac{R2\beta}{(1-\beta^2)}$$

DIAGRAM A2–11.

CHAPTER

A3

Transducers

The term *transducer* has become quite fashionable since the mid-1960s as a general term for a host of devices that link the physical world to the world of electronics. Of course, it is the possibility of making this interconnection that permits electronics to become so pervasive in science, technology, and even the household.

Various physical effects in themselves are often not electrical. For example, temperature may be monitored by a human observing the length of the mercury column in a conventional thermometer. But if temperature is converted into some proportional electrical effect, the power and versatility of electronic circuitry may be used to condition the corresponding *electrical signal* in a variety of ways. We may achieve greater sensitivity, or automatic recording over extended periods, or even analysis of information. If a computer is to be used for analysis of the measurements, it is often desirable for the computer to make the measurements automatically rather than to require a human intermediary. But this procedure requires conversion of the physical effect such as temperature to the electrical domain used internally by the computer.

For these reasons, the application of electronics to a problem may begin with a transducer, and the success of the application may depend on the selection of an appropriate transducer. In this chapter, we consider the general possibilities and limitations of a modest selection of commonly used transducers. These basically respond to one of the following effects: temperature, light intensity, strain in an object, movement of an object, position of an object, and ion concentration. However, it is useful to notice that these effects may often be used to monitor other phenomena; therefore some examples are briefly pointed out in the chapter.

One scheme for classifying transducers is through the electrical effect that is output by the device. Thus, the physical effect may produce a change in some resistance, capacitance, or inductance; or it may actually generate a voltage (emf). Since most transducers discussed in this chapter are of the resistive or the voltage type, two principal chapter sections are devoted to them. However, it is helpful to first put the general elements of an electronic system in perspective. This perspective is given in the first section of the chapter.

General View of Electric Systems

It is useful to our understanding of laboratory electronics to organize the typical electronic instrument or application into a few general blocks. This arrangement is illustrated in Figure A3–1. As we have mentioned, there is first some parameter that we want to study or monitor in some physical system. Because typically the parameter is not in itself electrical, a transducer is employed to generate an electrical effect that is related to the particular parameter. If the electrical effect is not already voltage, such a transformation is usually made, and this voltage is then sent to an *amplifier* that increases the typically small and feeble voltage level to some robust value. Finally, the latter voltage is put to an *output device* such as a meter, oscilloscope, stripchart recorder, loudspeaker, fluid valve, motor, and so on. At times the desired end result may be an effect back on the original system; this is typical of experiment control or industrial process-control situations.

We speak of input transducers and output transducers. An *input transducer* causes an electrical effect in consequence to some physical parameter at its input. The transducer block pictured in Figure A3–1 is actually an input transducer. Generally speaking an input transducer converts or *transduces* energy from some form in the physical world into electrical energy. Conversely, an *output transducer* converts electrical energy into some other form

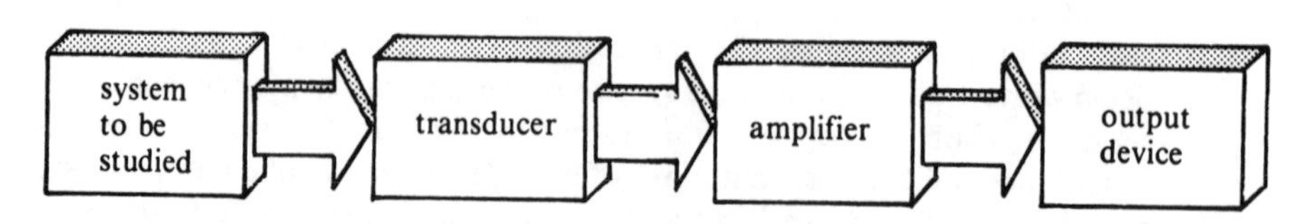

FIGURE A3–1. *Typical Electronic Application in Science*

of energy and causes a desired physical effect in the process. The output device block in Figure A3–1 is actually an output transducer. Only a few output transducers will be discussed in this chapter.

A simple example that uses both transducer types is a public-address (PA) system. The input transducer is the microphone into which someone speaks. The *air vibrations*—that is, sound energy—are converted to a small voltage by either a crystal or a dynamic microphone. The PA amplifier increases the voltage and power level to a magnitude (or degree) that is sufficient to operate or *drive* the output transducer, which is the loudspeaker system. In the output transducer, electrical energy is converted back into sound energy again--hopefully at a level heard throughout the auditorium. Of course, the additional power in the sound amplified must come from somewhere. It comes from the *power supply* or voltage source for the amplifier. This source could be a battery or, more likely, a nearby electrical power generating station connected through the 115-V line cord.

Resistive Input Transducers

Photoresistor

The *photoresistor* is a resistor made of a semiconducting material that changes its resistance in response to the intensity of *light* falling on it. Ordinarily, a semiconductor does not conduct electricity well because of the relatively small number of charge carriers that are free to move. However, when photons of light strike a semiconductor, electrons are kicked away from host atoms into a mobile condition and the conductivity of the material increases. Conversely, resistivity decreases.

The resistance R depends on the intensity I_* of the light according to

$$R = R_0 I_*^{-K} \tag{A3–1}$$

where R_0 is a multiplying constant and K is a constant in the range of 0.5 to 1.0. Thus the dependence is not linear but follows a power law. The change in resistance does *not* occur instantly with change in light level; delays of microseconds to milliseconds are typical, depending on the particular photoresistor type. Figure A3–2 shows a typical photoresistor.

The resistance of a typical photoresistor may vary from 50 megohms in the dark to 9 kilohms in an illumination of 2 footcandles or 20 lux.

NOTE: 1 footcandle = 1 lumen/foot = 10.7 lumens/meter^2; 1 lumen/meter^2 = 1 lux.

For comparison, the proper *illuminance* in a building hallway is about 30 lux; in a classroom, about 400 lux; direct sunlight is about 10^5 lux.

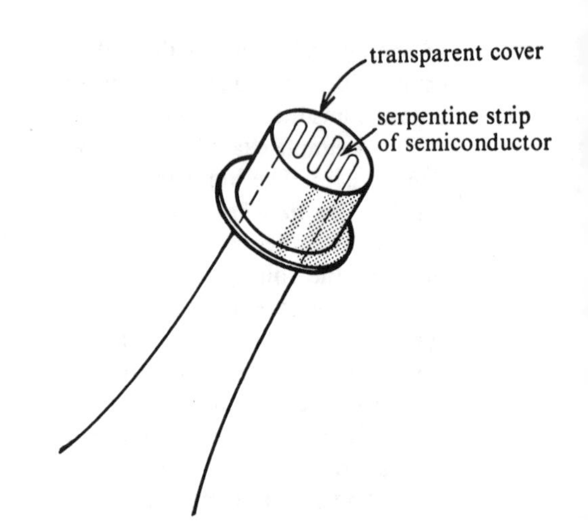

FIGURE A3–2. *Typical Photoresistor (Twice Normal Size)*

Photoresistors can be used in a number of ways, and a few applications are sketched here to give an idea of the possibilities. Some of the applications are rather novel; notice that light itself is often not the physical quantity of basic interest.

1. Photoresistors may obviously be used to monitor light intensity when this quantity is of direct interest. Thus, a typical contemporary light-meter for photographic applications is essentially a photoresistor connected to an ohmmeter. The ohmmeter scale is calibrated in appropriate photographic units.

2. The passage of an object can be detected. The passing object may block a light beam that normally illuminates a photoresistor; this event can be detected by monitoring the voltage variation across the photoresistor while it is connected in a voltage divider circuit, and the voltage signal may be passed to an electronic counting circuit. Thus, we may monitor the number of persons passing through a door, or the number of items passing on a conveyor belt, or the number of times an animal is in part of an enclosure.

3. Rather tiny movements of objects can be monitored. A movable opaque vane should only partially block a light beam that falls on the sensitive area of a photoresistor. Suppose that the vane is connected to a frog muscle; tiny movements of the stimulated frog muscle can be detected by the variation in resistance of the photoresistor. Similarly, finger tremor can be measured by using a person's finger as the vane.

4. The heart rate of a person can be monitored. A small light source and a photoresistor can be placed on opposite sides of the earlobe. At each heart beat, blood pumps into the capillaries and momentarily renders the flesh slightly more opaque. The light reaching the photoresistor then changes, and the change in resistance of the photoresistor may be detected and monitored by an electronic circuit. In fact, an electronic system for this application is described as an example system in Chapter A7.

5. The level of translucent fluid in a container can be monitored. A light beam can be arranged to shine from the top toward a photoresistor installed at the bottom of the container. As the fluid thickness increases, less light reaches the photoresistor. The fluid level could be indicated on the scale of an ohmmeter, or more elaborate automatic electronic recording circuitry could be used.

Thermistor

Like the photoresistor, the *thermistor* is a resistor made of a semiconducting material; but it makes use of the fact that *increasing temperature* rapidly decreases the resistivity of semiconductors. Increasing temperature generates more carriers by thermally shaking electrons loose from host atoms and consequently decreases the resistivity of the material. The resistance depends approximately exponentially on the temperature:

$$R = R_0 e^{+E/kT} \tag{A3-2}$$

where R is the resistance of the thermistor (at temperature T), R_0 and E are constants for the given unit, k is Boltzmann's constant (1.38×10^{-23} joules/molecule °K) and T is the absolute temperature. In one sense it is inconvenient to make a thermometer from a thermistor, because it is not linear. On the other hand, the exponential behavior implies a rapid change with temperature and therefore a sensitive element. At room temperature the resistance of a thermistor may decrease by some 5% for each degree Celsius rise in temperature.

Standard thermistors may be used over the temperature range −100°C to +300°C and have a resistance specification of ±10% at 25°C. However, selected (more expensive) types are interchangeable and accurate to ±0.2°C over a 100°C range. Note that very high *relative* accuracy is possible with even standard units; 0.0005°C changes are quite readily detectable with appropriate circuits.

Table A3–1 shows sample resistance-temperature data for a typical thermistor, the Fenwall GB32J2 thermistor. One may typically purchase a thermistor that has a certain resistance between 10 ohms and 10^5 ohms at 25°C. The thermistor will then have a resistance at other temperatures in nominal proportion as given in Table A3–1.

TABLE A3–1. *Resistance Vs. Temperature for a Typical Thermistor*

Temperature (°C)	0	25	50	100	150	200
Resistance (ohms)	5700	2000	810	185	59	25

Example: A thermistor is specified to have a resistance of 10,000 ohms at 25° C. What is expected to be its resistance $R(0)$ at 0° C? Assume the thermistor material is the same as that for the thermistor of Table A3–1.

Solution: By ratios from Table A3–1 we have

$$\frac{R(0)}{10{,}000\ \Omega} = \frac{5700\ \Omega}{2000\ \Omega} \quad \text{or} \quad R(0) = 28{,}500\ \Omega$$

Thermistors may be used in simple ohmmeter circuits or in a Wheatstone bridge circuit. The dial of the ohmmeter type of circuit can be calibrated to indicate the temperature directly. The bridge circuit is often used to detect small temperature variations. The bridge output may be amplified before being sent to a meter, and sensitive thermometers may be constructed in this way.

Thermistors come in various shapes and sizes to accommodate different experimental situations. Some are much smaller than a pinhead. Figure A3–3 illustrates some typical forms.

NOTE: There is one precaution in using thermistors. If a significant current is passed through a thermistor in the measuring process, it will *self-heat* from I^2R and indicate a temperature higher than the actual temperature.

The self-heating effect can, however, be put to use in a number of ways. Two examples are given here.

1. A self-heating thermistor can be used as the basis of a simple but effective *liquid-level sensor.* If a fluid is to be pumped into a tank, the pump should be stopped when the fluid reaches a certain level. Therefore, a yes/no (or binary) liquid-level sensor is required. We can exploit the fact that a self-

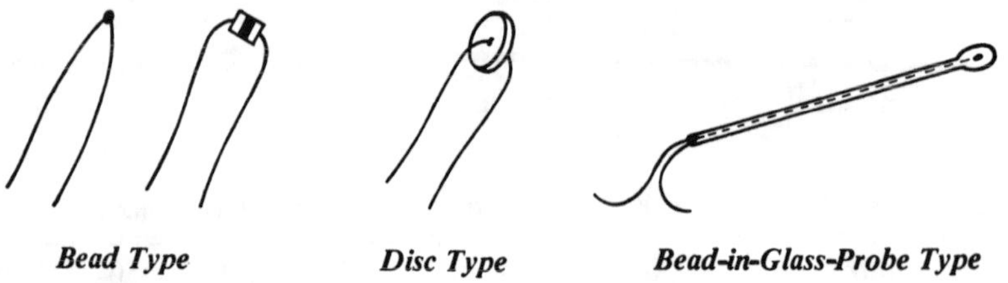

Bead Type *Disc Type* *Bead-in-Glass-Probe Type*

FIGURE A3–3. *Typical Thermistors*

heating thermistor is cooled by conduction and convection more effectively when immersed in liquid than in air. Consequently, when the liquid level reaches a thermistor suspended in the tank, its resistance increases significantly. The thermistor can be connected in a voltage divider circuit, and the change in voltage across the thermistor can be used to distinguish whether the liquid is above or below the particular level.

2. A self-heating thermistor can be the basis for a rugged *vacuum gauge* that measures intermediate vacuum pressures between about 10^{-4} and 1 torr. If the thermistor is placed in the vacuum to be measured, it will be cooled more effectively at higher pressure than at lower pressure. The corresponding change in its resistance can be used as a measure of the vacuum. Of course, it must be calibrated against an absolute vacuum measurement instrument such as the McLeod gauge. The Pirani gauge is a similar vacuum gauge technique that has been used for many years; however, the conventional Pirani gauge uses the self-heating of a *metal* resistor.

Platinum Resistance Thermometer

The resistivity of any metal increases in an approximately linear fashion from the small value it has when near absolute zero temperature. Therefore the resistance of a metal wire can be used as a temperature indicator; however, near 0°C the resistance can be expected to change by a fraction only on the order of 1/273 or about 0.4% per degree Celsius. For reasons of linearity and stability, platinum has become the metal of choice and in fact the *platinum resistance thermometer* is the world standard for temperature measurement from −270°C to +660°C. High-purity platinum wire with a resistance of the order of 100 ohms is typically wound on a ceramic core in a manner to relieve wire stresses. The size of the resulting sensing element can be expected to be larger than that for a thermistor, and instrumentation using the platinum resistance thermometer is generally more elaborate than for a thermistor. However, if performance approaching that of laboratory standards is required, this approach may be necessary.

Strain Gauge

If a metal or semiconductor wire is stretched, its resistance increases because of its longer length and smaller cross-section. Provided the stretch is not too great, the elasticity of the material causes it to return to the original shape and resistance if the force of stress is removed. The amount of stress (expressed as a fraction of its length) that a rod undergoes is called the *strain,* and fine wire used expressly to measure or "gauge" small amounts of elongation is called a *strain gauge.* One scheme is to bond fine wire to an elastic insulating sheet, as shown in Figure A3–4. The resulting strain gauge is

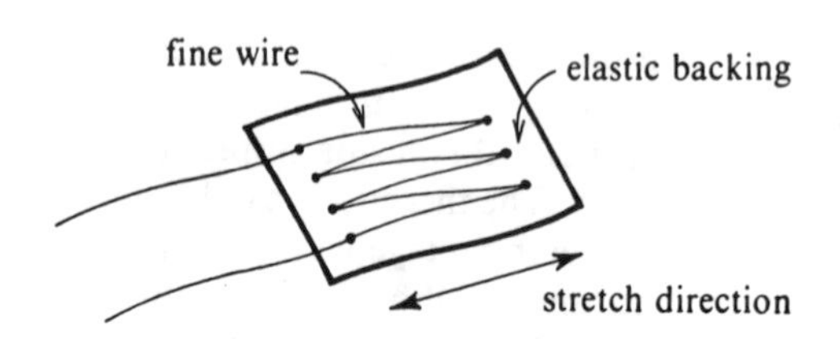

FIGURE A3–4. *One Form of Strain Gauge*

bonded to the object to be studied. As the portion under the gauge strains slightly, the wire in the strain gauge must stretch or compress by an equal strain. The resistance change of the gauge is then a measure of the strain in the object.

Unfortunately, the resistance change with strain is small and the resistance change with temperature can readily be large. A desirable approach to minimizing the temperature problem is to use two strain gauges—one *active* and one a *dummy*—in a Wheatstone bridge. The dummy strain gauge is placed as resistance R_2 in Figure A2–4 and is arranged to experience the same temperature as the active strain sensor, but not the strain. Then, as discussed in Chapter A2, the output of the bridge should change very little with change in temperature of the sensors.

Using strain gauges constructed from metal wire, we may expect bridge outputs on the order of 0.01 V for strains of 10^{-3}; that is, 10^{-3} centimeters stretch per centimeter of length. Amplification of this small voltage is typically required and quite possible, so that rather small displacements are accurately measurable. *Load cells* are available that use strain gauges in order to obtain transducers for force and weighing applications. For example, the weight of a container as it is being filled during a manufacturing process may be monitored electronically with the aid of a load cell. A second strain gauge application is used in electronic measurement of fluid pressure. The pressure transducer incorporates a thin diaphragm with a strain gauge mounted on it. The diaphragm bulges slightly under the pressure of the fluid, and the strain gauge resistance changes accordingly.

Conductivity Cell

Pure water has a very high resistivity and, conversely, low conductivity (because H_2O dissociates rather little into H^+ and OH^- ions). Ions in solution increase the conductivity rapidly and ion concentration can be obtained by measuring the resistance between a pair of conductors immersed in the solution. To be quantitative, a *conductivity cell* is constructed with plates and dimensions such that the resistance measured is related to the resistivity of the liquid (in ohms × meters) by a simple factor like 0.1, 1 or 10. Also, to prevent the ions coming out of solution at the respective electrodes, an ac current should be used in the measuring process. Figure A3–5 illustrates a typical

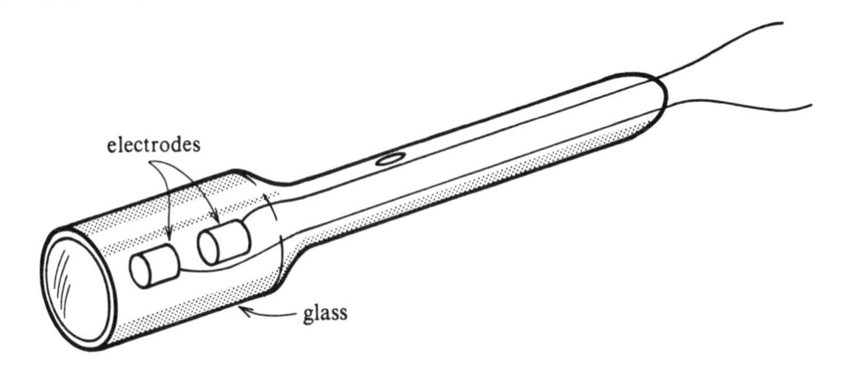

FIGURE A3–5. *Typical Conductivity Cell (Immersion Type)*

conductivity cell that is intended to be dipped into a liquid in order to measure its resistivity.

Voltage Input Transducers

Thermocouple

The *thermocouple* is a junction between two different metals, and represents a temperature-to-voltage transducer. For measurement purposes, the two metals are usually in the form of two fine wires that are welded or soldered together at one end to form a junction. The junction is effectively the sensor. The thermocouple combines the virtues of good accuracy, stability, low cost, broad temperature range, and small size; consequently, it has been widely used in science and industry for temperature measurement.

What may be called the *thermocouple effect* is actually the combination of two effects:

1. An emf or voltage develops between the ends of any wire that has its two ends at different temperatures. The voltage produced increases with increasing temperature difference between the two ends and is different for different metals.
2. An emf develops across the junction between two different metals. This emf depends on the metals used and on the temperature of the junction.

If a sensitive voltmeter is connected between the "free ends" of the thermocouple wire pair, it senses a small voltage that depends on the temperature of the thermocouple junction and also on the temperatures of the two junctions that are now formed between the voltmeter and the two thermocouple wires. To permit convenient use of the thermocouple, the two "free end" junctions should be kept at the same stable temperature. They are usually immersed in an ice water mixture, which becomes the *reference temperature.* The junctions immersed in the bath are called the *reference junctions.* See Figure A3–6 for a diagram of the resulting system.

The emf around the thermocouple circuit is approximately linearly proportional to the difference between the reference junction temperature and the temperature of the monitoring junction. Unfortunately, thermocouple effects can occur anywhere along the circuit path, including circuit connections within the sensitive voltmeter. To reduce these problems, copper wire should be used throughout the measuring circuitry, and temperature gradients should be avoided within this circuitry.

The thermocouple voltage developed is quite small. For example, the commonly used *copper-constantan* junction develops about 40 microvolts per Celsius degree. A potentiometer or good DVM is required for such measurements. Tables of thermocouple emf vs. temperature can be found in the *Handbook of Chemistry and Physics,* although thermocouples actually must be calibrated to be relied upon for accuracies better than 1C°. Copper-constantan thermocouples may be used in the temperature range of −183°C (boiling point of oxygen) to +400°C. *Chromel-constantan* units may be employed up to 1000° C, while some tungsten alloy units are usable to 2800° C.

Ion-Selective Electrode and pH Measurement

The rapid and convenient measurement of concentrations of ions in solution is important in many contexts; for example, the testing of fluoride ion content in fluorinated water. Scientists, as another example, are particularly interested in the concentration of hydrogen ions [H^+] or *pH measurement*

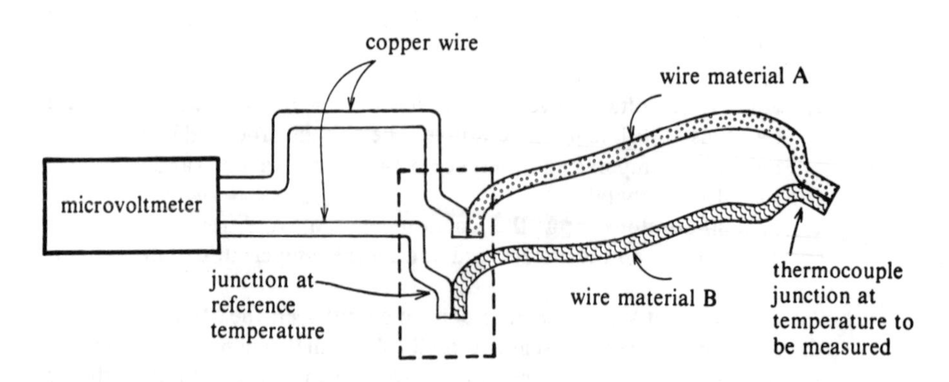

FIGURE A3–6. *Thermocouple Temperature Measurement*

because it measures the *acidity* or *alkalinity* of a solution. To review briefly, since the concentration of H^+ ion can range over many orders of magnitude, the pH measure of acidity/alkalinity is defined as

$$pH = -\log_{10} [H^+] \tag{A3–3}$$

where $[H^+]$ is the *molar concentration* of H^+ ion (in moles/liter). Pure water has $[H^+] = 10^{-7}$ molar so that its pH = +7. An *acidic solution* has a pH of less than 7, while an *alkaline solution* has a pH of more than 7.

Membranes have been developed that exhibit a voltage between the two sides when the concentrations of a selected ion type are different on the two sides of the membrane. The behavior is as though the membrane is permeable only to the one ion, so that diffusion from the high-concentration to low-concentration side by the selected ion would set up a potential difference that depends on the concentration ratios. The actual mechanism is generally much more complex, but in any case, charge transfer through the membrane material is involved, and the resistance of the membrane material is often quite high. For example, the selective membrane of the common *pH-indicating electrode* is made of a type of glass, and the resistance of the probe is typically on the order of 100 megohms.

Suppose then that the concentration of a certain ion in a solution is to be measured. The solution of unknown ion concentration is arranged to be on one side of the ion-specific membrane, while a *reference solution* of known ion concentration is placed on the other side. The emf or voltage V that develops between the two solutions must be measured, because this provides a measure of the relative ion concentrations. This voltage difference must be measured with a pair of electrodes. One electrode, the *reference electrode,* is placed directly in the solution with the unknown concentration, while the other electrode is placed within the reference solution that in turn is contained in the *indicating electrode,* or *selective-ion electrode.* The indicating electrode incorporates the selective membrane and is placed into the solution to be measured. Figure A3–7 is a diagram of the general configuration of a selective-ion measurement.

The voltage difference V is expected to obey Nernst's Law

$$V = V^\circ + \frac{kT}{ne} \ln\left(\frac{a_1}{a_2}\right) \tag{A3–4}$$

where V° is a constant depending on the particular ion, k is Boltzmann's constant, T is the absolute temperature, e is the charge of the electron, and n is the number of electrons in excess or in deficiency for the particular ion. The quantities a_1 and a_2 are the so-called *activities* of the ion in the unknown and the known solutions respectively. For dilute solutions, the activities approach the actual ion concentrations and can be expressed either in ions per unit volume or in moles/liter (molar concentration).

Suppose in particular we are interested in the pH-sensitive electrode. Then $n = 1$ and we have from Equation A3–4

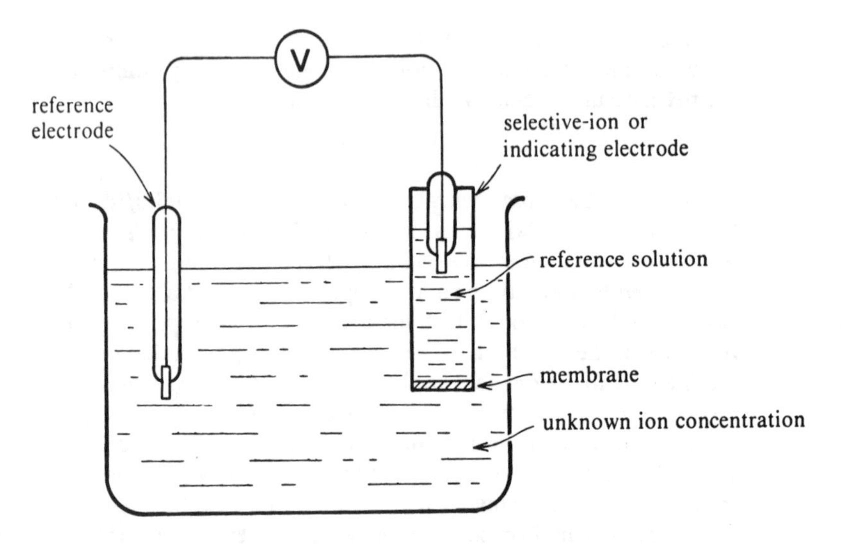

FIGURE A3–7. *Diagram of a Selective-Ion Electrode and Reference Electrode Placed in a Solution to Determine the Concentration of Ions in the Solution*

$$V = \left(V^{\circ} - \frac{kT}{e} \ln\, [H^+]_2\right) + \frac{KT}{e} \ln\, [H^+]_1$$

$$= K + \frac{2.3\, kT}{e} \log\, [H^+] \qquad \textbf{(A3–5)}$$

where we have grouped the constant quantities together as K and have changed to common logarithms as desired for the pH definition. Finally, at a temperature of 25° C, we have from Equation A3–3 and Equation A3–5 that

$$V = K - 0.05915 \left(\frac{\text{volts}}{\text{pH}}\right) \text{pH} \qquad \textbf{(A3–6)}$$

We see that the voltage difference between electrodes in a pH meter changes by about 59 millivolts for each unit change in pH (or each order of magnitude change in hydrogen ion concentration). Accurate measurement of this small voltage through an impedance of 100 megohms is a difficult assignment. Thus, if a voltmeter with an impedance of 10 megohms is used (for example, a conventional digital voltmeter of good quality), a factor-of-ten error results because of voltage divider action. This practical difficulty will be addressed in Chapter A6.

Electromagnetic Induction Transducers

We reviewed in part of Chapter A1 the fact that a changing magnetic flux within a coil of wire induces an emf or voltage in the coil circuit. The value of the emf is given by Faraday's Law as Equation A1–41. A number of input transducers that are sensitive to movement make use of this effect and are called *electromagnetic induction transducers.* The magnetic field is typically produced by a small permanent magnet in proximity with the sensing coil, so that a movement of the magnet relative to the coil produces the change in magnetic flux and consequently the voltage in the coil circuit. Note that a voltage is induced only while the magnetic flux through the coil is changing, so that this transducer approach is most suitable for detecting oscillating movements; the voltage out of the coil is then an ac voltage at the frequency of oscillation.

Figure A3–8 shows schematically several applications of this transducer type. Figure A3–8a shows the principle of construction for a *dynamic microphone.* A coil is attached to a flexible diaphragm that is in turn held fixed at the edges. As sound waves strike the diaphragm, it flexes and vibrates; then the coil attached to the diaphragm also oscillates in the field of the fixed permanent magnet. Consequently, the magnetic field through the coil oscillates and an ac voltage proportional to the amplitude of the oscillation and having the same frequency is induced in the coil. Voltage on the order of tenths of volts may be expected, with a relatively low output resistance because of the relatively few number of turns of the coil.

Figure A3–8b shows one type of *magnetic phonograph cartridge.* As the stylus oscillates from side to side in the record groove (very slightly in modern microgroove records), a small permanent magnet attached to the stylus cantilever oscillates between the iron core tabs that pass through a stationary coil. The iron core serves to "gather" the magnetic field through the coil from the air in the vicinity. This is due to the low *reluctance* of the iron; that is, the magnetic field prefers to pass through it. Thus an oscillating magnetic flux occurs in the coil and an ac voltage is generated in response to the audio recording.

In prospecting for oil and mineral deposits, geophysicists routinely use seismic reflections from underground rock structures. These scientists set off a near-surface explosion or vibration of some sort and measure the time for the seismic wave thus generated to return to the earth's surface. Transducers to measure these vibrations are termed *geophones,* and the type commonly used on land relies upon electromagnetic induction. Figure A3–8c shows that a movable permanent magnet is suspended by springs from the case of the geophone. As the earth vibrates, the magnet is set into oscillation relative to the coil (which is fixed to the geophone core) so that a voltage is induced in the coil.

Piezoelectric Transducer

When certain crystals (for example, quartz or Rochelle salt) or ceramics (for example, barium titanate) experience a mechanical strain or pressure, a voltage develops between the opposite faces of the crystal or ceramic during

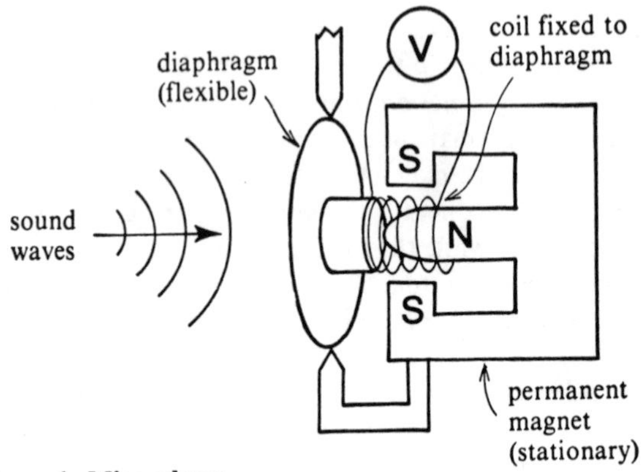

(a) Dynamic Microphone

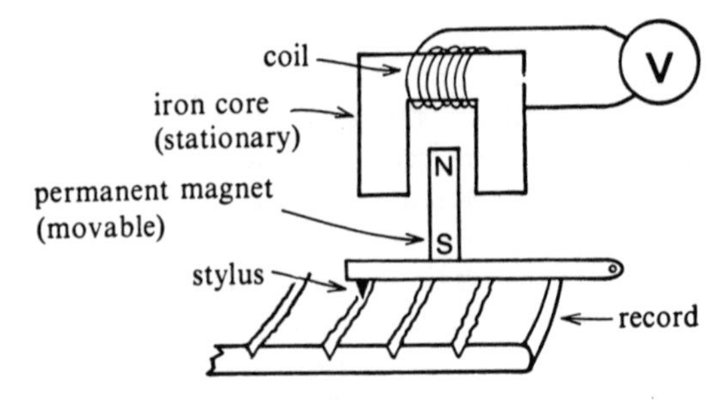

(b) Magnetic Phonograph Cartridge

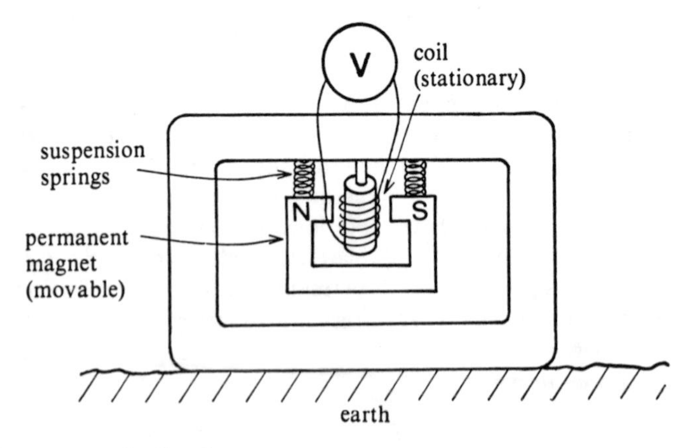

(c) Electromagnetic Geophone

FIGURE A3–8. *Schematic Illustration of Three Transducers That Use the Electromagnetic Induction Principle*

the distortion process. Materials that exhibit this phenomenon are called *piezoelectric.* Metal plates can be attached to the opposite sides; the voltage measured between these plates can then be used as a measure of the force that is producing the distortion of the crystal or ceramic. A number of force-to-voltage transducers use this effect and may be classified as piezoelectric transducers.

In a *ceramic phonograph cartridge* the stylus is attached to a small piezoelectric slab. The output from this type of pickup can be in the range of tenths of volts and thus needs less voltage amplification than the magnetic type; however, the output impedance of the device is quite high: approximately 1 megohm. Also, the fidelity to input amplitude tends to fall above about 15 kilohertz. *Crystal microphones* are sound transducers that use a piezoelectric element and therefore operate on much the same principle as the ceramic phonograph cartridge and with similar advantages and disadvantages. Similarly, geophones for use underwater in offshore explorations generally use piezoelectric elements to sense the water pressure variations associated with a seismic signal.

Photovoltaic Cell and Photodiode

The *photovoltaic* cell can be used to convert light energy directly into electrical energy and thus has aroused considerable interest during this energy-conscious era. Space satellites have used these *solar cells* to generate their electrical energy requirements. Scientists are further interested in them as a light transducer that generates a voltage or current in relation to the light irradiating the cell. The *photodiode* is essentially a small photovoltaic cell that is especially constructed to be useful as a light sensor.

The construction of a photovoltaic cell or photodiode is basically the same as that of a normal diode. The *semiconductor diode* is discussed in some detail in Chapter A4, but the *silicon photovoltaic cell,* invented at Bell Telephone Laboratories in 1953 by D. Chapin and C.S. Fuller, is described here. The description uses some of the terms from semiconductor physics that will be clarified in Chapter A4.

A thin layer of P-type silicon is formed near the surface of a slab of N-type silicon, thereby developing a *PN junction* as shown in Figure A3–9a. A silicon semiconductor diode is therefore constructed. When light photons strike the PN junction region, charge carriers (holes and electrons) are produced that move under the electric field that exists in the junction. Electrons move to the N region and the positive-charge-acting holes move to the P region so that a voltage difference is produced in the contacts to the cell. That is, metal contacts to the opposite sides of the cell exhibit this voltage and carry current sustained by the process to an external circuit. If very little current I is drawn from the cell—that is, *no load*—the voltage V of the cell increases with the light intensity but approaches a *saturation level* of about

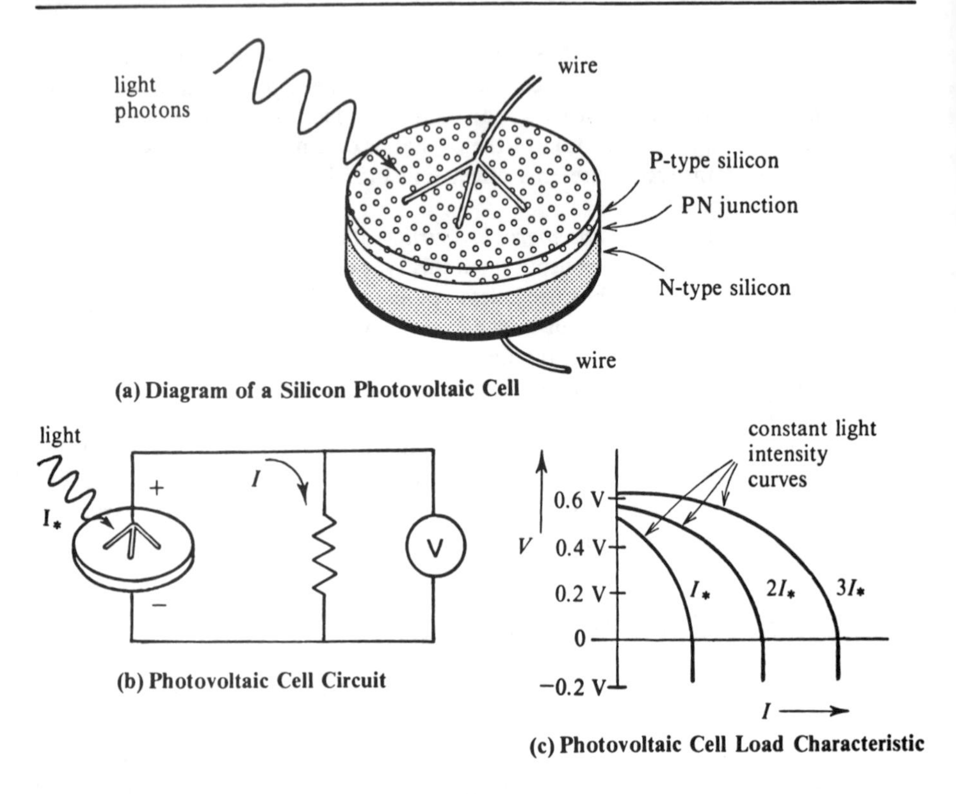

(a) Diagram of a Silicon Photovoltaic Cell

(b) Photovoltaic Cell Circuit

(c) Photovoltaic Cell Load Characteristic

FIGURE A3–9. *Silicon Photovoltaic Cell*

0.6 V that cannot be exceeded. Thus the voltage increases with light intensity I^* but in a very nonlinear fashion. This voltage versus current behavior is illustrated in the *load characteristic plot* of Figure A3–9c by the intersection points on the V axis of the various curves of constant light intensity. (Note the crowding together near 0.6 V.) However, the *short-circuit output current* is directly proportional to the light intensity, I_* (note the even spacing of the intersections along the I axis). That is, the current through a very low resistance (such as a good ammeter) connected across the cell is directly proportional to the light intensity striking the cell and provides a convenient linear light transducer.

However, it should be realized that the cell is responsive to light only over a certain range of light wavelengths. For silicon cells this *spectral response* ranges from about 400 nanometers (4000Å) to about 1100 nanometers (11,000Å). The amount of current depends upon the area of the cell and its efficiency of light energy conversion; therefore, no units are placed along the I scale of Figure A3–9c. However, when a cell is placed in full sunlight, an output current on the order of 20 milliamperes per centimeter^2 of cell area can be expected.

The PN junction may also be conveniently used in a different mode. An *external* voltage may be applied to *reverse bias* the junction: positive applied to N-type side (the cathode) and negative applied to P-type side (the anode). Then the current that flows is again directly proportional to the light intensity falling on the diode junction. This behavior is illustrated in Figure A3–9c by the constant intensity curves as they extend to the negative quadrant in terms of voltage. Generally only a tiny chip of semiconductor material is used for devices designed to be operated in this mode, and they are termed *photodiodes*. An important feature of photodiodes is their rapid response times to changes in light intensity. Thus the Texas Instruments TIED80 is rated with a 15 nanosecond (15×10^{-9} seconds) response time, while special *P.I.N.* photodiodes can respond in a fraction of a nanosecond.

Other Transducers

Snap-Action Switch

Detecting the presence of an object (perhaps a rat's paw or some moving mechanism) can sometimes be accomplished quite simply by a little switch that is activated by a small force acting through a small distance. For example, a psychologist may want to measure the number of times that a rat presses a bar and obtains a portion of food; or a physicist may want a small motor to be turned off when a zone refiner heating element passes a point in its travel. The small switch that is convenient for application in both cases is called a *snap-action switch*, although the term *microswitch*, which is a trade name, is also commonly used.

Pressing a small button on the snap-action switch either causes the switch to close a pair of *normally open* (NO) contacts or open a pair of *normally closed* (NC) contacts. Both types are available and operate with a *common* contact in the single-pole–double-throw (SPDT) type of switch. Often a hinged lever is attached to increase the mechanical advantage. An operating force in the range of 10 grams (0.35 ounces) to 100 grams is typical, with a current handling capability on the order of 5 amperes at 115 V. Removing the force causes the switch to snap back to its normal contact state. Figure A3–10a illustrates a typical switch of this type. Figure A3–10b shows a circuit that might be used with an electromechanical counter to monitor the number of times the switch has been pressed. Counters requiring a pulse of current (perhaps 300 milliamperes) at 24 V to turn the number wheels up one unit are commonly available, and such a counter is assumed in Figure A3–10b.

Magnetic Reed Switch

Another simple but useful switch that has many applications is the *reed switch;* it is especially useful in situations where we want to sense the

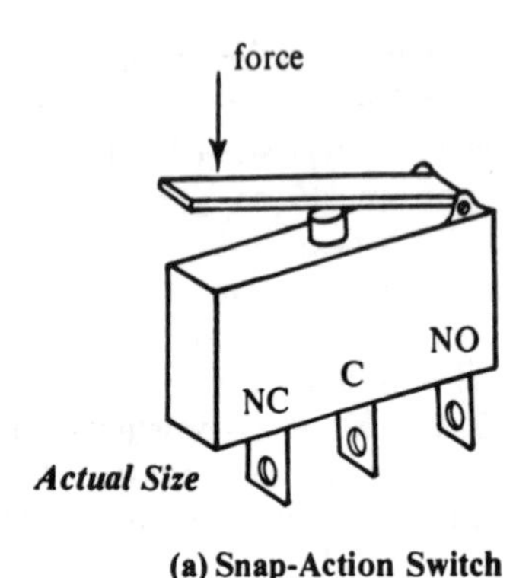

(a) Snap-Action Switch

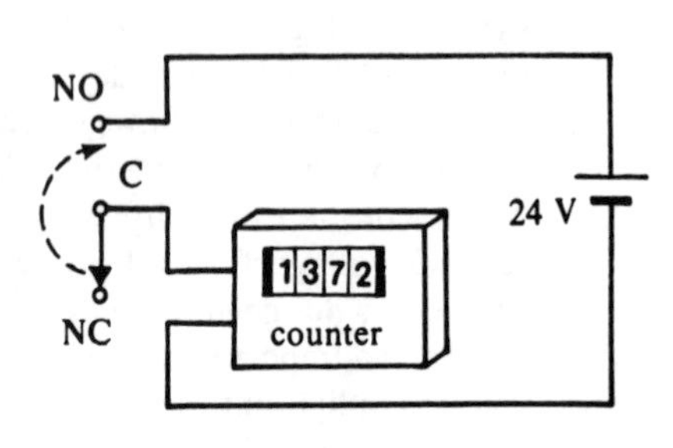

(b) Snap-Action Switch Used in a Simple Counting Circuit

FIGURE A3–10. *Snap-Action SPDT Switch*

proximity of an object without physical force or contact. The switch consists of overlapping reeds constructed of a ferromagnetic material that are placed inside a hermetically sealed glass tube perhaps 3 centimeters long. If the reeds are normally slightly separated (normally open), bringing a small permanent magnet near the reeds causes them to snap together and close the switch; removing the magnet will permit the reeds to flex back to their normally open configuration. These switches will work reliably for millions of operations at rates up to several hundred operations per second and can typically control currents of the order of a hundred milliamperes at up to 50 V. (Various ratings are available.)

Figure A3–11 illustrates a typical reed switch in a simple circuit in an application that requires counting the number of revolutions of a shaft. As the shaft turns, the magnet closes the switch and increments the counter.

A *reed relay* is constructed by placing a wire coil around the reed switch. Passing a current of perhaps 20 milliamperes through the coil pro-

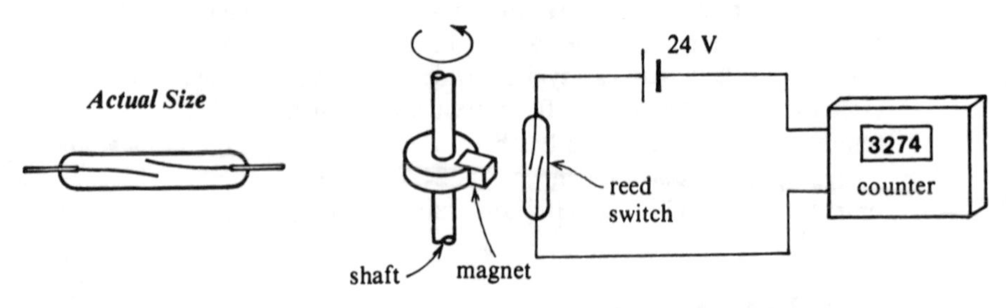

(a) Reed Switch

(b) Reed Switch Used in a Simple Circuit

FIGURE A3–11. *Reed Switch*

duces a magnetic field within the coil that closes the reed switch. Thus, an electrically operated switch is obtained; this is what is meant by the term *relay*. (Relays are discussed in the next section.)

Electromagnetic Relay

The *electromagnetic relay,* or simply *relay,* can be considered a type of output transducer. It is simply an electrically operated switch and has been a staple of electrical circuits for many years. In recent years, the *solid-state relay* (discussed in Chapter A8 and presented in Figure A8–29) has replaced the electromagnetic relay in a number of applications. However, the "old-fashioned relay" still enjoys a great deal of usage. It is a simple solution to the situation that calls for a small current to control—that is, turn on and off—a large current in some circuit.

The schematic symbol for the electromagnetic relay strongly suggests its construction (see Figure A3–12). The *relay coil* is in fact an electromagnet. When a current greater than the *pull-in current* for the particular relay passes through the coil, its magnetic field pulls the iron *armature* of the relay to a position nearer the coil. Conversely, when the coil current falls below the *drop-out current,* a spring attached to the armature snaps it back to the "dropped-out" position. But the *pole* of a switch is an integral part of this armature; often it is the pole of a *single-pole–double-throw* (SPDT) switch. Therefore, we have a SPDT switch that operates in response to an electrical input. The SPDT switch configuration is illustrated for the relay in Figure A3–12.

Sensitive relays may have a pull-in current of a few milliamperes in response to a coil voltage on the order of 12 V. Typically, the coil's resistance and voltage for pull-in are given by the manufacturer. A second important rating is that of the switch contacts. The contacts of a sensitive relay are typically capable of switching 1 ampere at about 120 V. Heavy-duty relays

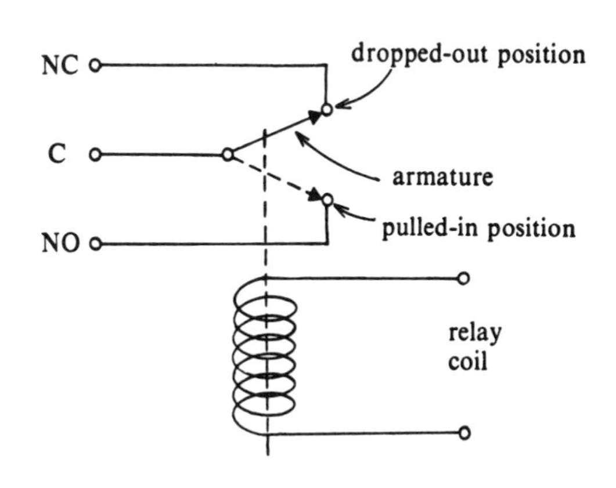

FIGURE A3–12. *Conventional Electromagnetic Relay*

often require coil currents of some hundred milliamperes at a coil voltage of 24 V and can switch tens of amperes at hundreds of volts. However, pitting of the switch contact points can be a problem. A significant resistance can develop between the closed points and thus inhibit current flow in the target circuit. Pitting results from the arcing between the points at the moments of "make" or "break." Life expectancy may range from 100,000 operations to several million, depending on the design.

Solenoid and Solenoid Valve

The pull-action *solenoid* is an output transducer that converts electrical energy to a mechanical movement. This device is useful when a small physical motion is desired as an output process. Push-pull–type solenoids are also available. These devices use an electromagnet in the form of a hollow solenoid to pull an iron plunger through a stroke that is typically about 1.5 centimeters (see Figure A3–13). The pull on the plunger exists as long as the solenoid (coil) receives sufficient current, which may be ac or dc depending on the type. The force exerted on the plunger ranges from a few ounces (~0.5N) for miniature units with exterior dimensions of about $1.5 \times 1.5 \times 4\,cm^3$, to three pounds (~15 N) for larger units with $3.5 \times 3.5 \times 4\,cm^3$ dimensions. The power required by the electromagnet coil ranges from 3 W to 24 W for these respective capabilities and may be from 12 V or 24 V_{dc} sources, or 120 or 240 V_{ac} mains.

For on/off control of fluid flow through tubing, a *solenoid valve* may be used. This output transducer is an electrically operated valve. It uses an internal solenoid to move a plunger which opens a flow path. 120 V_{ac} coil voltage at a current on the order of 0.2 A is a typical requirement to operate the valve, which may be normally open or normally closed. These valves are common fixtures in any automatic clothes washer or dishwasher machine.

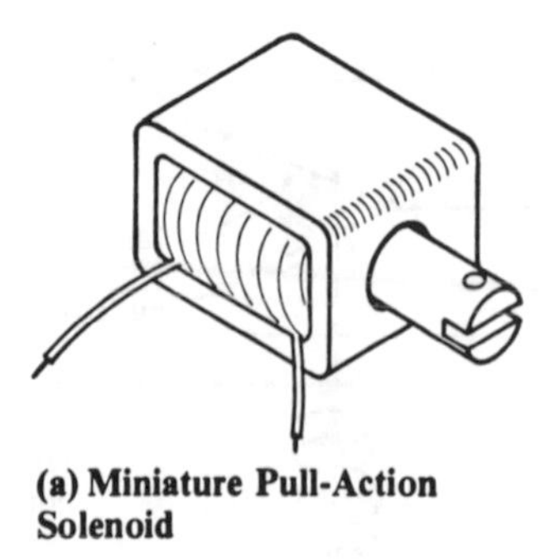

(a) Miniature Pull-Action Solenoid

stroke
coil
closed position
ferromagnetic plunger

(b) Diagram

FIGURE A3–13. *Pull-Action Solenoid*

REVIEW EXERCISES

1. (a) Does the resistance of a photoresistor increase or decrease as it is brought from a dark room to a well-lit room?
   (b) Explain briefly the direction of the effect.
2. Select from the choices given for a characterization of: A. a thermistor and B. a platinum resistance temperature detector (RTD).
   (a) Its resistance increases with temperature increase.
   (b) Its resistance decreases with temperature increase.
   (c) Its resistance has approximately exponential dependence on temperature.
   (d) Its resistance has approximately linear dependence on temperature.
   (e) It has high absolute accuracy.
   (f) It has high sensitivity.
3. Describe how a thermistor may be used to give a measure of wind speed or of fluid flow in a tube. It thus becomes the essential element in a flow meter.
   NOTE: Consider wind chill.
4. Diagram a Wheatstone bridge with one active strain gauge, and one dummy gauge. Label these gauges. Describe the purpose for this arrangement and, in qualitative terms, how it achieves the purpose.
5. What is the difficulty expected with using a conventional ohmmeter with a conductivity cell?
   NOTE: This difficulty holds true even for high-accuracy, wide-range ohmmeters found in many quality DVMs.
6. What is the difference in the electrical output from a thermocouple compared with that from a thermistor?
7. The voltage output from a pH indicator/reference cell pair is 59 mV/pH near room temperature. The majority of DVMs on the market can measure voltage with 0.1-mV resolution or better. Nevertheless, the majority cannot be used as the voltage measuring component in a pH meter. Why not?
8. Why are electromagnetic induction transducers (which do not use an external ac voltage) suitable only for measuring an oscillating motion?
9. Briefly contrast the operating principle of the dynamic microphone to that of the crystal microphone.
10. The photoresistor and the photovoltaic cells are sometimes confused. What are the essential differences between these devices?
11. Some computer circuits are being interconnected with fiber-optic ("light-pipe") cables for high-speed, electrical-interference-immune communications. The light intensity into a cable can be modulated (turned on and off) at a rate of many megahertz. For use as a detector at the receiving end, we might consider a photoresistor or a photodiode. Which would we choose and why?
12. We wish to use a photovoltaic cell to obtain an output that is directly proportional to the light intensity falling on it. What should be the mode of operation of the cell?
13. Describe the principle of operation of a reed-type electromagnetic relay.
14. (a) What are meant by the NO, C, and NC contacts, respectively, on an elec-

tromagnetic relay? Make a diagram to illustrate your discussion.

(b) What are meant by the pull-in and drop-out currents of an electromagnetic relay?

PROBLEMS

1. Suppose a photoresistor has a resistance of 2000 ohms at an illuminance (light intensity) of 400 lux. What would be its resistance in direct sunlight, where the illuminance is about 10^5 lux? Assume that the photoresistor obeys a power law with an exponent of 0.7.
2. A simple calculation can be made to test how accurately the thermistor of Table A3–1 (in the chapter text) follows exponential behavior. For accurate exponential change, the device should show the same *percentage* change in any identical temperature intervals.
   (a) Evaluate the percentage change in resistance for the four 50°C temperature intervals in Table A3–1.
   (b) Is the thermistor accurately exponential?
   (c) Is there a trend in the data?
3. The thermistor of Table A3–1 is used in the bridge pictured in Diagram A3–1.
   (a) What is the magnitude of the voltage V_{out} at 0°C, 25°C, and 50°C?
   (b) What point (*A* or *B*) is more positive at each temperature in part (a)?

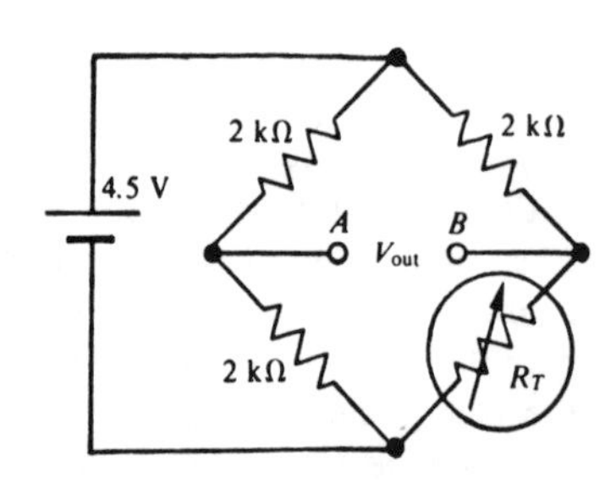

DIAGRAM A3–1.

4. A certain thermistor has a resistance at 25°C that is a factor of 19.8 more than its resistance at 125°C.
   (a) Assume exponential behavior over the temperature range. What is the value of *E* in Equation A3–2 for this unit? What are its dimensions?
   (b) Suppose the resistance for the thermistor is 2000 ohms at 25°C. What will the resistance be at 100°C?
5. Suppose a resistance thermometer is used in a Wheatstone bridge wherein the three other arms have equal resistances. The excitation voltage to the bridge is 6 V, and the bridge balances at 25°C.
   (a) If the resistance thermometer is a pure metal, the temperature coefficient for its resistance can be expected to be about $(1/300)(C^\circ)^{-1}$. What is the magnitude of the output voltage when the temperature of the thermometer is 26°C?

(b) Assume that the resistance thermometer is a thermistor with a temperature coefficient of $-5\%(C°)^{-1}$. What is the magnitude of the bridge output at 26° C?

6. The fractional change in resistance $\Delta R/R$ of a strain gauge is proportional to the strain $\Delta L/L$ that it experiences (for $\Delta L/L$ less than about 10^{-3}). We write $\Delta R/R = G\Delta L/L$, where G is the *gauge factor* for the particular gauge. For most metals, G is on the order of 2, but for semiconductors it can be on the order of 200.
   (a) Assuming a single active strain gauge in a Wheatstone bridge with equal arm resistances, find an expression for the output from the bridge in terms of bridge excitation voltage, gauge factor, and strain.
   (b) Argue that other things being equal, the output from a bridge with a semiconductor strain gauge is much greater than for the case with a metallic strain gauge. However, what rather serious problem can be expected with a semiconductor strain gauge?

   NOTE: Recall what a thermistor is.

7. Diagram A3–2 illustrates a force transducer that is based on strain gauges. The diagram shows a bar (originally straight) with strain gauges A and B mounted on the opposite sides (by use of an adhesive). One end of the bar is fixed, while a force applied to the opposite end causes the bar to bend, as shown in Diagram A3–2.

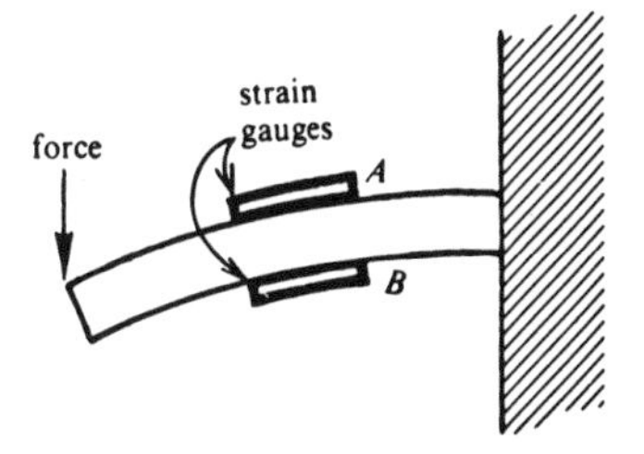

DIAGRAM A3–2.

   (a) Which gauge is in extension and which is in compression?
   (b) Discuss how this arrangement can be used to advantage in a half-active Wheatstone bridge.

   NOTE: When a strain gain is made to compress, its resistance decreases.

   (c) Discuss how a platform (weighing) scale with electrical readout could be constructed using this force transducer. (This is a form of load cell.)

8. Assume that a copper-constantan thermocouple is used to monitor the temperature of a sample. Approximately what voltage resolution is required to observe 0.1C° variations in temperature?
9. Given the definition of pH (Equation A3–3) and Nernst's Law (Equation A3–4), justify the numerical value: 59 mV per unit change of pH at 25° C.
10. A displacement transducer based on the variable capacitance of two metal plates can readily be imagined. We could consider having a capacitor with the displacement changing the distance between the plates, or changing the overlap area of two parallel plates. Select and justify the case:
    (a) Which would be most sensitive to small displacements?
    (b) Which would have capacity approximately proportional to displacement?

11. Suppose that a ferrite rod or a laminated iron rod is free to slide into a solenoid.
    (a) Describe how this arrangement could be a transducer for measuring linear position.
    (b) What property of the solenoid could be measured? Can this be a dc measurement?
12. A certain sensitive electromagnetic relay requires 5 V at a coil current of 20 mA for pull-in operation.
    (a) What is the coil resistance of the relay?
    (b) What is the input power required to operate the relay? (This is a valid measure of relay sensitivity.)
13. Electromagnetic relays that require less than 5 V at perhaps 20 mA to operate the relay are rare.
    (a) Why is it not possible for a single photovoltaic cell (solar battery) to operate a relay?
    (b) How might several photovoltaic cells be interconnected to operate a relay directly? What would be an approximate minimum number, assuming silicon photovoltaic cells and a sensitive relay as described?
    (c) Why is the area of each photovoltaic cell also of importance?
14. We wish to regulate the temperature of a warm box using a thermistor R_T, a sensitive electromagnetic relay, and a 100-W heater operated from 110 V. (The heater could be a 100-W bulb.) The relay's characteristics and the circuit are shown in Diagram A3–3. Of course, the thermistor and heater are both placed in the warm box.
    (a) Justify qualitatively that if the temperature of the box and thermistor are sufficiently high, the system behavior is such that the heating will stop. However, as the temperature then falls, heating will resume at some temperature.
    (b) We desire the system to stop heating at 30° C. What should be the thermistor resistance at 30°C to achieve this behavior?
    (c) As the temperature falls, what is the thermistor resistance when heating again resumes?
    (d) Assume that the thermistor resistance changes by −5%/C° near 30°C. At what temperature does the unit begin heating again? Note that the temperature is maintained only within a certain interval, often called the hysteresis of the regulator. The regulation is *not* particularly good for this simple circuit.

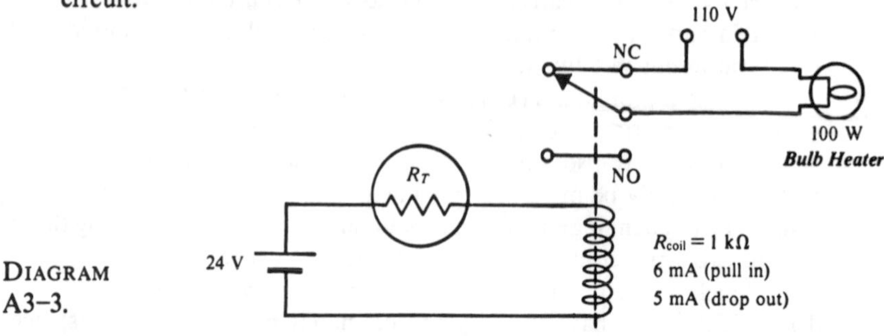

DIAGRAM A3–3.

CHAPTER A4

Diodes and Power Supplies

We begin the study of electronic devices and their utilization in this chapter. The principal actor in this study is the diode—a device that permits electrons to flow through it in only one direction. Since the diode does not actually achieve amplification of electrical signals, it is generally placed in the class of the passive devices that were studied in Chapter A1. Yet an understanding of the mechanism by which diodes operate is a direct precursor to the understanding of amplifying devices. Thus, in this chapter we study briefly the vacuum tube diode; this study permits the essential behavior of the vacuum tube triode to be quite readily understood in Chapter A8. Then our study moves to the solid-state diode.

The term *solid state* has quickly become a part of everyday language. Consider for example advertisements for the "all solid-state television." Whereas electrons are controlled as they move through a vacuum in the vacuum tube, it is their rather special behavior in *semiconducting solids* that is exploited in the solid-state diode and in various forms of transistors. In this chapter a basic account of semiconductor behavior is given that is then used to understand the solid-state diode; semiconductor behavior is later used to understand the essential features of transistor action (Chapter A8). Since transistors and diodes are fundamental constituents of integrated circuits (ICs), the story of modern miniaturized electronics properly begins in this chapter.

Although the diode itself has a surprising number of applications, in this chapter the diode will be applied only to the process of *rectification;*

that is, any process that changes an alternating current to a direct current. Rectification is central to the theme of the chapter: techniques for making an electronic battery from an ac power source. The formal term for an electronic battery is *power supply.* Power supplies are important to the scientist because dc power is required, at least internally, for the operation of virtually all electronic equipment.

Ideal Diode

The *ideal diode* is a circuit element that passes current only one way. It is a "one-way street" for electrons. Figure A4–1a shows the circuit symbol for a diode, and Figure A4–1b and c show the voltage polarity that results in *current flow* and *no flow,* respectively. The arrow of the symbol points in the direction of permitted *conventional* current flow. If the anode of the diode is made more positive than the cathode, current will pass through and the diode is said to be *forward biased.* Figure A4–1b, therefore, shows a forward-biased diode, while the diode in Figure A4–1c is *reverse biased.*

The diode is a *nonlinear* device. The reason for this statement becomes clear when we look at the contrasting graphs of current versus voltage for an ideal resistor and an ideal diode. Figure A4–2 shows such graphs, which are called the *current-voltage characteristics* for the two devices. On the horizontal axis is the voltage across the device, while the current flowing through it is on the vertical axis. Notice that both negative and positive voltage (as well as current) possibilities are graphed; these correspond to one voltage sense or the other across a device similar to Figures A4–1b and c.

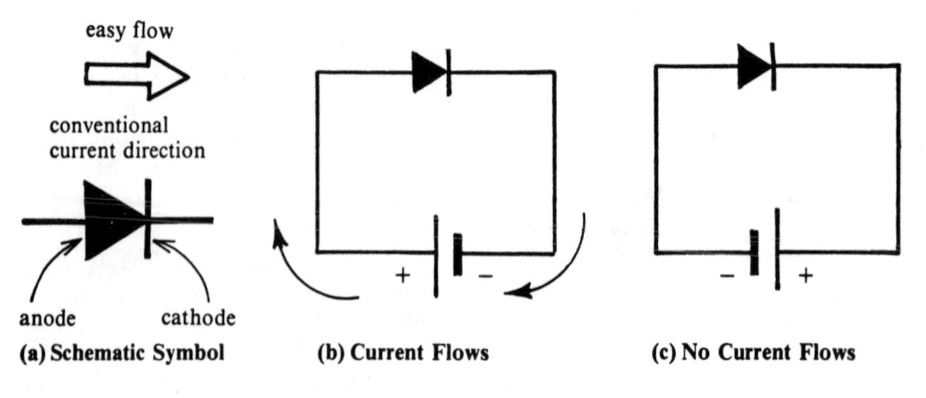

FIGURE A4–1. *Diode Symbol and Current Flow Sense*

The positive voltage axis in Figure A4–2b corresponds to a forward-biased condition for the diode. The current that then flows is called a *forward current*. Notice the large forward current that flows as soon as the *ideal diode* is even minutely forward biased. The negative voltage axis in Figure A4–2b represents a reverse-biased condition for the diode. Any current that then flows is called a *reverse current* or *leakage current*. Figure A4–2b shows the reverse current is zero for any reverse voltage across an ideal diode.

Real Diodes

Vacuum Diode

The *vacuum diode* has been in existence since about 1900 and for many decades was the diode most widely used. It is reviewed here very briefly because it is not used in new designs; however, it is still encountered in the lab in older equipment that uses vacuum tubes. Examples include vacuum tube voltmeters, oscilloscopes, and oscillators.

The vacuum diode consists of a filament, cathode, and plate inside an evacuated glass tube as shown in Figure A4–3a. The schematic symbol for a vacuum diode is shown in Figure A4–3b. The *filament* or heater is typically operated from 6.3 V or 12.6 V and glows red hot. (When the filament burns out or opens, similar to the filament of a light bulb, it is useless.) The very hot filament heats the metal cathode structure and literally boils electrons out of

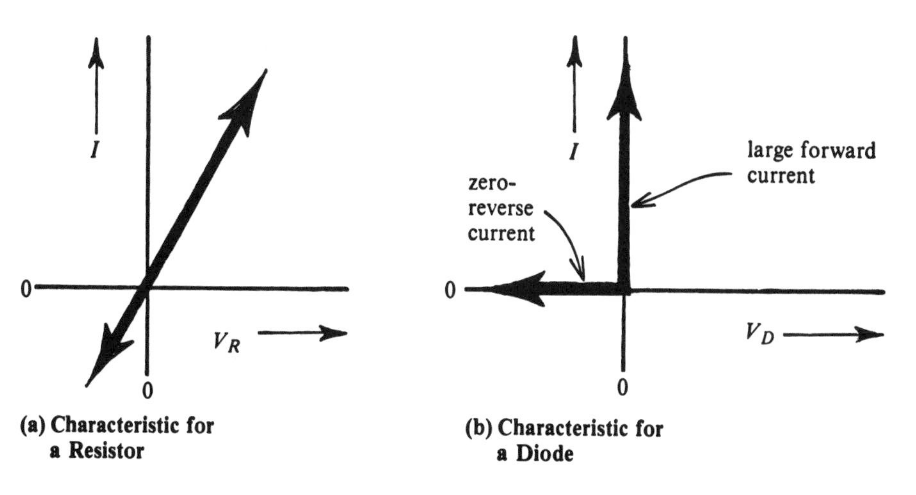

(a) Characteristic for a Resistor

(b) Characteristic for a Diode

FIGURE A4–2. *Current-Voltage Characteristic Graphs for an Ideal Resistor and an Ideal Diode*

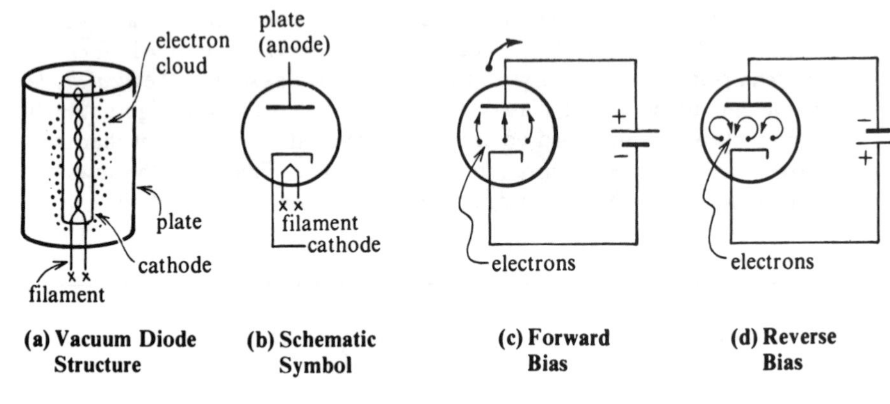

FIGURE A4–3. *Vacuum Diode*

it. The *cathode* is covered with a special oxide that promotes the emission of electrons at lower temperatures.

Surrounding the cathode cylinder is a second cylinder called the *plate*. If the plate is made positive with respect to the cathode, as shown in Figure A4–3c, the electrons are attracted toward it and current flows through the tube. If the plate is made negative relative to the cathode, the electrons are simply repelled back nearer the cathode and no sustained flow of charge or current results. The device acts like a one-way street for electrons.

NOTE: The term *diode* derives from the fact that just two elements—plate and cathode—are principally involved.

Semiconductor Diode

Like the vacuum diode, the *solid-state diode* has actually also been around since about 1900. The "cat whisker" of an old-time crystal set is a solid-state diode; it was a *point-contact diode* that utilized the curious one-way conducting behavior of the contact between a wire and a semiconductor crystal, but this was a temperamental diode that was not easily mass produced.

When pure semiconductor behavior and fabrication became well understood in the 1940s, the semiconductor diode became possible. *Silicon* and *germanium,* which both have 4 valence electrons, are the commonly used semiconductors of electronics. Each electron is shared with a neighbor's electron to form covalent bonds as diagrammed in Figure A4–4; a group of 8 electrons is thus in the outer shell of each atom. The simple picture, sufficient for our purposes, is that these electrons are quite tightly bound in place and not free to move around; thus, the charge carriers are not available and

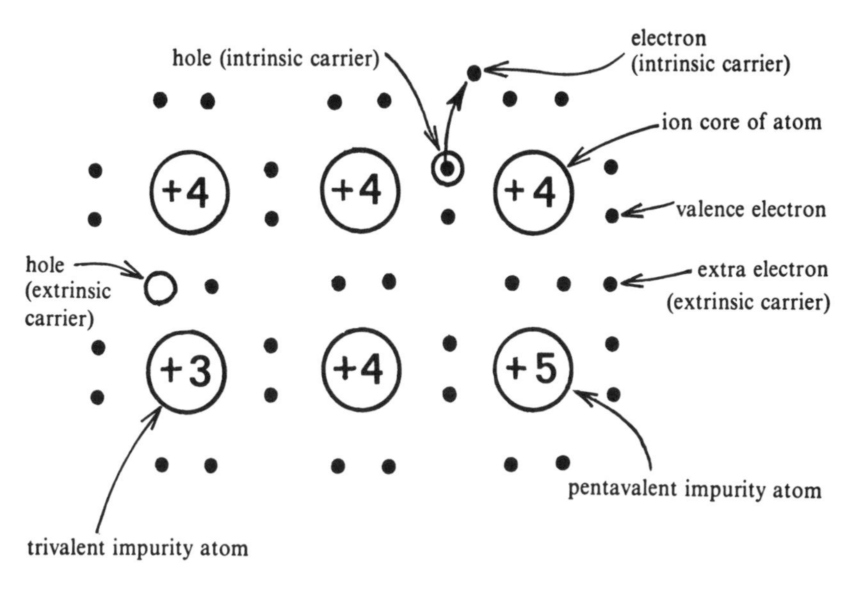

FIGURE A4–4. *Representation of Silicon or Germanium Atoms Showing Bonding Electrons and Impurity Atoms*

an insulator material would result if it were not for the mechanisms discussed next.

Thermal agitation will shake a few electrons free; these electrons can move under an electric field and thus be charge carriers. One such electron is circled at the top of Figure A4–4 and labeled an *intrinsic carrier;* because it is negatively charged, it is called an *N-type charge carrier.* A *hole* is left where the electron was, and neighbor electrons are inclined to jump into it. When one does, the hole moves to a different atom and charge motion is involved. The charge motion behaves as though a positive charge is jumping along with the hole; thus, we may think of a positive or *P-type charge carrier* as having also been broken loose by thermal agitation.

Not many of these intrinsic carriers are present at room temperature in pure silicon or germanium (about one per 10^{12} atoms in silicon and one per 10^{9} atoms in germanium), therefore, they are poor conductors, or *semiconductors.* However, even tiny amounts of an impurity can change the situation. If a pentavalent impurity such as phosphorus is present, the fifth electron is not needed for bonding and becomes quite free to roam around the lattice. The fifth electron is a negative or N-type carrier, and such an atom is diagrammed in the bottom segment of Figure A4–4. On the other hand, if a trivalent impurity like boron is present, a missing bonding electron or hole automatically exists. A neighbor electron can quite easily jump into it and, as the process continues, a hole or positive-acting P-type carrier hops about the lattice. Either situation will increase the conductivity dramatically. Carriers introduced through impurities in this way are called *extrinsic carriers.*

The magic of semiconductor electronics comes about when one type of impurity is added *on purpose* to produce mostly N-type carriers and thus produces an N-type semiconductor or else a different impurity is added to produce a predominance of P-type carriers and thus produces a P-type semiconductor. The adding of impurities is called *doping.* If a block of P-type semiconductor is joined to a block of N-type semiconductor, the resulting *PN junction* exhibits diode action and becomes the curious one-way street for current.

Figure A4–5a shows a picture of the features important to an understanding of how a PN junction works. The N-type semiconductor block is pictured as having – charges (electrons) free to move about; the positive host ions are rigidly fixed in the lattice and are not free to move. This is indicated by drawing circles around the +. On the other side, the P-type semiconductor block is pictured as having + charges (actually electron holes) free to move about and rigidly fixed negative ion cores.

When the two blocks of oppositely doped semiconductors are joined, there is a significant tendency for the – carriers in the N material to wander or *diffuse* over into the P block, and also for the + carriers in the P material to diffuse into the N block. This tendency is very similar to the tendency in nature for different gases in a volume to distribute themselves uniformly throughout the volume. Suppose two adjoining rooms are first filled with pure oxygen and nitrogen gas, respectively, and then a door between the two rooms is opened. The oxygen molecules will diffuse into the room with no oxygen concentration, and the nitrogen molecules will diffuse in a similar manner. After a short time both rooms are filled with an oxygen and nitrogen mixture.

We may similarly think of the – and + carriers in these semiconductor blocks as charged gas-like particles bouncing about in rooms. When the

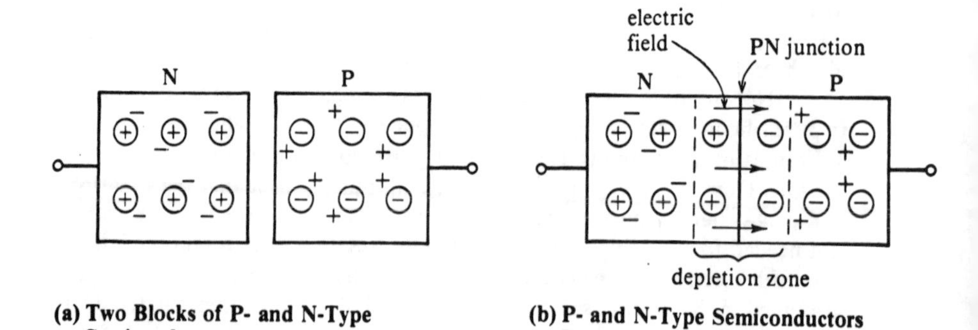

(a) Two Blocks of P- and N-Type Semiconductors

(b) P- and N-Type Semiconductors Placed in Contact

FIGURE A4–5. *P- and N-Type Semiconductors*

blocks are joined, the − carriers in the N block will wander into the P block for some distance. There each soon encounters a + carrier (a hole); the electron drops into the vacant bond represented by the hole, and both carriers disappear as a normal bonding arrangement results. However, the − carriers do *not* achieve a uniform distribution throughout the two blocks by any means, because they leave behind bare or *uncovered* + stationary host ions in the N block near the junction, and these ions set up an electric field tending to pull the wanderers back. Similarly, some of the + carriers in the P block diffuse into the N block, but they leave behind uncovered − host ions in the P block near the junction that inhibits other + carriers from diffusing over. The result is a *depletion zone* of primarily host ions—that is, a zone depleted of carriers—near the junction, and this situation is shown in Figure A4–5b.

An electric field develops in this depletion zone between the + and − ions near the junction (much as between the plates of a capacitor), and the push of this field will keep most of the − and + carriers on their respective sides of the junction. This field is sketched in Figure A4–5b. Because of this field there is a voltage difference V_0 that develops between the two blocks of semiconductors after they are joined; this voltage difference is called the *barrier voltage.*

Now the remarkable diode action of this device can be explained. Consider a battery connected with its plus terminal to N and minus terminal to P as shown in Figure A4–6a. The − carriers in the N material will initially be attracted toward the plus terminal and move some distance toward it. However, more uncovered + ions then appear near the junction, and soon an equilibrium tension develops between the attractions to these two positive elements. Similarly the + carriers in the P block are attracted initially toward the minus battery terminal, but soon stop due to the attraction back toward the additional − ions that appear near the junction. The end result as shown

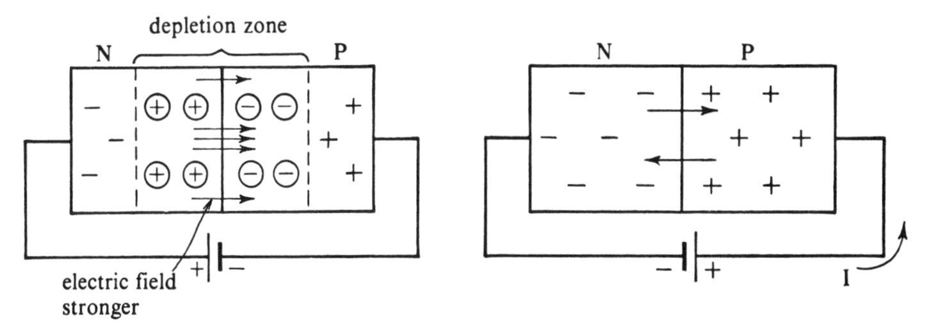

FIGURE A4–6. *Reverse- and Forward-Biased PN Junctions*

in Figure A4–6a is a widening of the depletion zone (and an even greater electric field barrier to diffusion across the junction) but no sustained current flow. We have a *reverse-biased diode.* Actually, a tiny *reverse-leakage current* flows because the very few minority intrinsic − carriers in the P block diffuse over to the junction and are swept across by the field there (the field direction does *not* inhibit minority carriers) so that they continuously neutralize some of the retarding + host ions and permit a like number of − carriers to flow out the end of the N block to the plus battery terminal. A similar flow of minority intrinsic + carriers out of the N block occurs in the opposite direction. This current is essentially constant no matter what the battery voltage is.

On the other hand, if the plus terminal of the battery is connected to the P material (and minus terminal to the N material) as shown in Figure A4–6b, the carriers in each block will initially be repelled from the battery terminals and propelled toward the junction. The width of the depletion zone is decreased and the barrier presented by the electric field there is lessened. Now many carriers have the energy to pass through the depletion zone and into the other side (where they soon neutralize with carriers of the opposite type). The continuous neutralization of the carriers in the vicinity of the junction causes more carriers to be supplied by the battery at the ends of each block, where they then move toward the junction. (Electrons enter from the wire at the N end; they depart from the P-end wire, but leave + holes in the process that then move toward the junction.) Thus, a continuous current flows, and we have a *forward-biased diode.* In fact, if the applied battery voltage is as large as the barrier voltage (about 0.3 V for germanium diodes and about 0.6 V for silicon diodes), the barrier is removed and the current becomes very large; that is, limited only by the actual resistance of the semiconductor blocks.

Characteristic of a PN Junction: Real Devices

The current-voltage characteristic of a PN junction is not the perfectly sharp angle graph of an ideal diode, where even an infinitesimal forward-bias voltage results in a very large current. However, it is much closer to the ideal than the old vacuum diode. When forward biased, the current does not begin rising appreciably until the battery voltage is of the order of the barrier voltage V_0, and V_0 depends on the semiconductor used in construction. As was just mentioned, the barrier voltage is about 0.3 V for germanium, and about 0.6 V for silicon.

A graph of the semiconductor diode characteristic is shown in Figure A4–7. It will be observed that the current appears to increase steeply after the *knee* of the diode curve is reached. This knee is at about 0.2 to 0.3 V for diodes constructed from germanium and at about 0.6 V to 0.7 V for silicon

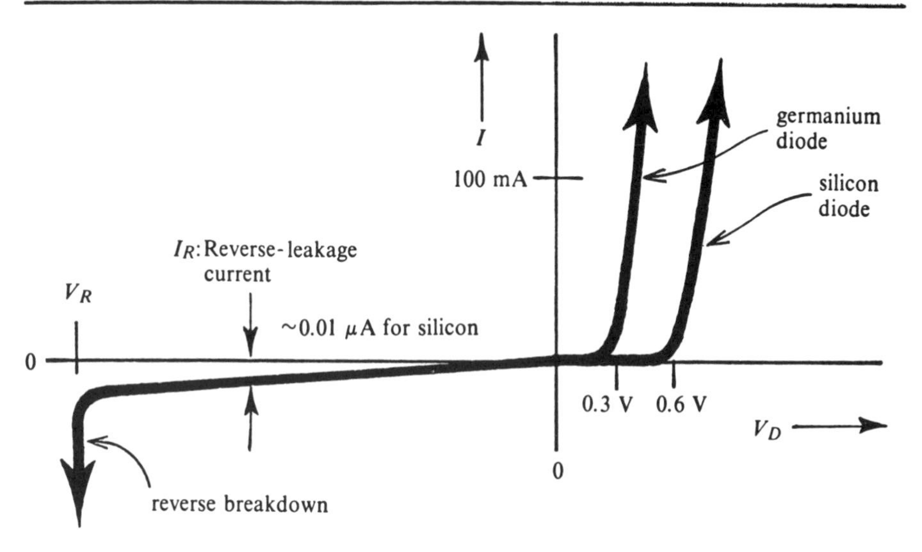

FIGURE A4–7. *Semiconductor Diode Characteristic*

diodes. In fact—especially for applied voltages below the knee—the curve follows a theoretical exponential curve quite well (derived in solid-state physics). The theoretical relation is often called the *diode law:*

$$I = I_R\left(e^{\frac{|e|V_D}{kT}} - 1\right) \qquad \textbf{(A4–1)}$$

Here I is the current passing through the diode, I_R is the reverse-leakage current, $|e|$ is the magnitude of the charge on the electron, V_D the voltage across the diode, k is Boltzmann's constant, and T is the absolute temperature. The reverse-leakage current is a few hundredths of a microampere for silicon diodes; for germanium it is much larger, perhaps 50 microamperes. This leakage current increases with temperature because of the increase in thermally generated intrinsic carriers and increases more rapidly for silicon than for germanium. The ideal diode would have zero reverse-leakage current, of course.

In addition, the characteristic graph of real semiconductor diodes in Figure A4–7 shows that they do *not* withstand a reverse voltage or back-bias voltage to arbitrarily high values. There is a limit, called the *reverse-breakdown voltage, V_R*. At this limit, reverse current then begins to flow, and if it is not limited in some way, the diode can be destroyed. The value at which this breakdown occurs can fluctuate from diode to diode in even one production batch. However, the limit is sharply defined for a given diode and put to good use as a voltage reference in *zener diodes* that are manufactured to have this reverse knee at a low-to-moderate voltage. (Zener diodes are briefly discussed within the Problems at the end of the chapter.)

Diodes intended for forward / reverse operation are usually available in two categories: signal diodes and rectifiers. *Signal diodes* are designed to be able to switch from the forward-conducting to the reverse or nonconducting state very rapidly—in nanoseconds (10^{-9} seconds)—so they can be used at radio frequencies and beyond. However, they are usually limited to perhaps 100 milliamperes in their forward-current capability. *Rectifiers* are intended to pass currents of the order of amperes in the forward direction and are typically used to construct power supplies (which are discussed in the remaining sections of the chapter).

It follows from the above discussion that diodes are typically rated and priced on the basis of three and sometimes four parameters:

1. Maximum safe forward current,
2. Maximum reverse (or inverse) voltage,
3. Reverse-leakage current,
4. Conducting to nonconducting switching time.

Figure A4–8 shows some typical diode shapes and identification features.

A longtime standard germanium signal diode is the 1N34, which has a maximum average forward-current capability of 50 milliamperes (though 1-second surges of 500 milliamperes are withstood), a maximum continuous-reverse voltage of 60 V, and a maximum reverse current of 50 microamperes at −10 V. A popular silicon switching diode (a *computer diode*) is the 1N914, which has nearly the same maximum characteristics as the 1N34 except that the reverse-leakage current is less than 0.025 microamperes. The 1N4001 to 1N4007 family of epoxy-case silicon rectifiers are widely available for use in power supplies and other applications where high currents must be passed. The maximum average forward current for these devices is 1 ampere, although 0.1-second surges of 10 amperes are possible. The maximum continuous reverse-voltage ranges from 50 V for the 1N4001 to 1,000 V for the 1N4007. Notice that the prefix 1N occurs with all these diodes and indicates a *JEDEC-registered diode.* Many semiconductor manufacturers produce them. The prefix *2N* indicates a *JEDEC-registered transistor.* JEDEC is the abbreviation for the Joint Electron Device Engineering Council.

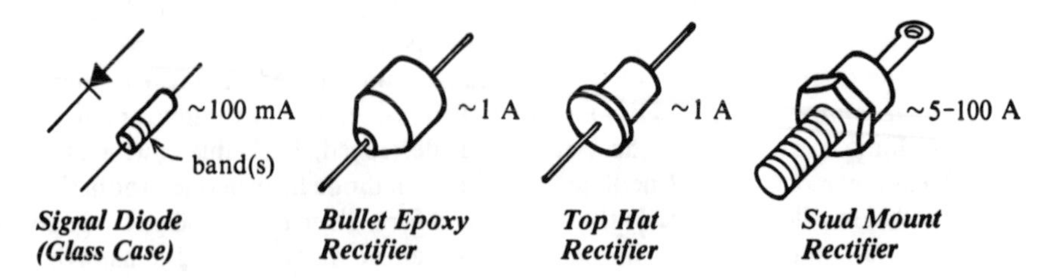

FIGURE A4–8. *Common Diode (Rectifier) Types*

Rectification and Filtering

Perhaps the most common application of diodes is in *rectification* of ac current/voltage to produce dc current/voltage. In many cases, rectification is done in a module that amounts to an "electronic battery"; that is, an ac voltage goes in and a dc voltage, such as from a battery, comes out. However, an appropriate filter circuit is required to approach a true dc voltage.

Half-Wave Rectification

First, the negative swings of the ac voltage must be blocked from passing to the output, and the diode is needed in order to do this. Consider the connection of transformer and diode shown in Figure A4–9a. To use a specific illustration, a transformer with a 12 V_{rms} secondary is illustrated. A time graph of the voltage between the leads of the transformer secondary is shown in Figure A4–9b. When the transformer lead at *A* swings positive in voltage compared to *G*, the diode becomes forward biased and conducts current easily, like a closed switch. But when the voltage at *A* goes negative with respect to *G* so that current would like to flow *into* terminal *A*, the diode will

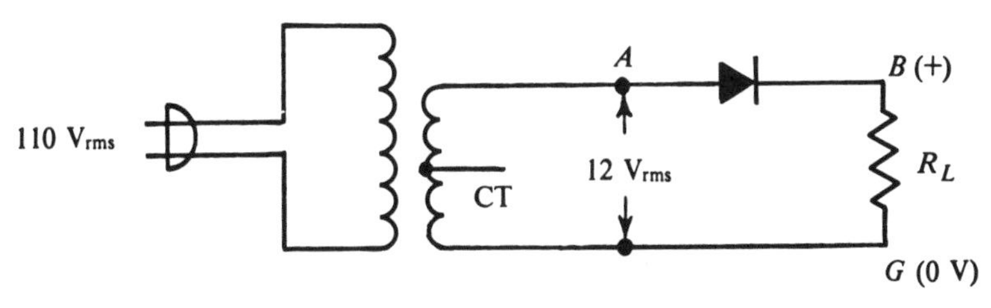

(a) Half-Wave Rectification Circuit with Load Resistor Attached

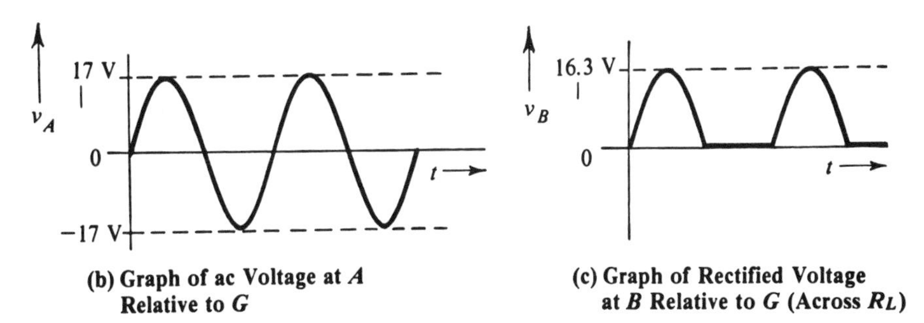

(b) Graph of ac Voltage at *A* Relative to *G*

(c) Graph of Rectified Voltage at *B* Relative to *G* (Across R_L)

FIGURE A4–9. *Half-Wave Rectification Circuit and Characteristic Graphs* not permit current flow in this sense. The diode reverts to acting like an open switch that disconnects the transformer at point *A* from the load at point *B*.

Thus, only the positive voltage swings appear across R_L, and current surges through R_L from *B* toward *G* only during these times. The current flows in only one direction through R_L, and point *B* is only made *positive* relative to *G*. The peak voltage appearing at *B* is the peak voltage from the transformer, which is 12 $V_{rms} \times \sqrt{2} = 17$ V in this case. Actually, a little less appears at *B* due to the nominal 0.7 V drop across the diode while conducting (assuming a silicon diode). Figure A4–9c shows a time graph of the voltage v_B that appears across R_L. The voltage curve v_B has the shape of half of the sine wave from the transformer; therefore the action of the circuit is called *half-wave rectification.*

Notice that during the half cycle that no current is flowing, no voltage appears across R_L; therefore the peak transformer voltage $V_p = 17$ V must appear as reverse bias across the diode. The existence of this reverse bias across the diode means we must use a diode with a minimum reverse breakdown of V_p in this circuit.

Although the voltage between *B* and *G* is unipolar, it is by no means steady and cannot yet be used as a battery replacement; it is satisfactory as a battery charger, however. If a battery of less than 16 V is connected between *B* and *G* as shown in Figure A4–10a, the current passes through the battery in only one sense (in surges) and the battery charges. Connecting the battery to just the transformer as shown in Figure A4–10b does *not* work since the current would pass in one sense just as much as the other. (Actually, the battery would discharge through the resistance of the secondary windings.) Charging current flows in Figure A4–10a whenever the instantaneous output voltage exceeds the battery voltage.

Full-Wave Rectification Using a Center-Tapped Transformer

Let us consider the battery charger of Figure A4–10a again. The current flows during half of each cycle; during the remaining time nothing is ac-

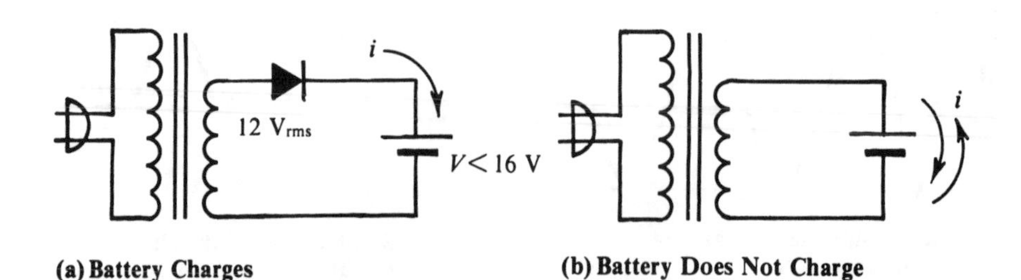

FIGURE A4–10. *Simple Battery Charger*

complished. The battery would charge more rapidly (in half the time) if we could arrange for current to flow from (+) to (−) through the battery during each half of a cycle. This can be done, and making full use of each cycle is called *full-wave rectification.*

One scheme for full-wave rectification requires a *center-tapped* (CT) *transformer* and two diodes. Figure A4–11a shows the diagram with only the secondary drawn. For illustration we assume that the same transformer as in Figure A4–9a is used (notice it was diagrammed as a CT transformer, although not needed there). The voltage between each end of the transformer and the CT ranges up to 8.5 V_{peak}, and the polarities alternate. First assume the "top end" (labeled *T*) of the transformer is (+) and the "bottom end" (labeled *B*) is (−) relative to CT as shown in Figure A4–11a. Then diode (1) is forward biased and conducts current, while diode (2) is reverse biased and blocks any current flow through it. During this half cycle, the current flows from the top (+) end through diode (1) and R_L to the center tap as shown by the dashed line. A peak voltage of about 8.5 V − 0.7 V = 7.8 V develops across the R_L load resistor as shown by the dashed half sine wave in Figure A4–11b. During the next half cycle, the polarities of the top and bottom ends of the transformer reverse relative to the midpoint (the top end goes negative compared to the bottom end). Now the bottom diode (2) is forward biased, since positive voltage is applied to its anode. Current flows from end *B* through diode (2) and the load resistance R_L to the CT as shown by the dotted line. Notice that the current flow is through R_L in the *same* sense as for the

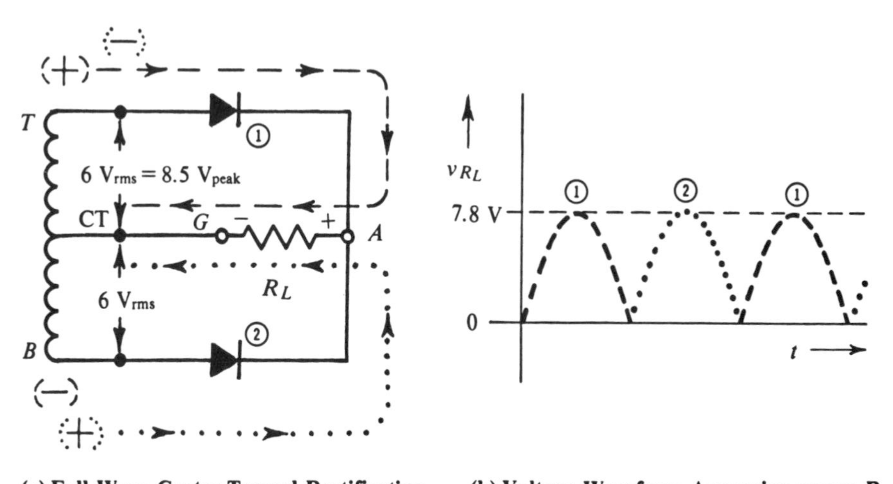

(a) Full-Wave Center-Tapped Rectification **(b) Voltage Waveform Appearing across R_L**

FIGURE A4–11. *Full-Wave Center-Tapped Rectification Circuit Showing Alternate Voltage Polarities and Current Loops along with the Voltage Waveform Appearing across R_L*

first half of the cycle, and the voltage across R_L follows the dotted half sine wave (2) in Figure A4–11b. A full-wave rectified waveform is pictured there. Notice that the peak voltage is only half of what was obtained in the half-wave case. However, the peak reverse voltage that the diodes must sustain is just as much as if the transformer were used in a half-wave circuit. This fact can be seen by noting that when point *A* is positive at +7.8 V, the voltage at the bottom of the transformer (point *B*) is at −8.5 V (relative to the CT); thus nearly 17 V develops across diode (2).

Full-Wave Rectification Using a Bridge of Rectifiers

It is possible to obtain full-wave rectification without a center-tapped transformer, but at the expense of using four diodes. Today diodes are so cheap that using four diodes is the common approach.

Four diodes are arranged in a square with the load resistance connected across one diagonal and the transformer connected across the other as shown in Figure A4–12a. The name of the scheme—full-wave bridge rectification—derives from this bridge arrangement (like the Wheatstone bridge).

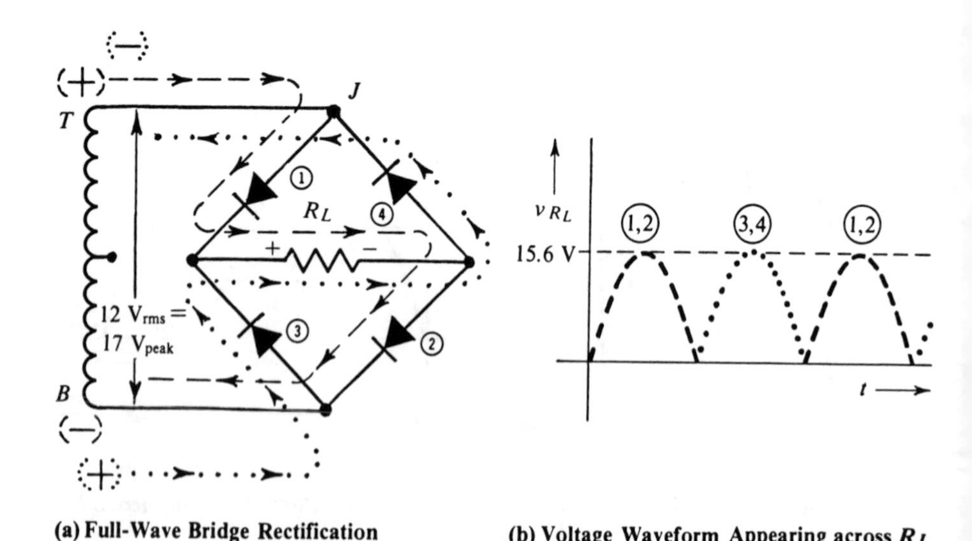

(a) Full-Wave Bridge Rectification

(b) Voltage Waveform Appearing across R_L

FIGURE A4–12. *Full-Wave Bridge Rectification Showing Alternate Current Paths along with the Voltage Waveform Appearing across R_L*

Again assume the top of the transformer secondary is (+) and the bottom is (−) first. Current will attempt to flow out of *T* of the transformer to the junction *J*. Diode (1) will permit the current to progress through it while (4) will not (note the arrow senses of these two diodes). The dashed-line current then comes to a second junction with a choice of diode (3) or the load resistance R_L. Diode (3) says *no* (note its arrow) and the current passes left to right through the load resistance. Finally the current passes through agreeable diode (2) down to the negative bottom terminal of the transformer *B*. Note that the current would not go back "uphill" in voltage through diode (4).

Half a cycle later the secondary winding polarity has reversed to the bottom (+) and top (−). The student should justify that the current now follows the dotted path from *B* through diode (3), R_L, and diode (4) to the negative top *T*. Again note that the current passes in the same direction through the load resistor so that the left end is more positive and the right more negative. The voltage across R_L is full-wave rectified as illustrated in Figure A4–12b, with a peak value of approximately 17 V − 0.7 V − 0.7 V = 15.6 V. With the same transformer, the voltage in the full-wave case is virtually the same as the voltage in the half-wave case. The student should justify why 0.7 V is subtracted twice in this case. The peak reverse voltage appearing on the diodes is also about 16 V.

Filtering to Obtain dc

The rectified waveform obtained in the circuits just described works fine for charging batteries, but such a device is a poor battery eliminator. Using one on a transistor radio, for example, would result in a loud 60-hertz or 120-hertz buzz from the radio. The peaks need to be "smoothed out." The easiest way to manage this need is with a simple *capacitor (C) filter*. A large capacitor (typically electrolytic) is connected across the rectifier output and thus is in parallel with whatever load resistance R_L is connected, as shown in Figure A4–13a.

To understand the smoothing process, assume the transformer is plugged in at instant $t = 0$ in Figure A4–13b. Then the voltage swings positive and the diode is forward biased; it conducts current like a closed switch to the load R_L and capacitor *C*. The capacitor charges and the voltage across it follows the waveform input up to point (1). At this time the voltage from the transformer begins dipping, and the capacitor should like to discharge through the diode (as well as through R_L). However, this current direction through the diode is the wrong way. The diode will not permit it and becomes back biased as the transformer voltage drops toward zero and beyond. Therefore, between points (1) and (2) in Figure A4–13b, the diode behaves like an open switch, and the capacitor must discharge (we hope, slowly) through the load resistance R_L. The voltage decreases exponentially with characteristic

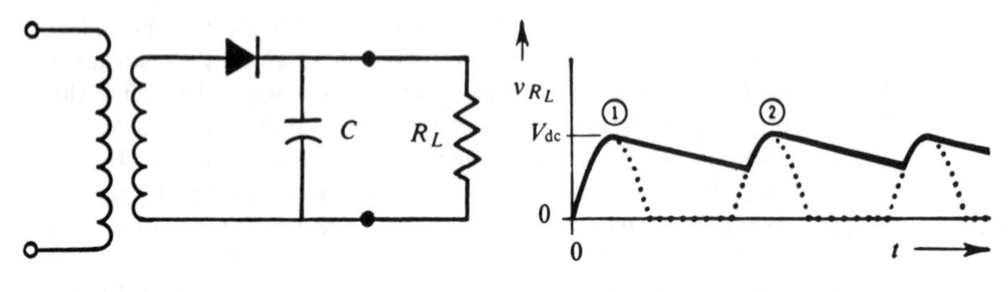

(a) Filtered Half-Wave Rectifier

(b) Smoothed Voltage Appearing across R_L

FIGURE A4–13. *Half-Wave Rectifier with Simple C Filter and Smoothed Voltage Appearing across R_L*

time RC between times (1) and (2). When the transformer voltage again rises to more than the voltage across the capacitor, the diode becomes forward biased, and the capacitor again charges to the peak voltage from the transformer. This process continues, and if the time to discharge (roughly RC) is much more than the time between voltage pulses, a quite steady voltage is sustained across the load resistor.

Notice that the dc voltage that develops is just the same as the peak voltage of the rectified waveform. There is always some *ripple* with a sawtooth-like shape riding atop the dc voltage. The ratio of the peak-peak voltage V_r of this sawtooth to the dc voltage is a measure of the quality of the dc voltage (ideally it would be zero). To calculate this ratio, it is sufficient for a rule-of-thumb design to imagine that the capacitor discharges linearly toward zero over a time RC as shown in Figure A4–14a. In fact, this approximation becomes precise in the case of small ripple.

The triangles with which we shall work in Figure A4–14b are obtained from the ripple and rectified ac waveforms (see Figure A4–14a). From the similar triangles we see that

$$\frac{V_r}{V_{dc}} = \frac{T}{RC} = \textit{ripple ratio} \qquad \textbf{(A4–2)}$$

The time T in this expression is approximately the period between half sinusoids. If the transformer used in the power supply operates from the United States power lines, this time is 1/60 seconds (see Figure A4–15a).

Now we are able to appreciate the advantage of smoothing out a full-wave rectified form. Again a capacitor may be placed directly across R_L for the job. The discharging now takes place over only 1/120 seconds however, as shown in Figure A4–15b, and this value for T should be used in Equation A4–2. The ripple is reduced by a factor of two (for a given C and R_L) because the capacitor has only half as long to discharge.

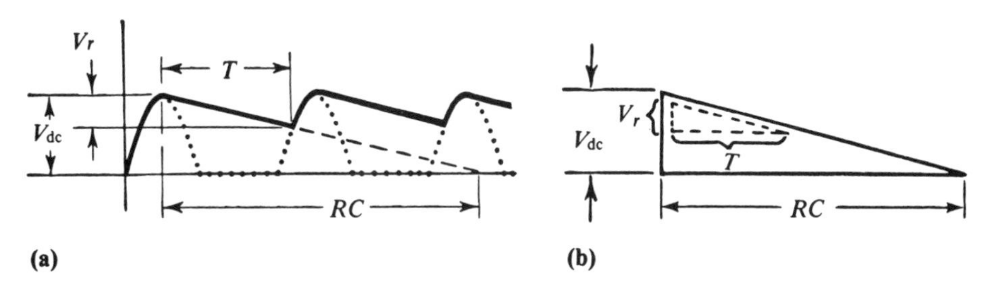

FIGURE A4–14. *Approximate Calculation of the Peak-Peak Ripple Voltage V_r*

Figure A4–15c shows a dc *power supply* or battery eliminator using a simple C filter in conjunction with a full-wave bridge rectifier circuit.

Example: Suppose in Figure A4–13 a 60-hertz, 12-V_{rms} transformer and capacitance of 500 microfarads are used. What is the dc voltage and ripple if this power supply is used with a load resistance of 100 ohms? What dc current is drawn from the supply?

Solution: The dc voltage will be $V_p = \sqrt{2} \times 12\ \text{V} = 17\ \text{V}$ and the dc current drawn is 17 V/100 ohms = 170 milliamperes; next

$$V_r = \frac{V_{dc}T}{RC} = \frac{17\ \text{V}(1/60\ \text{s})}{100\ \Omega \times 500 \times 10^{-6}\,\text{F}} = 5.6\ \text{V}$$

This is a lot of ripple (33%) and a larger capacitor should be used. Using the bridge rectifier of Figure A4–12 with this transformer would reduce the ripple to 2.8 V (16%).

Other Filter Schemes

The problem at hand is to reduce the ac voltage and retain the dc component of the voltage coming from the rectifier output. Clearly, a low-pass filter is an appropriate tool. We can use a resistor and capacitor—that is, an *RC filter*— as shown in Figure A4–16a, or better (though more expensive and bulky), an inductor and capacitor—that is, an *LC filter*—as shown in Figure A4–16b. Because of the ┐ shape of the schematic drawing of the filter they are called *L-section filters.*

We examined the action of a filter in Chapter A1. Again, think of the ac ripple as being reduced by ac voltage divider action. We have impedance elements shown as boxes in Figure A4–17. The "dc part" of the rectified ac

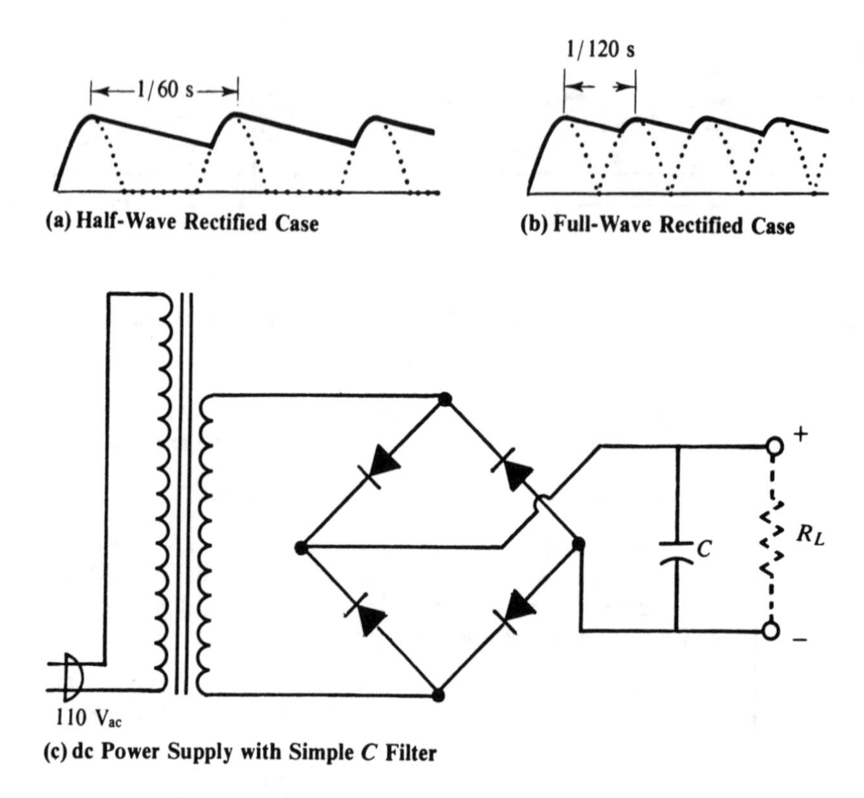

(c) dc Power Supply with Simple C Filter

FIGURE A4–15. *Reduced Ripple from a Filtered Full-Wave dc Power Supply*

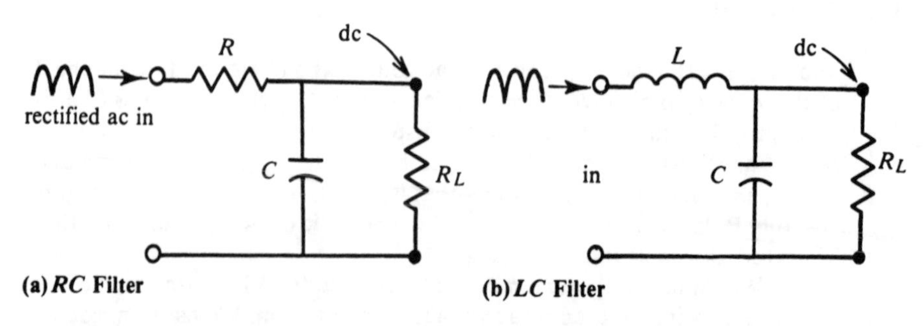

(a) RC Filter (b) LC Filter

FIGURE A4–16. *L-Section Filters to Reduce Ripple*

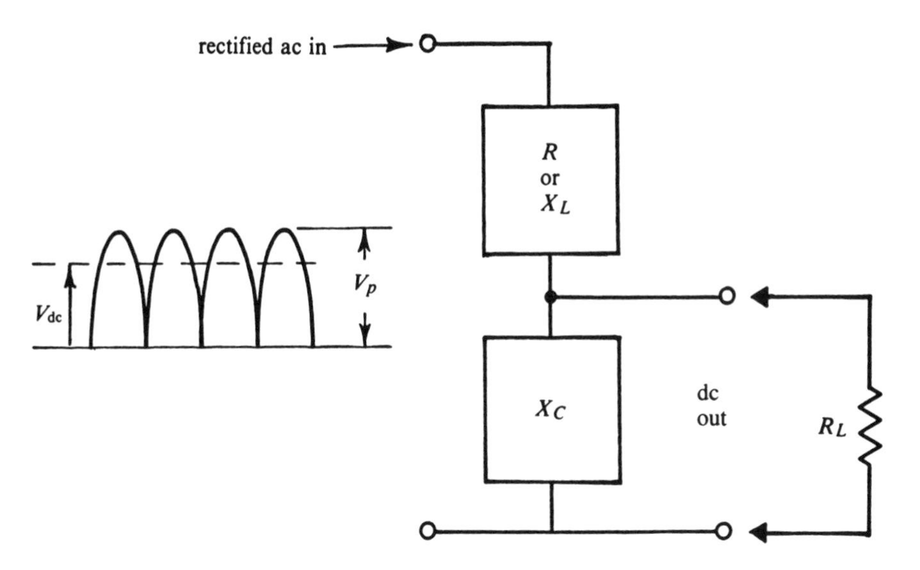

FIGURE A4–17. *Reduction of Ripple by ac Voltage Divider Action in the L-Section Low-Pass Filter*

voltage waveform—that is, the average value of the rectified ac waveform—passes to the output easily. The dc voltage from the output of the filter works out to be about $(2/\pi)V_p$ or 0.64 V_p for the full-wave rectified input. The ac component is reduced greatly if $R >> X_C$ or if $X_L >> X_C$. Then the ripple voltage out is of the order of X_C/R times $V_{ac,in}$ ($V_{ac,in}$ is approximately V_p) for the resistor L section; or X_C/X_L times V_p for the inductor L section.

The resistor input L-section filter is not as desirable if there are variations in the load current because there will be variations in the dc drop that develops across it. The dc voltage out then varies also.

Electronic Voltage Regulators

The ripple remaining after filtering may be in excess of what is tolerable in a scientific application. There are two other problems:

1. *Line voltage variations:* If the power company voltage fluctuates by 10%, for example, this fluctuation also appears in the transformer output and ultimately in the dc voltage.

2. *Load current variations:* When the load resistance changes, changes occur in the current drawn from the transformer and through the diodes and filter. Because of resistance in these elements, increased current results in a drop in the dc voltage (and also typically results in an increase in ripple).

Electronic circuits may be used to greatly reduce the fluctuations in dc voltage due to ripple, line voltage variations, and load current variations; these circuits are called *voltage regulators.* The general voltage regulator can be adjusted to provide the desired dc voltage as output. However, many applications simply require one or more of the standard voltages used with integrated circuits. Therefore, we will confine the discussion here to a group of widely used fixed-voltage regulators, also called three-terminal regulators.

Three-Terminal Regulators

Electronic techniques to regulate against variations in line voltage or load changes have long been available but were relatively expensive. However, integrated circuits (ICs) have now made this elegant approach as cheap as a banana split! One handy type of unit—the *three-terminal regulator*—takes as its input any voltage a few volts higher than the design dc output voltage and *reduces* it to a certain steady dc voltage out (see Figure A4–18a). There is also an upper limit to the voltage that the device can take as input voltage.

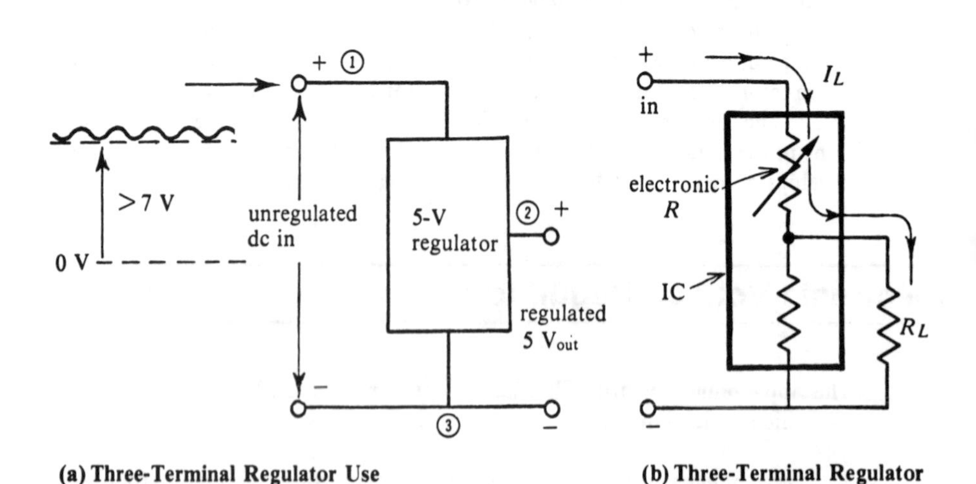

(a) Three-Terminal Regulator Use

(b) Three-Terminal Regulator Operation

FIGURE A4–18. *Three-Terminal Electronic Regulator*

Manufacturers' specifications sheets should be consulted for the capability of a particular device.

A simple view of how the task of reducing the voltage to the steady dc voltage out is accomplished is shown in Figure A4–18b. An electronic variable resistor (a transistor in the IC) is continuously adjusted by a sensing circuit in the device so that the voltage drop across it is just right to subtract from the instantaneous voltage in and yield the desired voltage out. Even the ripple is canceled out this way, as long as the voltage in does not drop below the output voltage. This eliminates the need to purchase very large electrolytics to produce ultra low ripple dc.

NOTE: Heating will occur in this series resistance due to the load current through it, and regulators capable of handling currents over 100 milliamperes typically must be *heat sinked;* that is, they are bolted to aluminum fins to give off the heat. Most units will automatically shut down if overheating occurs due to excessive current (and thus save themselves).

Figure A4–19a illustrates the LM109/209/309 positive 5-V, 1-ampere regulator in the TO–3 metal case that has now been available for a decade. The LM prefix is often used by the National Semiconductor Corporation to designate their various *Linear Microcircuit* (integrated circuit) products. Many other manufacturers also produce the LM309. The 1XX, 2XX, 3XX numbering system is used by National to designate *military temperature range* (−55°C to +125°C), *industrial temperature range* (−25°C to +85°C)

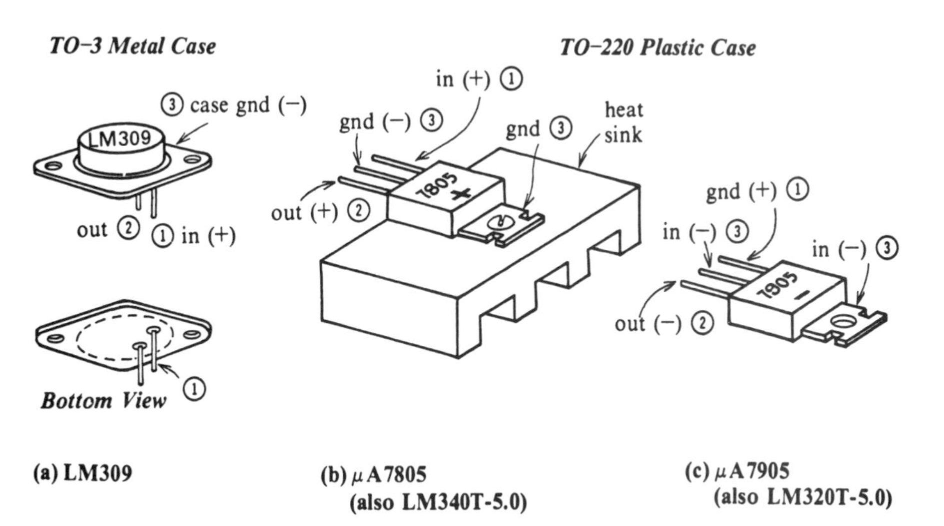

FIGURE A4–19. *Packages of Three Widely Used Fixed Voltage Regulators*

and *commercial temperature range* (0°C to 70°C) devices. As might be expected, the larger temperature range commands a higher price. The in (1), out (2), and ground (gnd) (3) pins are labeled in Figures A4–19a, b, and c and the corresponding connections are numbered in Figure A4–18a. Notice that the case is ground or negative, while pin (1) is positive; the pin usage establishes the device as a positive voltage regulator (unless the case is insulated from the ground or chassis of the power supply; then it may be used in a negative regulated supply).

Figure A4–19b illustrates the μA7805 5 V, 1.5 ampere *positive voltage regulator* in the TO–220 plastic case with heat-sink tab, which is in turn screwed to a heat sink. A variety of fixed voltages are available in the μA78XX group. Thus the 7806 is a +6 V unit, 7815 is a +15 V unit, and so on. There is a corresponding fixed *negative voltage regulator* family, designated μA79XX. The μA prefix is used by Fairchild Semiconductor to designate their microcircuit devices, but again many other manufacturers now produce this family of devices.

NOTE: In Figure A4–19c that the pin functions are defined differently on the negative regulator. Thus, the tab is NOT grounded!

Regulated Power Supplies

As we will discuss in the following chapters, most electronic circuits require a source of dc power (dc voltage and current) for operation; in some cases the dc voltage must be very stable. Portable equipment must obtain this power from batteries (or possibly photovoltaic cells), which are already dc sources of course. Although mercury batteries may be used because of their excellent stability, they are expensive. Rather, an electronic voltage regulator may be used with some battery that has a less desirable voltage discharge characteristic (and also may be rechargeable).

To eliminate batteries altogether in the laboratory, most electronic apparatus use a *regulated dc power supply* that plugs into the ac mains. The constituents for a regulated power supply have already been discussed. They are brought together in the form of a block diagram in Figure A4–20. Although a fixed-output-voltage regulator (e.g., +5 V) would suffice for a specific application, general purpose power supplies often employ adjustable-output-voltage regulators.

A regulated dc power supply is often built into electronic equipment, but sometimes the dc voltages must be provided by one or more exterior supplies. The latter case is often true of "home built" electronic equipment for the lab. When external dc power is required, the most effective approach is often to purchase one of the many power-supply types that are available com-

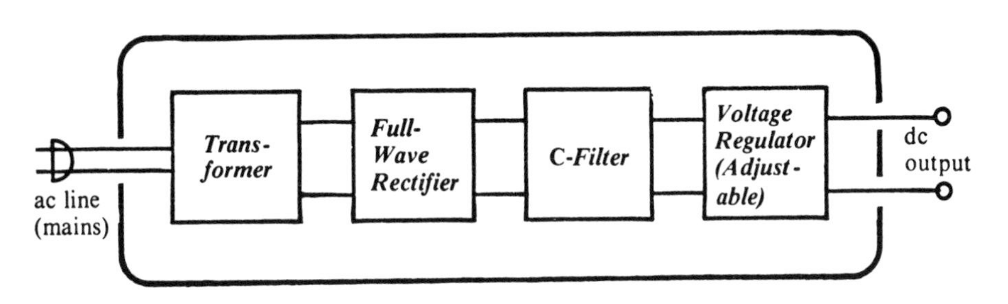

FIGURE A4–20. *Block Diagram of a Regulated dc Power Supply*

mercially. These range from *open frame* units designed to be incorporated into equipment, to more expensive general purpose *laboratory bench supplies.* The latter generally provide one or more adjustable output voltages and meters to monitor output voltage and current.

A modern bench supply also often provides adjustable *current limiting* at the dc output, which protects both the supply and the electronic equipment from catastrophic large currents in the event of a short circuit. This current limit should be adjusted to a value perhaps 30% greater than is expected to be required by the electronic circuit and is done by first actually connecting a shorting lead across the power-supply output. (This sight could have provoked heart failure in a laboratory in 1960!) Then the current-limit control knob should be adjusted for the desired current reading on the supply's current meter.

Because of the low cost of integrated circuits, almost all modern scientific power supplies are regulated. The quality of the regulation is specified relative to the three problems mentioned in the beginning of this section. They are:

1. *Ripple regulation:* Generally the maximum percent ripple is specified. This is the ratio of the rms-ripple voltage to the dc voltage present at the output.
2. *Line regulation:* Generally the maximum percent change in the output voltage is specified for a 10% change in the ac line voltage into the supply.
3. *Load regulation:* Generally the maximum percent change in the output voltage is specified for a change from zero current to the maximum rated output current for the supply.

In addition, stability of the output voltage against changes in ambient temperature and long term stability (or voltage drift) are specified for quality supplies.

REVIEW EXERCISES

1. Diagram the electrical characteristic for an ideal diode.
2. (a) Which element in the vacuum diode is the source of electrons within the device?
   (b) What should be the polarity of the voltage of the anode relative to the cathode for current to flow through the vacuum diode? Why?
3. (a) What is the difference between an intrinsic carrier and an extrinsic carrier in a semiconductor?
   (b) Which is a better insulator: a doped semiconductor or a pure semiconductor?
   (c) What is the difference between N-type semiconductor and P-type semiconductor?
4. (a) What is the depletion zone in a semiconductor junction diode? (Basically, why does it form?)
   (b) Why is there a barrier voltage in a semiconductor junction diode? What is it between, and what does it come from?
   (c) Which material in a PN junction should be more positive to produce significant current flow? Briefly explain why.
5. Consider the PN junction shown in Diagram A4–1. Reproduce the little drawing and name the two terminals. Also draw the corresponding diode symbol beside it.

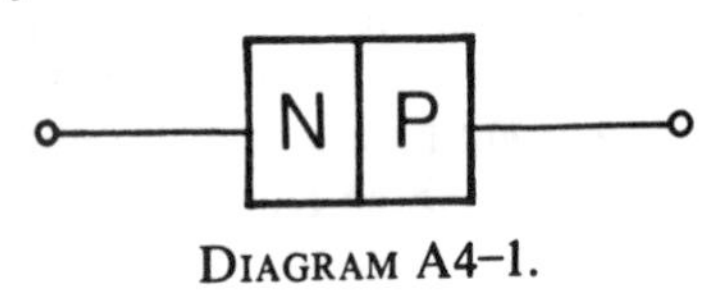

DIAGRAM A4–1.

6. (a) What is the reverse-leakage current for an ideal diode?
   (b) Compare the typical reverse-leakage currents for silicon and germanium diodes.
7. (a) What is meant by the knee of a real diode characteristic?
   (b) Give approximate values for this knee for silicon and for germanium diodes.
8. Name the three important electrical parameters that are commonly specified for diodes by their manufacturers.
9. Which of the circuits shown in Diagram A4–2 will certainly ruin the real diode, and why? In each of the other cases, explain why the diode is not likely to be ruined.

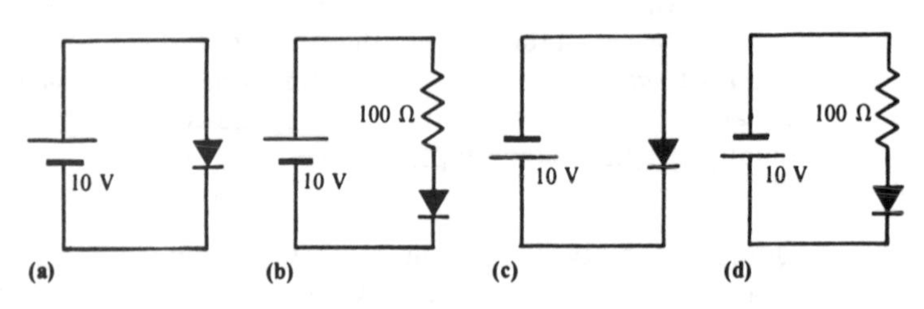

DIAGRAM A4–2.

10. Consider the two diodes connected in series to a voltage source as shown in Diagram A4–3.
    (a) If the two diodes are ideal, does any current flow? Why?
    (b) For real diodes, there is certainly a tiny current flowing, and there is the possibility of a large current. Explain what diode characteristic is involved in each case and to which diode in the circuit it applies.

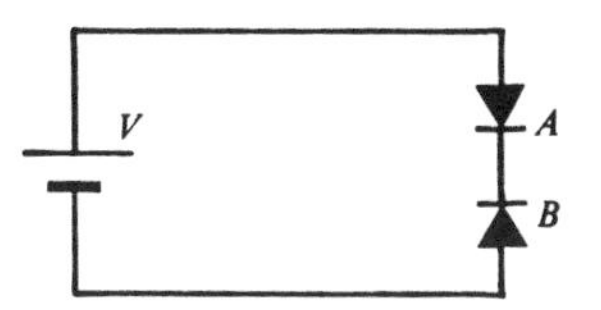

DIAGRAM A4–3.

11. (a) What specification is typically given by manufacturers for signal diodes, but not for rectifiers?
    (b) Give order-of-magnitude forward-current capabilities for signal diodes and for epoxy-case rectifiers.
    (c) Three widely used diodes are the 1N34, 1N914, and 1N400X. Designate each as silicon or germanium and signal diode or rectifier.
    (d) What is the difference between the 1N4001 and the 1N4007?

12. A signal diode is placed in two circuits as pictured in Diagram A4–4. In which case will a significant current flow?

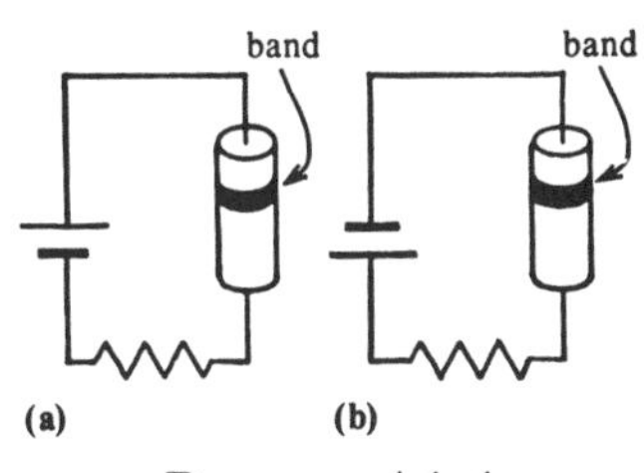

DIAGRAM A4–4.

13. Sketch the voltage output from a half-wave rectifier and a full-wave rectifier.

14. Consider Diagram A4–5.
    (a) Which diode connection will correctly produce rectification of the polarity indicated on the load resistor?
    (b) Which connection(s) will probably burn out all diodes?

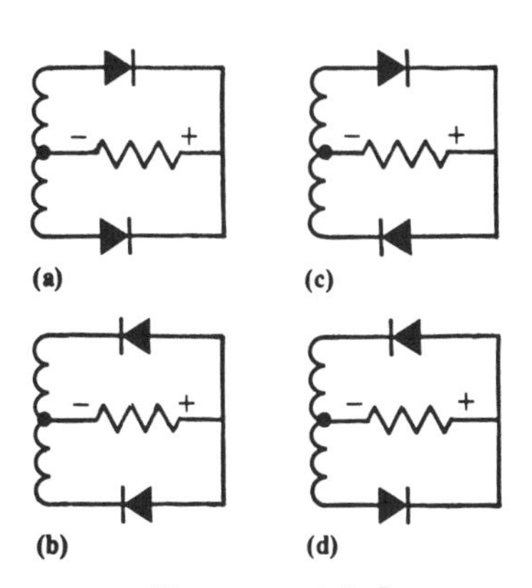

DIAGRAM A4–5.

15. In the bridge rectifier shown in Diagram A4–6, draw in diodes to produce the indicated polarity on the resistor.

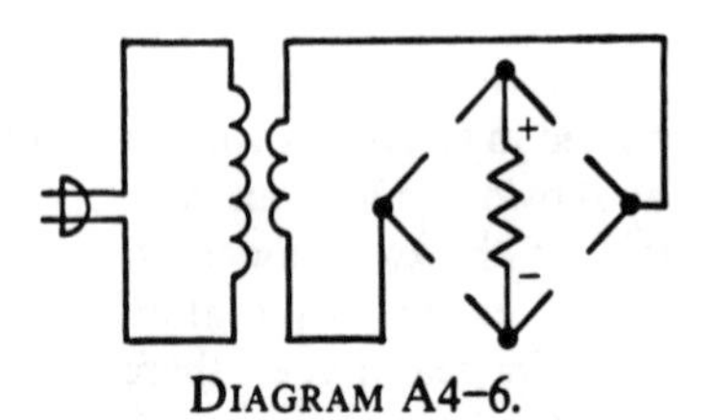

DIAGRAM A4–6.

16. What is the difference between a battery charger and a battery eliminator?
17. Describe how a simple shunt-capacitance filter is able to produce an "almost dc" voltage from a rectified ac voltage.
18. (a) Draw an *L*-section filter constructed from a capacitor and an inductor, as well as from a resistor and a capacitor.
    (b) What type of filtering action is desired, high pass or low pass?
    (c) Which *L*-section filter displays better load regulation, the *LC* or the *RC*? Why?
19. Which power supply is superior, one with 1% line and load regulation, or one with 5% line and load regulation?
20. Briefly, tell the purpose of each of the following elements of a power supply:
    (a) transformer
    (b) rectifier
    (c) filter
    (d) regulator

PROBLEMS

1. Consider the circuit drawn in Diagram A4–7. What is the current drain of the battery? (Include approximate diode drop for silicon.)

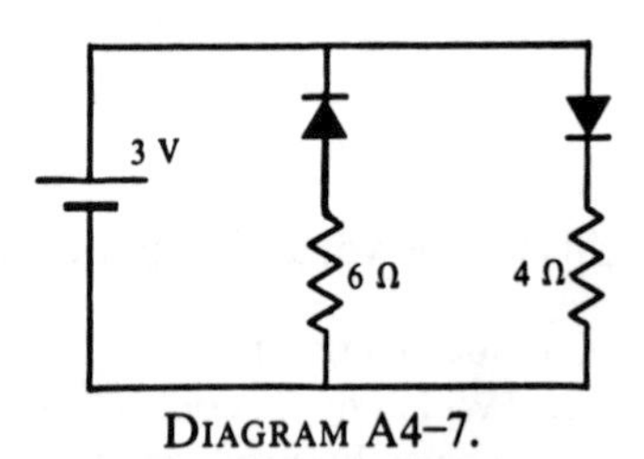

DIAGRAM A4–7.

2. (a) Consider the diode law. Show that the quantity $|e|/kt$ is equal to 39.6 V^{-1} at room temperature (20° C). Justify both the number and the units.
   (b) Assume a reverse-leakage current appropriate to silicon of 0.01 μA. Evaluate the diode law to find the currents for respective forward biases of 0.1 V and 0.45 V.
   Note that the change in current certainly does not follow Ohm's Law.
3. (a) Suppose the ac voltage $v_{in}(t)$ shown in Diagram A4–8(a) is input to the circuit diagrammed in (b). Sketch the graph of the output voltage

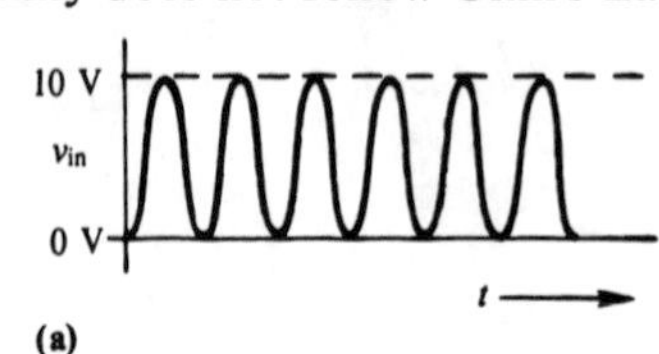

(a)

$v_{out}(t)$. Include a voltage scale.

(b) Suppose instead the voltage $v_{in}(t)$ is input to the circuit diagrammed in (c). Sketch the graph of the output voltage $v_{out}(t)$. Include a voltage scale.

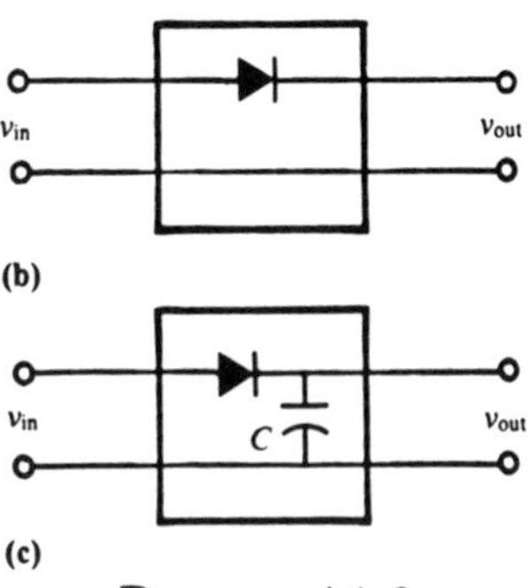

DIAGRAM A4–8.

4. (a) Why must we be concerned about the reverse-voltage rating of diodes to be used in a rectifier circuit?
   (b) What minimum reverse-voltage rating is required for the diode used in a half-wave rectifier circuit that is used in a 20-V_{dc} power supply?
   (c) What minimum reverse-voltage rating is required for the diodes used in a full-wave rectifier circuit with center-tapped transformer, if the circuit is used in a 20-V_{dc} power supply?
5. Draw a complete half-wave battery charging circuit. If the battery to be charged is 6 V, what is the minimum rms voltage from the transformer output?
6. Consider Diagram A4–9: What is the dc voltage out of each power supply?

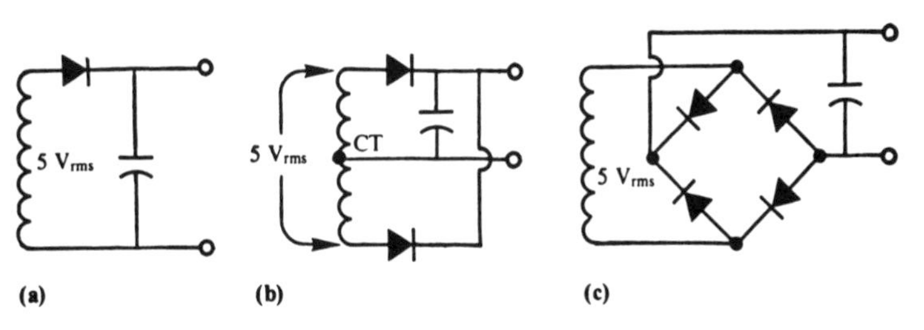

DIAGRAM A4–9.

7. (a) Consider the power supply shown in Diagram A4–10. Sketch the voltage appearing across R_L, $v_R(t)$ versus t, if the switch S is open. Include a scale on your two axes and use a dashed line for this $v_R(t)$.
   (b) Suppose $R_LC = 0.01$ s. Sketch with some accuracy $v_R(t)$ when the switch is closed. Use a solid line for this $v_R(t)$.

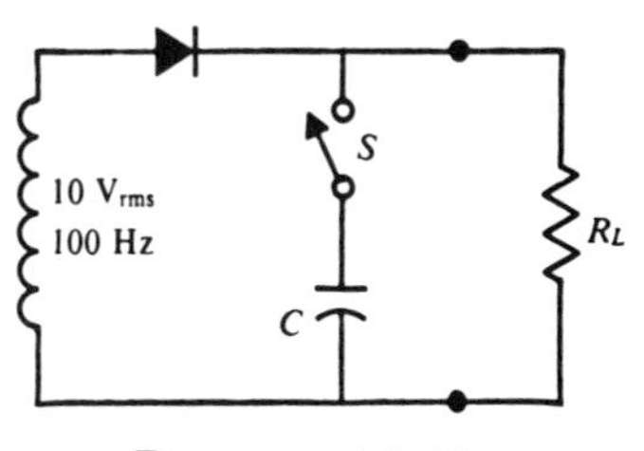

DIAGRAM A4–10.

   (c) Comment on the performance of this power supply.

8. Suppose you want to build a battery eleminator for a small portable radio. It operates from 4 AA (penlight) cells in series. With a VOM you measure the battery current drain to be 30 mA when the radio is on. You accept a ripple voltage (peak-to-peak) of 0.1 V and plan to use a bridge rectifier scheme.
   (a) What voltage transformer (rms) should you purchase?
   (b) What size filter capacitor should you obtain for simple C filtering? What type of capacitor should it be?
9. Consider the L-section filter shown below in Diagram A4–11.
   (a) What dc voltage is expected out?
   (b) What is the approximate value expected for the ripple voltage?

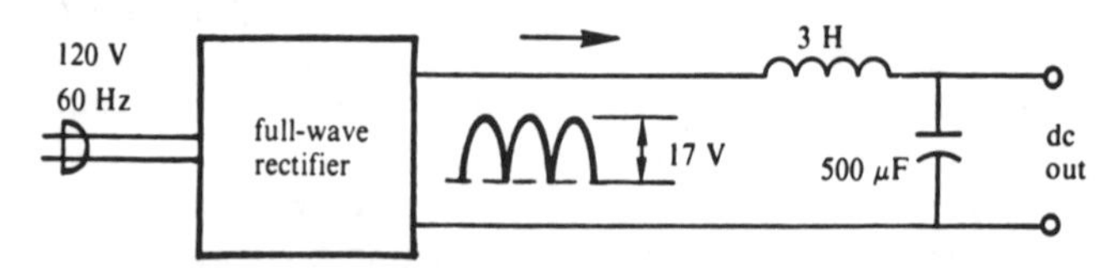

DIAGRAM A4–11.

10. (a) An LM309 three-terminal regulator is operated from a 12-V automobile battery to produce a +5 V_{dc} supply for a portable computer circuit. Suppose a 10-Ω load is connected as shown in Diagram A4–12. What is the power (heating) dissipated within the LM309?

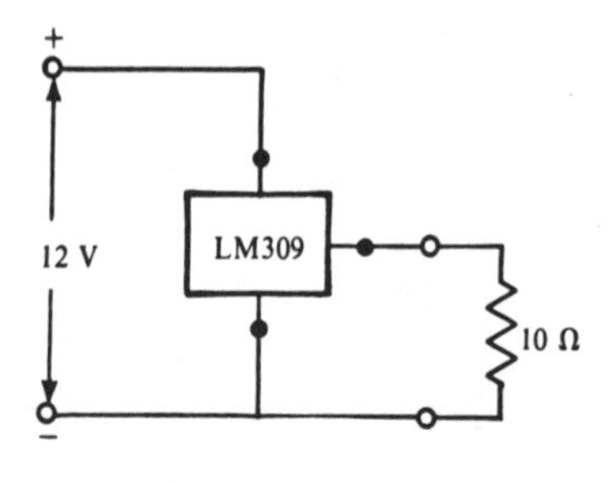

DIAGRAM A4–12.

   (b) Why would a 6-V automobile-type battery in principle be a better choice? (In practice, the input voltage should be about 2 V greater than the regulated output voltage, however. This rules out the 6 V battery.)
11. The load regulation of a power supply can be characterized by the power supply *output impedance*. This is simply the Thevenin equivalent resistance, R_{out} from the output terminals. Suppose the output voltage from the supply changes by an amount ΔV_{out} when the load current changes by ΔI_{out}. Show that $R_{out} = \Delta V_{out}/\Delta I_{out}$.
12. An *inverter* is a device that converts from one dc voltage level to a higher dc level. This is done with the aid of an oscillator that operates from the lower dc voltage and a transformer.
   (a) Make a block diagram of a possible inverter design. Why is an oscillator required?

(b) The frequency of the oscillator is usually chosen to be well above 60 Hz (for example, 400 Hz). Why does this permit the use of small-capacitance filtering?

13. Zener diodes are manufactured specifically to be operated in reverse-breakdown mode and to take advantage of the well-defined voltage at which it occurs for a given diode. Provided the power dissipated in the diode does not exceed the power rating of the zener, the diode is quite safe and operates with a nearly constant *zener voltage* across it.

Note that in normal operation the zener diode cathode is positive. A zener diode may be operated in a simple voltage-regulating circuit, typically providing about 1% line and load regulation.

(a) Consider the zener regulator circuit shown in Diagram A4–13 (which illustrates the zener diode symbol). If the zener voltage is 6 V, find the current in the 1-kΩ resistor, the 6-kΩ load resistor, and finally the zener diode. What is the power dissipated in the zener diode?

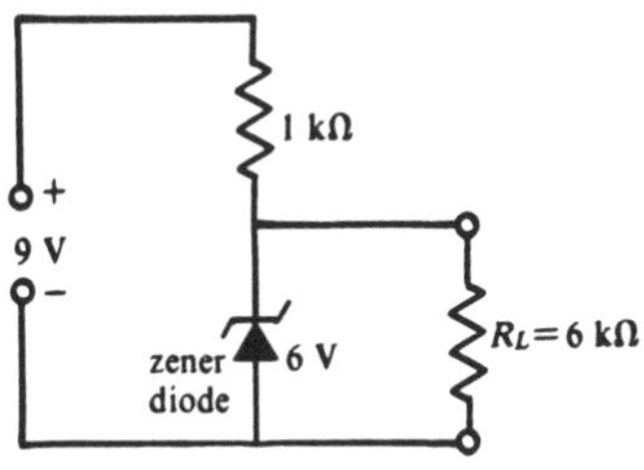

DIAGRAM A4–13.

NOTE: This problem can be worked out in your head.

(b) Suppose the load resistor R_L is reduced from 6 kΩ to 2 kΩ. Again find the current in the 1-kΩ resistor, the load resistor, and finally the zener diode.

(c) The load used in part (b) is the minimum value (and draws the largest load current) for which the voltage across the load resistor is nearly constant. For smaller R_L, the voltage across it is simply determined by voltage-divider action between the 1-kΩ and the R_L resistors. Justify this statement.

CHAPTER

A5

Amplifier Behavior

Fundamental Amplifier Properties

In an electronics application we are often faced with a source of small voltage that is not capable of producing the desired end results; therefore, an amplifier of some sort is needed. Actually, it is not only the voltage capability of the source that is of concern, but also the current capability. At times a source produces a quite satisfactory voltage but can supply only a small amount of current. These features must be considered in the system design, as we see shortly.

In dealing with amplifiers there are a number of amplifier properties that are always the same, no matter the design or manufacturer of the unit. These properties should be appreciated before examining specific amplifier designs in later chapters; they can be indicated in a fairly simple *black-box model* of the amplifier. Also, black-box models are useful for the input and output devices that connect to the amplifier. Note that this amounts to putting some equivalent circuits into the boxes of the system model in Figure A3–1.

Black-Box Models: Input Transducer, Amplifier, and Output Transducer

By Thevenin's Theorem we expect typical *input transducers* to behave like a voltage source (in some cases dc, while in others ac) that is in series with a certain resistance called its *output impedance*. The input transducer may then be represented by a box with contents *source voltage* $v_s(T)$ and source resistance R_s (see Figure A5–1). The source voltage depends on the parameter of the system being studied; we use T for temperature, but the parameter could be any measurable quantity.

Next the *amplifier* has two terminals called the *input port* that appear to have an *input resistance* R_{in} (also called an *input impedance*) between them as far as any external circuit is concerned. The amplifier monitors whatever voltage appears across the input resistance R_{in} and generates a voltage that is larger by the factor A_v. A_v is the *open circuit voltage gain* or *voltage amplification* of the amplifier. This voltage is represented by a voltage generator with voltage $A_v v_i$ that in turn is in series with an *output resistance* R_{out}. The voltage generator in series with the resistance R_{out} is connected to the two output leads—that is, the *output port* or output of the amplifier. Note the Thevenin equivalent nature of the amplifier output circuit.

Finally the amplifier is connected to an output transducer, which is the *load* for the amplifier. This load has a certain resistance R_L. A voltage develops across this resistance and power is dissipated in it. In fact, the name of the game is to begin with a low power P_{in} produced by the input transducer into the input of the amplifier and produce a large power P_{out} into the output transducer. Generally (but not always) this involves an increase in the voltage level of the signal. The *power gain* or *amplification of power* A_p is defined through the equation:

$$A_p = \frac{P_{out}}{P_{in}} \tag{A5–1}$$

Expressed in decibels it is equal to 10 log A_p.

To predict the voltage that appears across R_L, the source voltage v_s, the four resistors R_s, R_{in}, R_{out}, R_L, and the voltage gain A_v must be known. The situation is perhaps best discussed by example.

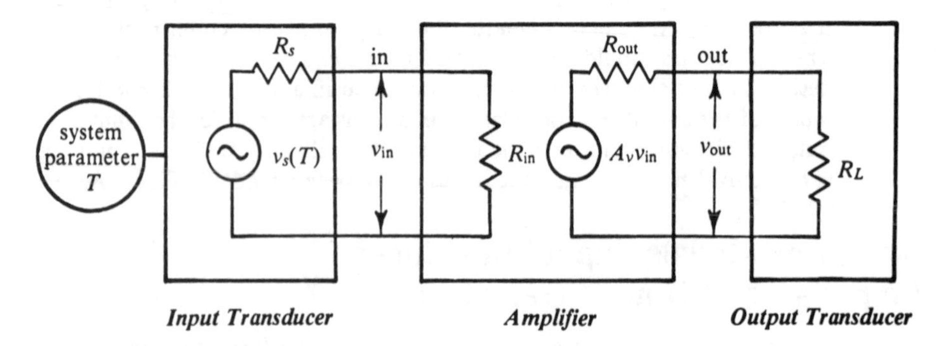

FIGURE A5–1. *Black-Box Model of an Electronic System*

Example: Suppose a microphone with an output impedance of 10,000 ohms is connected to an amplifier with an input resistance of 1,000 ohms. The voltage gain of the amplifier is 2,000 and its output impedance is 200 ohms. The amplifier is connected to a loudspeaker with a resistance of 8 ohms. If the microphone produces an *open circuit* voltage of 0.2 V_{rms}, what voltage appears across the loudspeaker and what power is dissipated in it? What is the overall voltage gain and power gain of the system? The situation is diagrammed in Figure A5–2.

Solution: We must work from input to output in the system. Although the voltage from the microphone is a time-varying or ac voltage, we may work with the rms value of the microphone voltage. Consequently, the other voltages throughout the electronic system are also developed as the respective rms values, and the uppercase letter V is used for V_{in} and V_{out} rather than the lowercase v of Figure A5–1.

1. A voltage V_{in} appears between the input terminals of the amplifier that is less than 0.2 V because of the voltage divider action. We see

$$V_{in} = 0.2\ V_{rms}\left(\frac{1\ k\Omega}{1\ k\Omega + 10\ k\Omega}\right) = 0.0182\ V_{rms}$$

2. The signal generator within the amplifier generates a voltage

$$2000\ V_{in} = 2000 \times 0.018\ V_{rms} = 36.4\ V_{rms}$$

3. The voltage out of the amplifier that appears across the loudspeaker is less than this, however, because of voltage divider action between the 200-ohm and the 8-ohm resistances. We see

$$V_{out} = 36.4\ V_{rms}\ \frac{8\ \Omega}{8\ \Omega + 200\ \Omega} = 1.4\ V_{rms}$$

4. The power developed in the loudspeaker is

$$P = \frac{V^2}{R} = \frac{(1.4\ V_{rms})^2}{8\ \Omega} = 0.245\ W$$

5. The power generated by the voltage source in the microphone is

$$P = \frac{V^2}{R} = \frac{(0.2\ V_{rms})^2}{(10{,}000 + 1000)\ \Omega} = 3.6 \times 10^{-6}\ W$$

6. The voltage gain of the *system* is the voltage 1.4 V_{rms} across the speaker compared to 0.2 V_{rms} in the microphone, for a ratio of only 7.

7. The power gain of the *system* is

$$\frac{0.245 \text{ W}}{3.6 \times 10^{-6} \text{ W}} = 68{,}000$$

In decibels it is

$$10 \log 68{,}000 = 48 \text{ dB}$$

8. The power gain of the amplifier is much more than that in (7). The power into the amplifier itself P_{in} is

$$P_{in} = V_{in}I_{in} \text{ or } \frac{V_{in}^2}{R_{in}} = \frac{(0.0182 \text{ V}_{rms})^2}{1000\ \Omega} = 3.31 \times 10^{-7} \text{ W}$$

(Note that this is only about 1/10 the power delivered by the voltage in the microphone, but it *is* the power delivered by the microphone into the amplifier.) The power delivered by the amplifier to the loudspeaker was obtained in (4) above. Then using Equation A5–1 we get

$$A_p = \frac{P_{out}}{P_{in}} = \frac{0.245 \text{ W}}{3.3 \times 10^{-7} \text{ W}} = 7.42 \times 10^5$$

Expressing this in decibels we have

$$\begin{aligned} \text{dB ratio} &= 10 \log (7.42 \times 10^5) \\ &= 10\,(5 + \log 7.42) \\ &= 10 \times 5.87 = 58.7 \text{ dB} \end{aligned}$$

We say the amplifier itself affects a 58.7 decibel power again.

NOTE: The power delivered by the voltage generator inside the amplifier is considerably more than the power delivered to the load—in fact, about 20 times more—but about 95% of this is lost in the 200-ohm resistor.

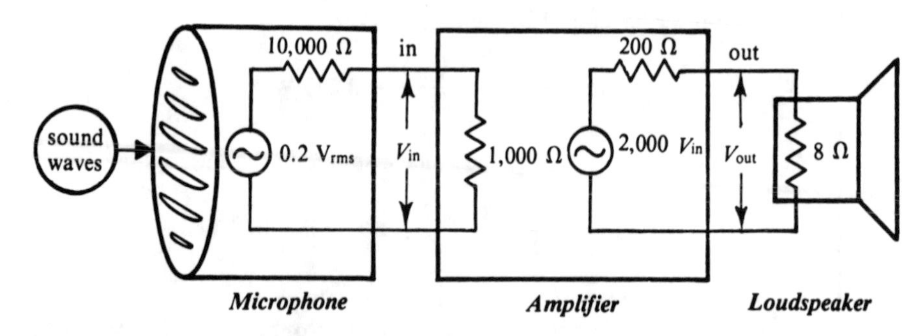

FIGURE A5–2. *Example of Black-Box Electronic System*

Several important features of electronic systems are evident in this example. First notice that the voltage V_{in} appearing at the input of the amplifier is greatly reduced by voltage divider action from the 0.2 V_{rms} developed *within* the microphone. And the amplifier can only amplify the signal V_{in} appearing between its input terminals, so that we have a case of "2 steps backward before making 4 steps forward." This is an example of *impedance mismatch*.

Instead of having R_{in} be much less than R_s (here 1,000 ohms and 10,000 ohms, respectively), R_{in} should be much greater than R_s. Then little voltage divider action would take place and the amplifier would develop an internal voltage of $2{,}000 \times 0.2\ V_{rms} = 400\ V_{rms}$ (rather than 36.4 V_{rms}). It should now be clear that *an ideal amplifier has an input impedance of infinity* (similar to a voltmeter).

The second feature to notice in this example is that the output voltage across the loudspeaker is greatly reduced because of the voltage divider action that takes place between the 200-ohms R_{out} and the 8-ohms R_L of the loudspeaker. We say the loudspeaker *loads down* the amplifier output. If the output impedance of the amplifier were very low (approaching zero) the full voltage would appear across R_L irrespective of its value. Therefore, we can see that *an ideal amplifier would have zero output impedance.*

The third feature to notice is that although the voltage amplification in this example is only 7, the power amplification is very much more: 68,000. This is because the impedance of the microphone and loudspeaker are so different. In fact, the voltage generated in the microphone is rather substantial as transducers go, but because of its high output impedance, the microphone is capable of delivering only a small current and, therefore, power to the outside world.

To demonstrate this: Suppose the microphone were connected directly to the loudspeaker. Then the current flowing from the microphone and through the loudspeaker is

$$I = \frac{0.2\ V_{rms}}{10\ k\Omega + 8\ \Omega} \simeq \frac{0.2\ V_{rms}}{10\ k\Omega}$$

The power delivered to the speaker is given by I^2R or

$$P_{speaker} = \left(\frac{0.2\ V_{rms}}{10\ k\Omega}\right)^2 8\ \Omega = 3.2 \times 10^{-9}\,W$$

This is indeed a tiny amount compared to that delivered to the loudspeaker with the help of the amplifier (0.234 watts calculated in part 4). We say there is a *large impedance mismatch* between the microphone and the loudspeaker.

Power Gain with an Ideal Voltage Follower

The fact that there was a large power gain with only a small voltage gain in the microphone/amplifier/loudspeaker system may have been surprising.

Actually, there can easily be power gain when there is no voltage increase; this accounts for the utility of that rather odd "amplifier," the voltage follower. The *voltage follower* is an amplifier with an open circuit voltage gain A_v of unity. Then the output voltage (at least if it is unloaded) is just the same as the input voltage. The output voltage *follows* the input voltage in that at every instant the voltage between its output terminals is a copy of the input voltage.

Let us consider the case where an ideal voltage follower has been interposed between the microphone and loudspeaker of Figure A5–2. Recall that by ideal we mean its input impedance R_{in} is very large ($\approx\infty$) and the output impedance R_{out} is very small (≈ 0). This system is diagrammed in Figure A5–3.

Because of its very large input impedance, the voltage divider action at the input to the ideal amplifier is negligible, and V_{in} is just 0.2 V_{rms}. This voltage is copied by the voltage generator in the black-box model of the voltage follower and passes unattenuated to the 8-ohm loudspeaker resistance. The power expended in the loudspeaker is then

$$P = \frac{V^2}{R} = \frac{(0.2\ V_{rms})^2}{8\ \Omega} = 5 \times 10^{-3}\,\mathrm{W}$$

The result is power over a million times greater than the power delivered to the loudspeaker if the microphone is connected directly to it. Thus, although there is no voltage gain, there is a very great power gain, and the voltage follower deserves the title of amplifier. The voltage follower is developing a rather respectable power (relative to the world of electronics) by maintaining 0.2 V_{rms} across the rather small resistance of the loudspeaker.

NOTE: If the voltage follower can sustain this same voltage across a very small resistor, the power delivered becomes very large.

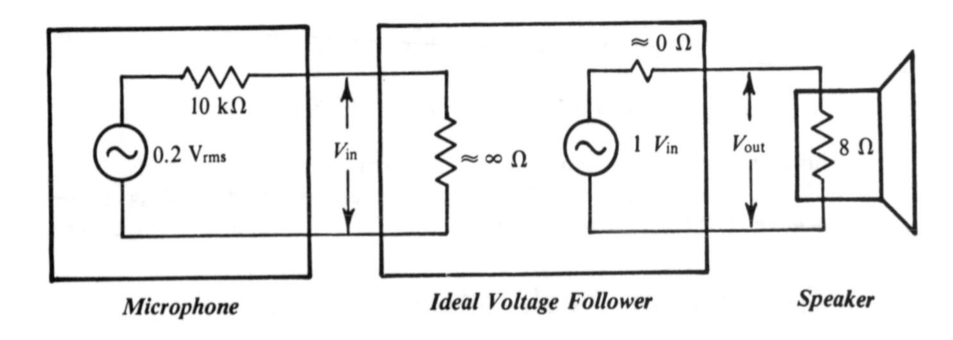

FIGURE A5–3. *Ideal Voltage-Follower Amplifier*

Impedance Matching and Maximum Power Transfer

Suppose we have a certain amplifier as a given—in particular, a certain output voltage and output impedance—but we have the choice of various impedance values for the load. For example, suppose that in the Figure A5–2 the amplifier's 200-ohm output impedance is something we must live with, but imagine we can choose the impedance of our loudspeaker. (This is not close to reality for loudspeakers, but it can be true for other output transducers.) The question now is: What value for load impedance maximizes the power delivered to the load? How do we obtain *maximum power transfer* to the load?

We first write down a general expression for the power delivered to the load and treat the load resistance R_L as a variable. Figure A5–4 illustrates the situation at hand. The power delivered to R_L (as a function of R_L) is

$$P(R_L) = \frac{V_{R_L}^2}{R_L} = \frac{[VR_L/(R_L + R_{\text{out}})]^2}{R_L}$$

$$= \frac{V^2 R_L}{(R_L + R_{\text{out}})^2} \qquad \textbf{(A5–2)}$$

Notice that in the limit of very small load resistance ($R_L \rightarrow 0$) the power goes to zero due to the zero R_L in the numerator. (The power is all dissipated inside the amplifier in the resistance R_{out}.) At the other extreme, as R_L becomes very large ($R_L \rightarrow \infty$), the power also goes to zero because of the infinity in the denominator. (No power is delivered to the load because the current flowing drops to zero.) There apparently is some R_L between these extremes that accomplishes the maximum power transfer.

From calculus we know that the slope of the function is zero at the maximum of the function. Therefore, to find the maximum of the function $P(R_L)$, we take the derivative of Equation A5–2 with respect to R_L, set the result equal to zero, and solve for R_L; that is,

$$\frac{d}{dR_L}\, P(R_L) = \frac{V^2}{(R_L + R_{\text{out}})^2} - 2V^2 R_L (R_L + R_{\text{out}})^{-3} = 0 \qquad \textbf{(A5–3)}$$

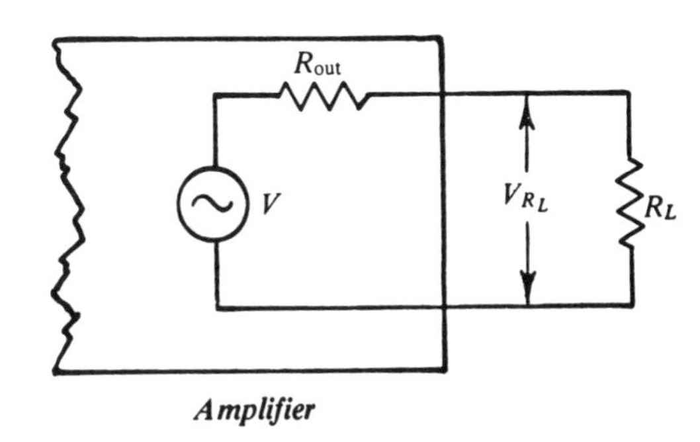

FIGURE A5–4. *Pertinent Circuit for Solving for Maximum Power Transfer*

Cancel the V's and multiply through by $(R_L + R_{out})^3$. Then we get

$$R_L + R_{out} - 2R_L = 0 \qquad \text{or} \qquad R_L = R_{out} \qquad \textbf{(A5-4)}$$

This is the condition for maximum power transfer: that the load resistance be equal (or match) the output impedance of the power source. It is a bit surprising since clearly half the power delivered by the voltage generator in the source of power is dissipated in the R_{out} of the source itself.

A common example of impedance matching occurred in the past for vacuum tube high-fidelity (hi-fi) amplifiers. The output of the amplifier was available through several impedance-value outputs: for example, 4 ohms, 8 ohms, 16 ohms. To obtain maximum power transfer to a 16-ohm speaker, for example, the speaker should be connected to the 16-ohm post. This impedance matching was done by using a step-down transformer with several taps on the secondary to provide various impedance levels.

Cascaded Amplifiers

Suppose we have several amplifiers, each with an *unloaded voltage gain* of 10, but there is need for a voltage amplification of 1,000. What can be done? The answer is quite clear: Connect the output of the first into the input of a second, and so on, in a sequence. The result is a *cascaded amplifier* as illustrated in Figure A5–5.

The output of the first amplifier becomes the input of the second, and so on. In general we must be concerned about the loading of the second amplifier on the output of the first amplifier in order to predict the output voltage of the first, which then becomes the input voltage to be amplified by the second. This concern carries on down the line.

If each amplifier were ideal ($R_{in} = \infty$; $R_{out} = 0$), then it is easy to see what the voltage out of the system should be, compared to v_{in}; that is, the overall voltage gain of the cascaded amplifier. The output voltage of the first

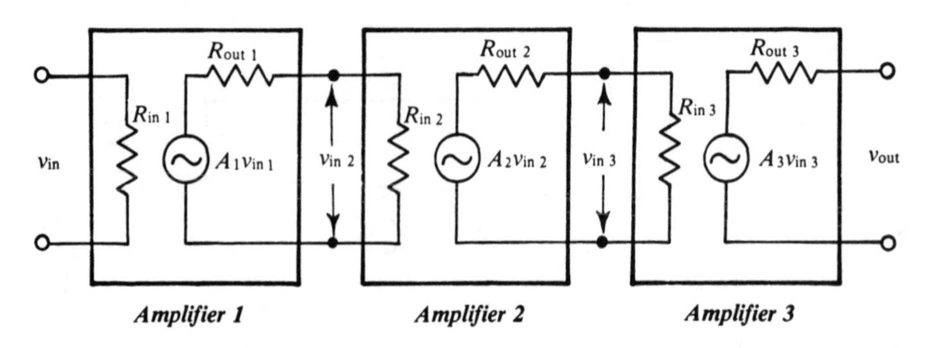

FIGURE A5–5. *Cascaded Amplifier*

amplifier would then be $v_1 = A_1 v_{in}$. This is amplified by the second amplifier to become $v_2 = A_2(A_1 v_{in})$. Finally, the third amplifier receives this as input and produces as output $v_3 = A_3(A_2 A_1 v_{in})$. We see the overall gain of ideal amplifiers would be the product of the individual amplifications:

$$A_{\text{total ideal}} = \frac{v_{out}}{v_{in}} = A_3 A_2 A_1 \qquad \textbf{(A5–5)}$$

The examination of a more realistic case is left to the problems. There the student is asked to derive a formula for overall gain if the three amplifiers are identical.

Fundamental Amplifier Types

There are a number of fundamental variations or types of amplifiers that are discussed in the remaining sections of this chapter. These types are introduced here and their properties and terminology are discussed. To aid this understanding we first consider a simple electrical model for the operation of a basic amplifier building block or stage.

Amplifier-Stage Model

The basic amplifier is constructed around one transistor or vacuum tube (as we see in Chapter A8) and can be characterized by a certain input impedance, output impedance, and voltage gain. In other words, a basic amplifier behaves like the amplifier black box that we have already examined. Typically the voltage gain from one basic amplifier is not enough for a task; thus several basic amplifiers must be cascaded as shown in Figure A5–5. Each basic amplifier in the overall chain of amplifiers is then referred to as an *amplifier stage.*

We see also in Chapter A8 that it is in the nature of amplifiers for the output voltage to tend to be midway between the negative and positive terminal voltages of the battery (or batteries) supplying the power to the amplifier. The reason can be understood from a voltage divider model of the amplifier as shown in Figure A5–6a. In an amplifier stage, the *active device*—that is, the transistor or vacuum tube—behaves like an electrically controlled variable resistor; it changes its resistance in response to an input voltage to a control input. The output voltage from the amplifier is then the output voltage from this voltage divider and can range between the voltage levels present at the top and at the bottom of the voltage divider.

Figure A5–6b emphasizes the connections that are made to the basic amplifier box of Figure A5–6a. Notice that there are actually just four wires

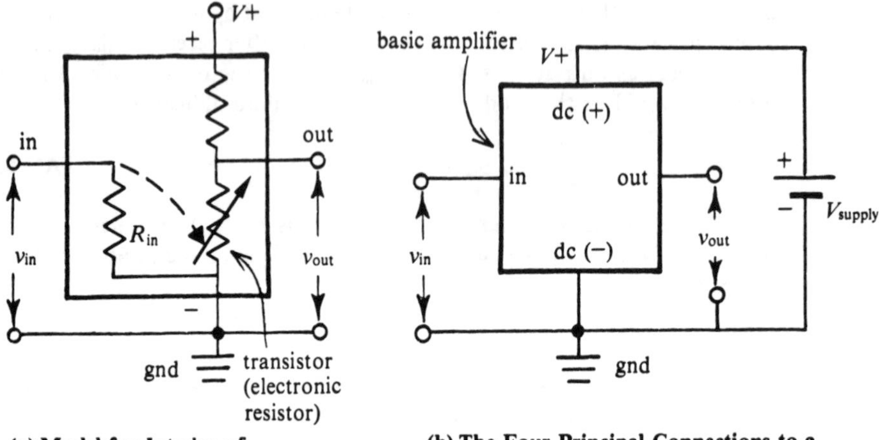

(a) Model for Interior of Basic Amplifier

(b) The Four Principal Connections to a Basic Amplifier

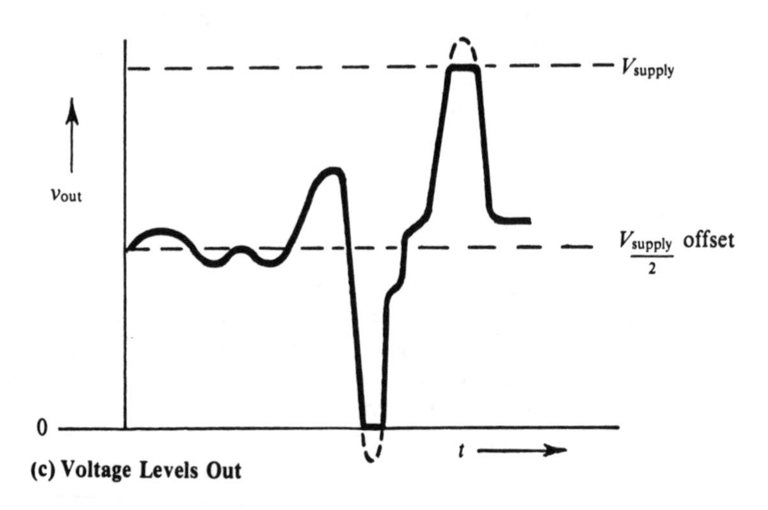

(c) Voltage Levels Out

FIGURE A5–6. *Connections and Behavior of a Basic Amplifier*

that lead from the box itself. There is a single signal input wire, a single signal output wire, a power supply wire V+, and a (bottom) wire that serves in both the input and output circuits. This last wire is often referred to as the *ground* (gnd) of the amplifier, although it may not actually be electrically connected to "true ground" (that is, to the earth). The term *common* would be a better term; it serves as a reference voltage level for both the input and the output. The negative terminal of the *power supply battery* is connected to this ground or common, while the positive battery terminal is connected to a dc (+) power-in or $V+$ terminal. The battery is drawn explicitly in Figure A5–6b, but to simplify the drawing the battery itself is usually omitted, and only the *V*+ and gnd connections are drawn, as in Figure A5–6a. For portable equipment, batteries may actually be used, but otherwise electronic power supplies (battery eliminators) are generally used. Some source of dc electrical power to the amplifier is always necessary.

The voltage out of the *electronic voltage divider circuit* must then vary between the V+ supply voltage and ground as shown in Figure A5–6c. Often what is of interest is the changing of the output signal, or what is called the *ac component* of the signal. The average value of the output signal is called the *dc component* of the signal. In Figure A5–6c the ac component signal is seen riding atop a dc voltage of about half the supply voltage. We see shortly why it is desirable that the design have this average voltage level (or *resting level* if there is no signal input) midway between the power supply voltages.

The ac component of the output signal is larger than the voltage signal in v_{in} by the factor of the amplifier voltage gain (provided the amplifier is not loaded down). Notice two places in the output voltage v_{out} of Figure A5–6c where the signal is distorted or *clipped off*. To faithfully follow the input voltage shape, the output signal should have gone higher and also lower in voltage. But it is physically impossible for the amplifier—that is, the voltage divider—to produce a voltage more positive than the + supply voltage or more negative than the − supply voltage. There is *no* source of voltage in the amplifier itself (for example, no batteries) to produce a higher voltage. It should be clear that to get an output with no clipping, the input signal peak-to-peak voltage must be smaller than the power supply voltage used by the amplifier by a factor equal to the amplifier's voltage gain. Also, for symmetrical input signals, a dc or resting-level output set midway between the supply voltages will accommodate the largest peak-to-peak output *swing*.

ac and dc Amplifiers

When we need to cascade two or more of these basic amplifiers in order to make an amplifier with higher gain, the large dc offset voltage of the output causes a severe problem in the next stage. The next stage will attempt to amplify this large dc input voltage, and its output will simply become stuck or *latch up* at the limiting supply voltage. A simple cure is to connect *blocking capacitors* as shown in Figure A5–7 between stages. These *interstage coupling capacitors* then pass or *couple* the ac signal on to the next stage to be amplified but *block* the offending dc. There is a penalty to be paid, however. The amplifier loses the ability to amplify dc voltages. It becomes an *ac-only amplifier* or as commonly stated, an *ac amplifier*. For example, the voltage produced by a copper-constantan thermocouple at 10°C is only about 0.4 millivolts; it needs to be amplified before being observable on most voltmeters. But it would do no good to connect this small dc voltage into an amplifier like the one shown in Figure A5–7. Connecting it into stage 1

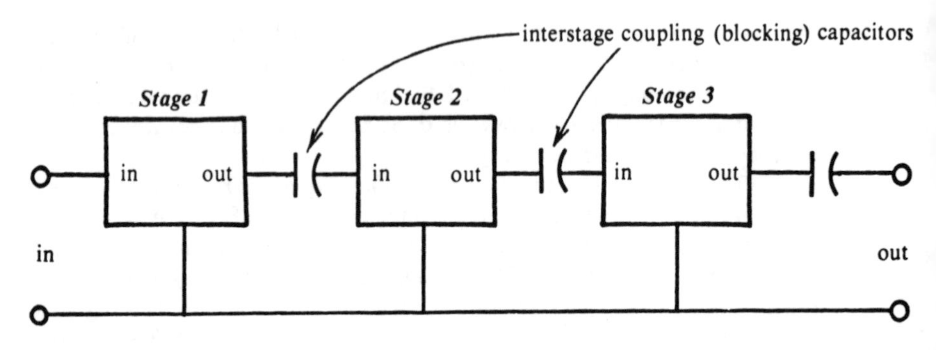

FIGURE A5–7. *Capacitively Coupled (ac) Amplifier*

causes that output to increase some amount, but this increase is blocked from influencing the input of the next stage by the capacitor (except for a brief charging interval).

A *dc amplifier* retains the capability of amplifying dc input voltages as well as ac voltages. In general, this is accomplished by operating the amplifier from two supply voltages that are equally plus and minus with respect to the ground of the amplifier. The situation is shown in Figure A5–8. Then the general rule that the output voltage be halfway between the two supply voltages puts it normally at—or at least near—the ground. The amplifier is also arranged so that the input voltage should be relative to the connection between the two supply voltages. This point is labeled gnd (ground) in Figure A5–8, and it is convenient to think of it as such, although *common* is

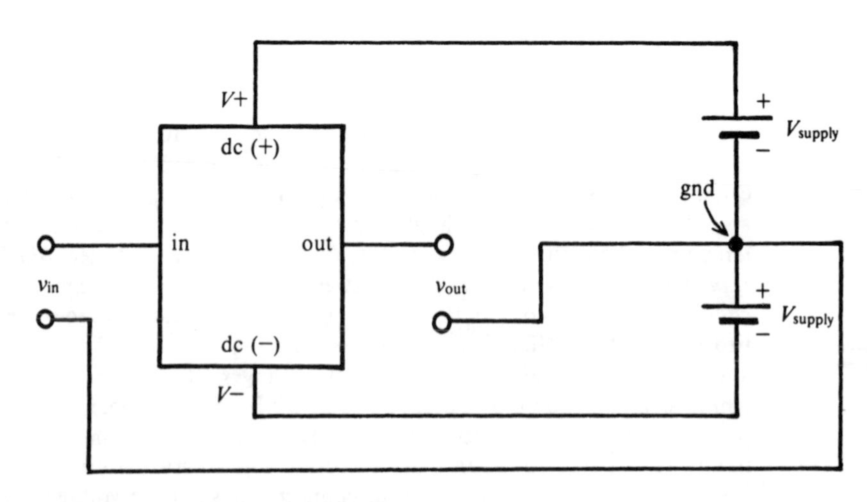

FIGURE A5–8. *Dual Voltage Supply, dc Amplifier*

again a better term. If v_{in} is zero ("in" lead shorted to ground), v_{out} should also be zero ("out" lead at same voltage as ground) for the perfect dc amplifier.

For example, if $V+$ is $+12$ V and $V-$ is -12 V, halfway in between is zero volts. The dashed output offset line of Figure A5–6c thus gets moved to zero volts. Now there is no dc component in the output signal, and several stages of this type of amplifier can be cascaded without blocking capacitors between stages. The overall amplifier then retains the ability to amplify dc voltage.

Because of this *dual supply voltage* arrangement, it is possible for v_{out} to go both above and below ground if the input voltage is positive or negative. This is often desirable. The output can typically go positive by nearly V, or negative by nearly V, where V is the voltage of either battery in Figure A5–8. Generally the output of the dc amplifier is *not* zero when the input is shorted for real amplifiers. This voltage is called the *output offset voltage* for the amplifier and is usually *referred to the amplifier input;* that is, the amplifier behaves as though a phantom tiny battery, called the *input offset voltage,* exists in the input wire to the amplifier. This input offset voltage is equal to the output offset voltage divided by the voltage gain of the amplifier.

Amplifier Bandwidth

Because of coupling capacitors, an ac amplifier is clearly limited with respect to the lowest frequency that it can amplify well. In addition, all amplifiers have some upper limit frequency beyond which they begin to fail. The high frequency failing comes about because capacitances in the tubes or transistors themselves begin to *short-circuit* the currents at high frequencies. The consequence is that the voltage gain A_v of the amplifier is a function of frequency. A plot of the logarithm of A_v versus the logarithm of signal frequency $\log f$ results in a fairly simple graph that tells the story (see Figure A5–9).

The voltage gain for this particular example is constant at 1000 over a frequency range of about 5 hertz to about 500 hertz. The frequency range of 500 Hz − 5 Hz = 495 Hz is called the *bandwidth* of the amplifier, while the value of 1000 is called the *midband gain* of the amplifier. The end points for the bandwidth are taken to be when A_v is decreased by the factor 0.707 (similar to the case for filters).

If the plot in Figure A5–9 were for a dc amplifier, there would still be some high-frequency limit, but the line would extend straight to the left to zero hertz or dc as shown by the heavy dashed line. The dashed line is precisely how a dc amplifier differs from an ac amplifier.

Noninverting and Inverting Amplifiers

We will define what is meant by noninverting amplifier and inverting amplifier. A *noninverting amplifier* produces an output in the same sense as its

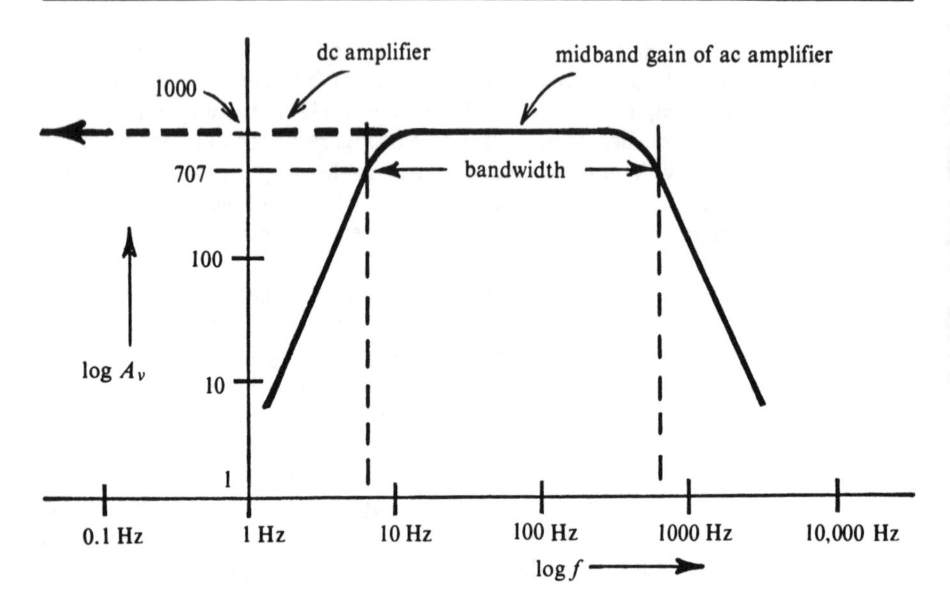

FIGURE A5–9. *Bandwidth of ac and dc Amplifiers*

input, as shown in Figure A5–10a. There the input signal is seen to change more positive, and the output changes in the positive sense, too, although it is greater by the factor A_v. If the input swings more negative, the output also swings more negative. On the other hand, an *inverting amplifier* has an output that changes in the opposite sense to any change in the input, as illustrated in Figure A5–10b. For example, if the input is made more positive, the output goes more negative. In both cases, the input and output voltages are between those respective leads and ground.

Several other conventions are introduced in Figure A5–10. First, rather than representing the amplifier by a rectangle (which reminds us of a black box), a *triangle* is used. The amplifier output extends from a triangle corner, while the amplifier input wire goes to the opposite triangle side; indeed, with this convention it is not actually necessary to label input or output. Second, if the amplifier is noninverting, the input is labeled with a +; if it is inverting, the input is labeled with a −.

NOTE: These lines are *not* + or − power supply connections. The power supply connections are actually left off the diagram for simplicity, although in fact power must be supplied, of course.

Difference or Differential Amplifiers

There are a number of situations, especially in chemistry and the life sciences, where it is desirable to have two inputs into an amplifier *besides* ground. For

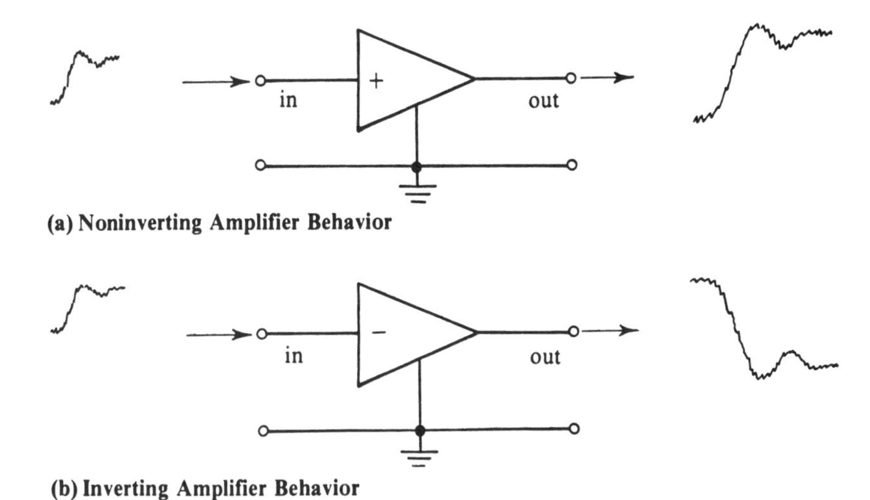

FIGURE A5–10. *Noninverting and Inverting Amplifier Behavior*

example, every electrocardiogram instrument requires these two inputs, and we see why in the next section of this chapter. Such amplifiers produce a nonzero output voltage if there is a *difference* in the voltages presented to the two inputs, and this output voltage is independent of whether or not both inputs are *off ground*—different in potential (or voltage) from ground potential—by any amount.

Thus the *difference or differential amplifier* has two signal inputs; they are inverting and noninverting inputs and are labeled with the − and + convention, respectively, in Figure A5–11. The behavior of a dc differential amplifier with a voltage gain of 20 is illustrated in the figure by showing the output resulting from various connections to the inputs. The student should study the various consequences.

First, Figure A5–11a shows that the output of an ideal difference amplifier is zero if both inputs are grounded, which is certainly no surprise. However, Figure A5–11b shows that the output is still zero if both inputs are brought off ground by the *same* voltage; that is, a *common-mode voltage.* The output is only nonzero if a voltage *difference* occurs between the two input terminals, as happens in Figures A5–11c, d, and e. Notice the output voltage is 20 times the input voltage difference of 0.1 V in these three cases, although a common-mode input voltage of 1 V occurs again in the case of Figure A5–11e. Also notice that the output voltage swings positive if the + input is more positive than the − input, but it swings negative if the − input is the more positive (see the case in Figure A5–11e). This behavior is in accord with the noninverting and inverting character of the inputs. We can write:

$$v_{out} = A_{vd}(v_{in+} - v_{in-}) = A_{vd}v_{id} \tag{A5–6}$$

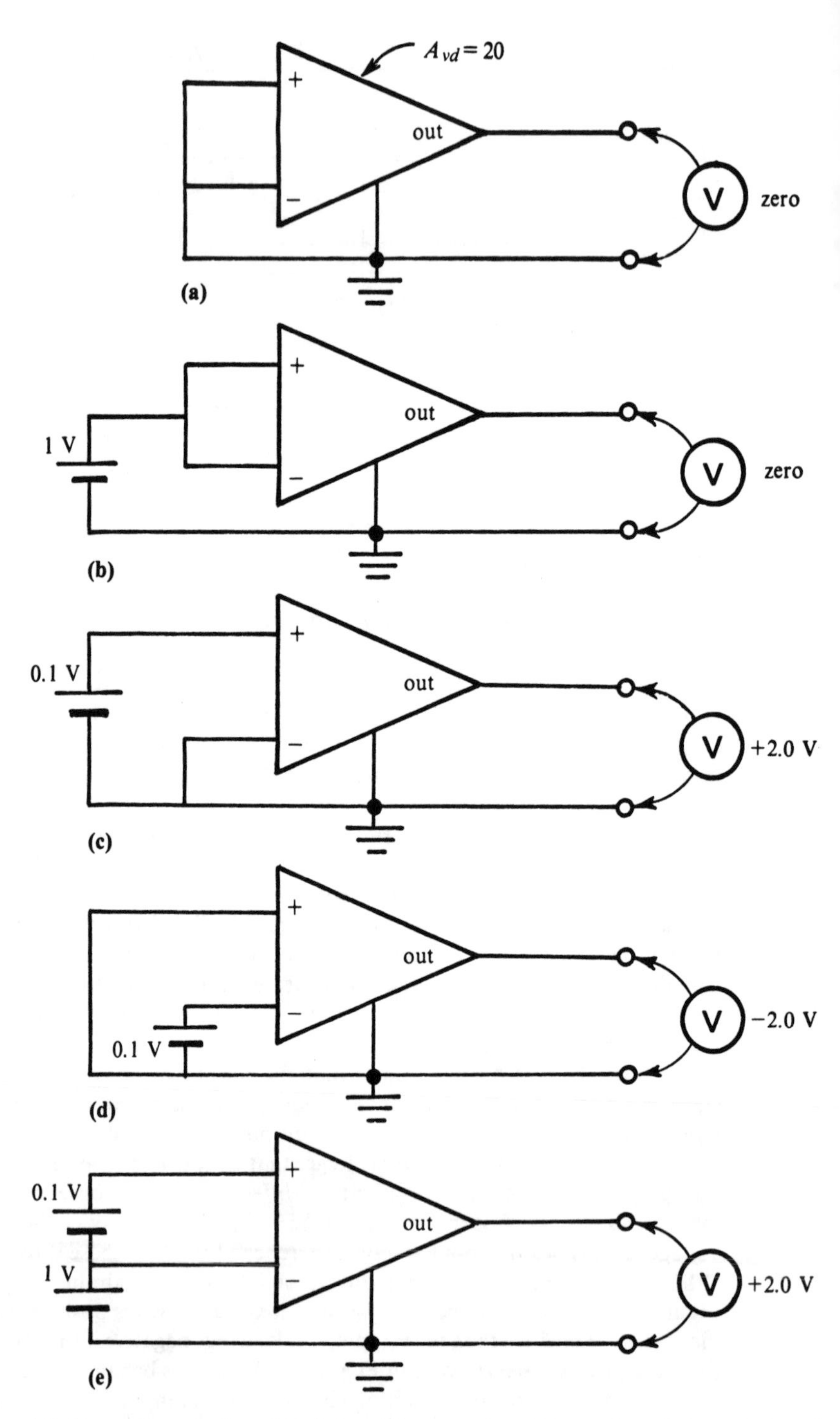

Figure A5–11. *Output of an Ideal dc Difference Amplifier with a Voltage Gain of 20 for Various Input Voltages*

where v_{out} is the output voltage; v_{in+} and v_{in-} are the voltages into the + and − inputs, respectively, with respect to ground; A_{vd} is the *difference voltage gain;* and we let the *input difference voltage* $(v_{in+} - v_{in-}) = v_{id}$. In the example of Figure A5–11, A_{vd} equals 20.

Real difference amplifiers would show a nonzero output voltage in Figure A5–11b, even though both inputs have a common (or the same) voltage. If v_{oc} is the voltage out for a common voltage v_{ic} in, the *common-mode voltage gain* A_{vc} is

$$A_{vc} = \frac{v_{oc}}{v_{ic}} \tag{A5–7}$$

We want this ratio to be small compared to the desirable difference voltage gain of the amplifier. If v_{od} is the voltage out for a voltage difference of v_{id} between the input terminals, the *difference voltage gain* is

$$A_{vd} = \frac{v_{od}}{v_{id}} \tag{A5–8}$$

The ratio of A_{vd} to A_{vc} is called the *common-mode rejection ratio* (CMRR). That is

$$\text{CMRR} = \frac{A_{vd}}{A_{vc}} \tag{A5–9}$$

It is quite common for this ratio to be 10^4 or more (the larger the better). Actually manufacturers usually quote the ratio in decibels. Since the ratio refers to voltage, we have 20 decibels for every factor of 10.

Example: Suppose for a realistic amplifier in Figure A5–11b, the connection gives an output voltage of 0.005 V. Find its CMRR.

Solution: We have

$$A_{vc} = \frac{0.005 \text{ V}}{1 \text{ V}} = 0.005$$

The difference voltage gain is clear from Figure A5–11c as

$$A_{vd} = \frac{2.0 \text{ V}}{0.1 \text{ V}} = 20$$

$$\text{CMRR} = \frac{A_{vd}}{A_{vc}} = \frac{20}{0.005} = 4000$$

In decibels we would have

$$\text{CMRR} = 20 \log 4000 = 20 \times (3.6) = 72 \text{ dB}$$

A Differential Amplifier Application

The differential amplifier is a very desirable part of the electrocardiograph that is routinely used in medical practice to diagnose certain heart diseases. Heart muscle action is initiated by an electrical pulse that passes around the heart. The body is a conductive volume, and the electric fields associated with each pulse support small currents and attending voltage drops out to the body surface. Voltage pulses on the order of 10-millivolts (10^{-2} V) amplitude and 0.1-second duration may be observed between the limbs of the body.

It is common to find oscilloscopes in science laboratories that can display 10-millivolt pulses quite satisfactorily. However, if a student connects a scope probe between his or her two hands, the scope display will almost certainly show a rather "noisy" appearing signal of perhaps 1 V in amplitude and 60 hertz in frequency. The 60 hertz is the telltale sign (in the United States) that the signal comes not from the heart but from the electric utility company! The problem is that the 60-hertz voltage in the wiring to lighting fixtures and so forth produces electric field lines that end on the person's body. In effect, there is a capacitive coupling C_1 between the wires and the person, as is diagrammed in Figure A5–12a. There is also capacitively coupling C_2 between the person's limb and ground, and the pair of capacitors constitutes an ac voltage divider that establishes a 60-hertz voltage between the limbs; this rather large voltage completely masks the electrocardic potential. Notice in the diagram that the one input signal lead for most oscilloscopes is connected to ground (true earth) through the ground prong of the three-prong power plug.

The solution is to connect the leads from the two limbs into a differential amplifier, where neither of the input leads is grounded. This is illustrated in Figure A5–12b. We can expect the noise voltages induced in the two parts of the body to be about equal and *in phase* with each other. The signals v_{arm} from the arm and v_{leg} from the leg are diagrammed in Figure A5–12c where we see the desirable bioelectric signals are an *out-of-phase modulation* (for clarity, shown here relatively large) on the constantly oscillating 60-hertz signal that is "in common" to both signals. The differential amplifier subtracts v_{arm} from v_{leg}; this largely eliminates the 60-hertz common-mode signal and the desired bioelectric signal emerges as output. The bioelectric signal may be displayed on an oscilloscope (electrocardioscope) or drawn on a fast-response strip-chart recorder (electrocardiograph).

Noise in Electronic Systems

When the magnitude of the voltage from a transducer is small—for example, less than 1 millivolt—electrical noise can be a problem. Several cases of electrical noise are familiar to almost everyone. For example, a hiss is heard

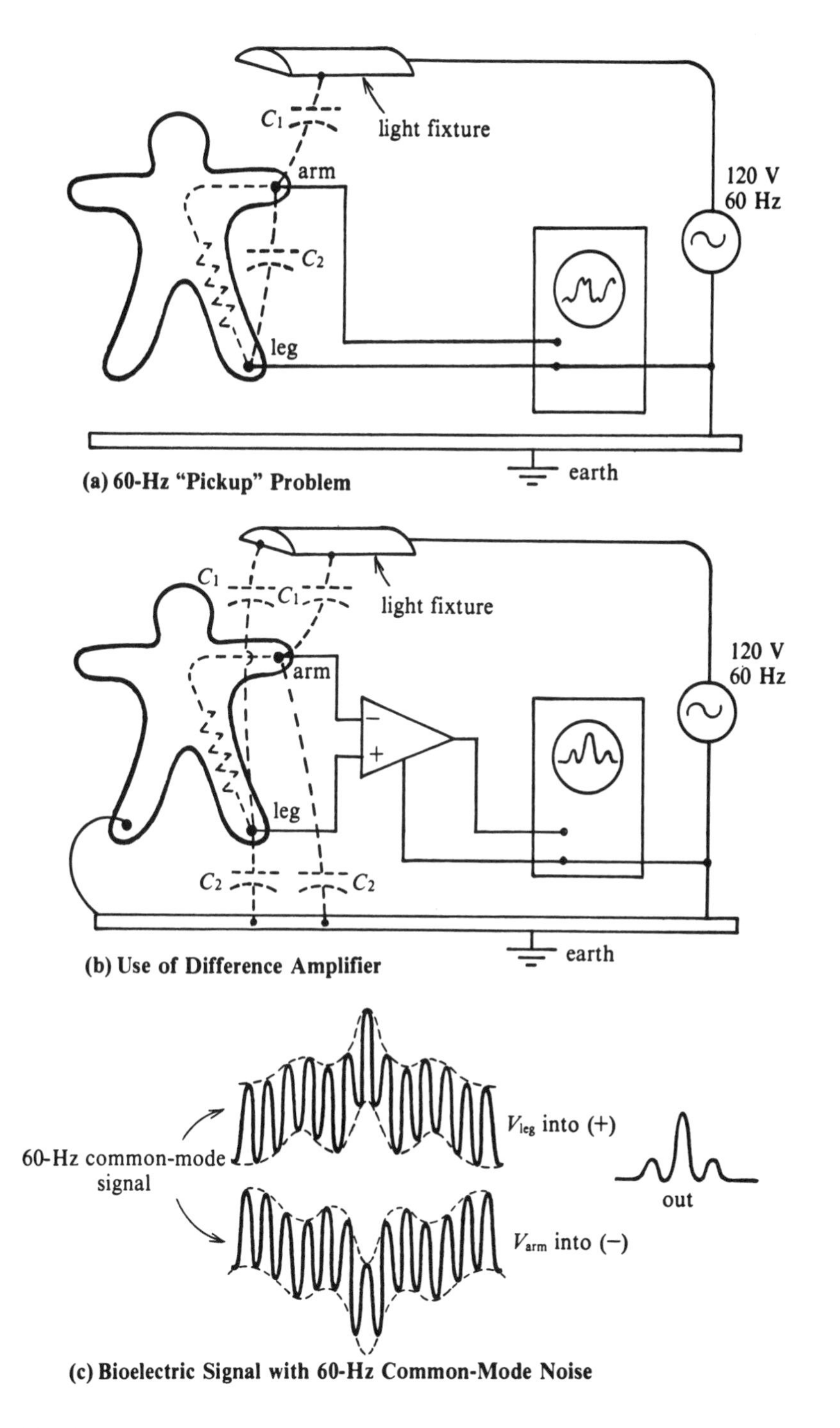

FIGURE A5–12. *Rejection of 60-Hertz Common-Mode Signal by Use of a Differential Amplifier in Electrocardiography*

when the volume control of a stereo amplifier is turned up while no music is being played; this background noise is annoying during quiet musical passages. Another example of electrical noise is literally seen as "snow" in a television picture when it is received in an area remote from the transmitter.

Electrical noise is unwanted or spurious electrical voltage or current that occurs in circuits. In general, electrical noise is of two types: It may be generated *within* the electronic system—that is, within the transducer as well as the amplifier—or it may be generated by some *external* source and be introduced into the system by electrostatic coupling or by electromagnetic coupling. Externally generated electrical noise is often termed electrical *interference,* and we have already discussed ways to reduce the problem of interference. Electrically shielded interconnecting cables—for example, coaxial cables—as well as shielding through use of a metal chassis for the electronic circuitry were discussed in Chapter A2. The use of difference or differential amplifiers to reduce common-mode interference was discussed in the previous section. In this section we want to introduce the subject of internally generated noise.

Thermal or Johnson Noise

Noise is always generated within the components of the electronic system—that is, within the transistors, resistors, inductors, capacitors, and even wires. A fundamental source of noise is the random movements of electrons as they travel within these devices. Each electron carrier possesses a thermal kinetic energy $(3/2)kT$, where k is Boltzmann's constant and T is the absolute temperature. The random motions due to this thermal energy are superposed on any uniform current motion and result in tiny fluctuations in the flowing current.

Of fundamental interest is the noise that occurs within an ideal resistor. An ideal inductor or capacitor does not generate noise; but a real inductor or capacitor does generate noise because of inherent resistance. We are interested then in the fluctuations that occur in the current flowing through a resistor. The random part of the current is called *thermal noise* current or *Johnson noise* current. This noise is described as *white noise* because it has frequency components spread evenly through the entire frequency *spectrum* from dc to about 10^{13} hertz.

Suppose we simply short the ends of a resistor. A tiny randomly alternating current that is caused by the random electron motions then flows within the circuit. In a given frequency band Δf, it is possible to show that the *noise power* P_N that is dissipated in the shorted resistance R due to the *noise current* $I_{N\text{rms}}$ that flows is given by

$$P_N = I_{N\text{rms}}^2 R = 4kT\Delta f \qquad \textbf{(A5–10)}$$

Notice that the power is proportional to the absolute temperature T and this is basically because the thermal energy of the electron is proportional to kT; further, the noise power is directly proportional to the frequency band Δf because a larger bandwidth includes more of the noise frequencies that are present across the entire spectrum. As we have stated, the noise current is a *randomly alternating* current, and some sort of average measure is required; therefore it is not surprising that the root mean square or rms measure of the noise current must be used. It is interesting that the value of the *power* does *not* depend on the resistance.

When dealing with amplifiers in electronics, it is the voltage present that is often of particular interest. We may think of an equivalent *noise voltage* $V_{N\text{rms}}$ that appears within any resistance R; the noise voltage is considered to cause the noise current that flows. The noise voltage may be modeled as a voltage generator $V_{N\text{rms}}$ that is in series with a noiseless resistance R, as shown in Figure A5–13. Let the terminals of the model resistor be shorted. Then the power dissipated in the resistance must be given by $V_{N\text{rms}}^2/R$. The noise power is given by Equation A5–10, and it therefore follows that the noise voltage must be

$$\frac{V_{N\text{rms}}^2}{R} = 4kT\Delta f \qquad \textbf{(A5–11)}$$

or

$$V_{N\text{rms}} = \sqrt{4kTR\Delta f} \qquad \textbf{(A5–12)}$$

Let us consider the implications of this result through an example.

Example: A 50 kilohm resistor at 20°C is connected to an ideal dc amplifier with a voltage gain of 3000; however, the amplifier has an upper frequency response limit of 1000 hertz. The system is diagrammed in Figure A5–14. What is the noise voltage expected at the amplifier output?

Solution: Since the amplifier amplifies voltage over a bandwidth of 1000 hertz, this is the bandwidth for which noise from the resistor is relevant. The equivalent noise voltage present in the resistor is then given by

$$V_{N\text{rms}} = (4 \times 1.38 \times 10^{-23} \times 293 \times 50 \times 10^3 \times 1000)^{1/2}\ \text{V} = 0.9\ \mu\text{V}_{\text{rms}}$$

Since the amplifier is ideal—the resistor is connected to the amplifier—there is no loading on this noise voltage source. (Note that the output impedance of the resistor is its own resistance R.) The rms noise voltage $V_{N\text{out}}$ appearing at the amplifier output is then

$$V_{N\text{out}} = A_v V_{N\text{rms}} = 3000 \times 0.9\ \mu\text{V} = 2.7\ \text{mV}_{\text{rms}} \qquad \textbf{(A5–13)}$$

NOTE: If an oscilloscope is set to 5 mV/cm and is connected to the output of the amplifier here, the scope trace shows a hash of jiggles like those in Figure A5–13. The scope picture of thermal noise looks like grass, and "grass" is one slang term for it. Due to its random nature, the peak-to-peak voltage of the noise signal is *not* a well-defined quantity. Nevertheless, it is possible to make a visual estimate on the scope display of the *p-p* noise, and we can expect it to be on the order of three times the rms voltage value, or about 9 millivolts in this case.

Several features of thermal noise that are of particular interest to the scientist practitioner of electronics may be rather quickly inferred from this example. First, if the 50 kilohm resistance were the output impedance of a transducer, we may expect a noise output from the transducer of at least 0.9 μV_{rms}. If the true output signal of interest from the transducer were on the order of one microvolt, then the desired signal would be almost hidden by the noise signal. In fact, we speak of the ratio of the signal voltage to the noise voltage as the *signal-to-noise ratio* (S/N) from the source. The ideal amplifier would amplify both the desired signal as well as the noise, so that the desired signal is still obscured when it immerges from the amplifier output.

Second, a larger resistance produces a larger noise voltage. Therefore, a transducer with a large output impedance has two strikes against it: The large resistance implies a larger inherent noise value, as well as a possible problem with excessive loading due to the input impedance of a nonideal amplifier. Actually, the Johnson noise given by Equation A5–12 represents a lower bound on the noise that may be expected from any real resistance. For example, metal wire or metal film resistors may approach this theoretical limit, but carbon composition and carbon film resistors are significantly more noisy. This noise is related to the granular structure of the conducting path in the carbon-type resistors.

Third, lowering the temperature of the resistor lowers its thermal noise output. In some scientific applications where the signal-to-noise ratio

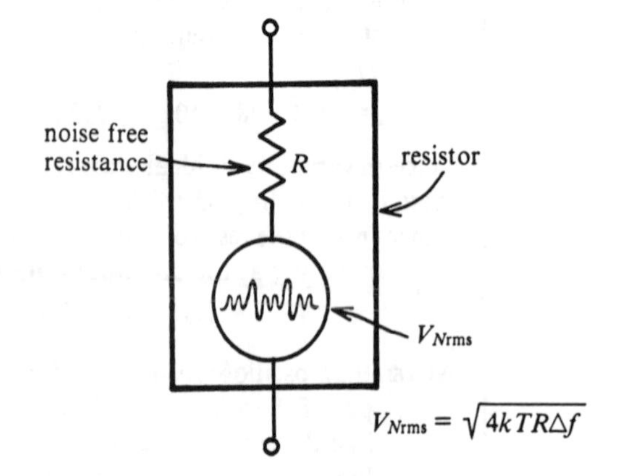

FIGURE A5–13. *Model for Thermal Noise or Johnson Noise from a Resistor*

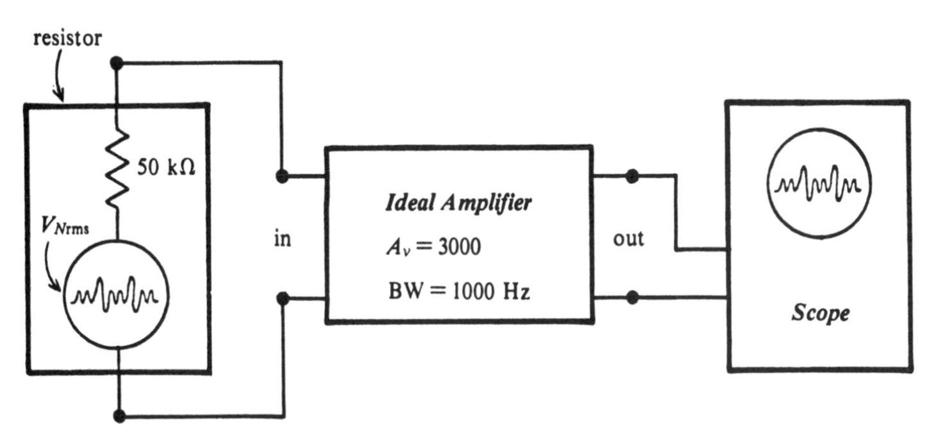

FIGURE A5–14. *Resistor Connected to Ideal Amplifier*

is critical, the transducer temperature is lowered to the temperature of liquid nitrogen or even liquid helium in order to improve the S/N.

Fourth, the effective input noise voltage depends on the size of the bandwidth that passes the signal—the *pass bandwidth.* An important technique that is used to improve the signal-to-noise ratio is to use an amplifier with a bandwidth that is as small as possible. For example, if the bandwidth in the previous example were reduced by a factor of 64 (to 15 hertz), the noise voltage would be reduced by the factor $\sqrt{64} = 8$. However, the bandwidth of the amplifier should be chosen large enough to pass without attenuation the range of frequencies that is present in the desired signal. If the transducer signal is a dc or low-frequency signal, a low-pass *RC* or *LC* filter may be placed at the amplifier output in order to reduce the effective amplifier bandwidth and therefore the noise.

NOTE: What are called *tuned amplifiers*—such as those present in radio and television receivers—use the narrow bandwidth (or filter) character of resonant (or tuned) inductance-capacitance circuits. In this way the receiver can achieve a narrow bandwidth for the radio frequencies that are of interest in this application. However, in this book we avoid the topic of resonant *LC* circuits since they are most appropriate for circuitry operating at radio frequencies; we are generally not considering the subject of radio-frequency circuitry. At audio frequencies it is possible to use *bandpass* filters that are constructed from resistors and capacitors. For a discussion of these more advanced *RC* networks, refer to the texts by Brophy, Jones, or Malmstadt, Enke, and Toren that are listed in the Reference section.

Amplifier Noise

The previous example assumed an ideal noiseless amplifier. But any real amplifier introduces noise of its own into the signal that immerges from the amplifier output. The most important noise sources in an amplifier are usually those present in the first stage of amplification in an amplifier. The quality of the first stage is important because the noise generated in later stages is generally well below the amplified noise that they receive as input from the first stage.

One useful measure of the noise generated within an amplifier is the *equivalent input noise voltage* V_{Nin} that may be modeled as present at the amplifier input. This amplifier noise voltage appears added to the signal (and the noise) that is input by the transducer to the amplifier. It should be quite clear that for a good—that is, high—signal-to-noise ratio from the amplifier, we desire that the equivalent input noise voltage V_{Nin} be much less than the signal voltage from the transducer.

Since the bandwidth of the amplifier may be adjustable, the equivalent input noise voltage is often quoted as an input noise–voltage *density,* with units of $V_{rms}/\sqrt{\text{hertz}}$. This reflects the fact that for frequencies above about 100 hertz, the internal amplifier noise is generally quite uniform (or white) in terms of its frequency dependence, similar to the noise from a resistor. From Equation A5–12, we see that the rms-noise voltage is proportional to the square root of the frequency bandwidth. To obtain the actual equivalent noise voltage at the amplifier input, we must multiply the noise-voltage density value by the square root of the amplifier bandwidth.

The input noise–voltage density for an amplifier was introduced as being uniform above approximately 100 hertz. Unfortunately, below this

Example: The OP–07 type of operational amplifier is quoted to have a typical input noise–voltage density of $10\,nV/\sqrt{\text{hertz}}$. An OP–07 is used to construct an amplifier with a voltage gain of 50 and a bandwidth of 10 kHz. What is its equivalent input noise voltage?

Solution: Although operational amplifier design is discussed in the next chapter, we may nevertheless use the quoted values of noise-voltage density and bandwidth to compute

$$V_{Nin} = \text{noise density} \times \sqrt{\text{BW}} = 10\,\frac{nV}{\sqrt{\text{Hz}}} \times \sqrt{10^4\,\text{Hz}} = 1.0\,\mu V_{rms}$$

Note that the voltage gain is actually not needed here.

NOTE: If a noise-free 1-microvolt signal is input to this amplifier, the signal out of the amplifier will be almost obscured by noise, with a poor signal-to-noise ratio of 1.

frequency the amplifier-noise density generally increases approximately in proportion to the quantity $1/f$, where f is the frequency of interest. This noise regime is termed *1/f noise* or *flicker noise* and originates in the transistors or the tubes of the amplifier. For example, in the frequency range of 0.1 to 10 hertz the OP–07 operational amplifier mentioned in the previous example is specified to have a typical peak-to-peak noise voltage of 0.35 μV. Notice that this is significantly more than what would be obtained from the input noise–density value for the operational amplifier, 10 $n\text{V}/\sqrt{\text{hertz}}$, since from the latter value we obtain

$$V_{N\text{rms}} = 10\ \frac{n\text{V}}{\sqrt{\text{Hz}}} \times \sqrt{(10\ \text{Hz} - 0.1\ \text{Hz})} = 30\ n\text{V}_{\text{rms}}$$

The *approximate* peak-to-peak noise voltage is then about $3 \times 30\ n\text{V} = 90\ n\text{V} = 0.09\ \mu\text{V}$, which is only about one-fourth the flicker noise that appears in this frequency interval.

We see that in practice, noise voltages of several origins are effectively present simultaneously at an amplifier input. But the magnitude of the total noise voltage present is *not* obtained by merely adding the noise voltages together. The total noise voltage is calculated by finding the square root of the sum of the squares of the contributing noise voltages. (This rule is characteristic for the probable value of the sum of random variables and evolves quickly here from the fact that the total noise *power* is the simple sum of the contributing noise powers.)

Example: A transducer with a noise voltage of 0.5 μV_{rms} over the frequency range of dc to 100 hertz is input to an amplifier constructed from an OP–07 operational amplifier. The amplifier has been designed with a bandwidth of dc to 100 hertz. What is the total effective input noise to the amplifier?

Solution: We must properly add the transducer noise, the OP–07 flicker or $1/f$ noise, and the broad-spectrum noise of the OP–07. The transducer noise is given as 0.5 μV_{rms} in the total frequency interval. The broad-spectrum noise in the frequency interval of 10 hertz to 100 hertz is given by

$$V_{N\text{in}} = \left(\frac{10\ n\text{V}}{\sqrt{\text{Hz}}}\right) \sqrt{(100 - 10)\text{Hz}} = 95\ n\text{V}_{\text{rms}}$$

The rms value of the OP–07 internal noise in the interval 0.1 hertz to 10 hertz including $1/f$ noise is approximately one-third of the peak-to-peak value, or $0.35\ \mu\text{V}/3 = 0.12\ \mu\text{V}$. Then the total rms noise that effectively appears at the amplifier input is given by

$$\text{total } V_{N\text{in}} = \sqrt{(0.5)^2 + (0.095)^2 + (0.12)^2}\ \mu\text{V} = 0.52\ \mu\text{V}_{\text{rms}}$$

Summary

Basic properties of amplifiers and principles for their use have been considered in this chapter. Especially with the miniaturization of electronics, it is important to appreciate that although the circuit within the chassis or the integrated circuit may be very complex, the scientist is often primarily interested in a relatively few properties of the box for its application in the laboratory. The black-box model for an amplifier is very useful for predicting and understanding amplifier performance. The notion of amplification or gain enters through this model, and the utility of impedance matching also becomes apparent.

The behavior of ac, dc, and difference amplifiers are important. It is useful to appreciate the essential nature of an amplifier stage through the voltage-divider model for it and the process by which the consequent dc offset problem in the stages of a cascaded amplifier can be solved with coupling capacitors. However, the capacitors limit the amplifier to ac performance, and indeed the general notion of amplifier bandwidth entered there.

The simple voltage-divider model for an amplifier stage also illustrates why the power supply voltage(s) limit the peak-to-peak voltage that can be output by an amplifier and why dc amplifiers usually require two supply voltages for their operation. The general consideration of input, output, and power supply connections to an amplifier are also illustrated in this model. The nature of the difference or differential amplifier, including its property of common-mode rejection, is an important final consideration.

REVIEW EXERCISES

1. Draw the black-box model for an amplifier. Label the ports and resistors. In what quantity does the voltage amplification appear?
2. (a) Define mathematically the open-circuit voltage gain A_v of an amplifier.
   (b) Define mathematically the power gain A_p of an amplifier.
3. (a) What should the input impedance be for an ideal amplifier? Why?
   (b) What should the output impedance be for an ideal amplifier? Why?
4. Suppose we connect an audio generator to one input of a stereo amplifier and an 8-Ω (power) resistor to the corresponding output. Then we measure the input voltage to be 0.01 $V_{p\text{-}p}$ and the voltage across the 8-Ω resistor to be 4.0 $V_{p\text{-}p}$.
   (a) Explain why the open-circuit voltage gain for the stereo channel is not 400.
   (b) Should it be more or less than 400? Why?
5. (a) What is a voltage follower?
   (b) Can its voltage gain be greater than one?

(c) Can its power gain be greater than one?

6. A high quality voltage follower is often inserted between a voltage source and a load. It is then often called a *buffer amplifier*. Comment on the reason for the term *buffer* here.
7. Suppose we decide to heat a cup of water for tea by connecting a power resistor to the output of the stereo in Question 4 and dropping it into the cup. (This is not an exaggeration of their power.) Is there an optimum value for the resistor? Explain.
8. In general, it is not possible for the peak-to-peak voltage from an amplifier output to exceed a certain voltage. What is the determining voltage? Explain briefly. (However, if inductors or transformers are used, this rule can be broken for ac signals.)
9. (a) What is meant by the clipping of the signal out of an amplifier? What is its cause?
   (b) Describe how to make a square wave, given a large-amplitude sinewave source, and an amplifier with substantial voltage gain.
10. (a) What is the reason (in terms of amplifier construction) that a high-gain ac amplifier cannot amplify dc voltages?
    (b) Why can't these offending items simply be replaced by wires within the amplifier?
11. (a) What is meant by the offset voltage of a dc amplifier? Describe how it can be measured.
    (b) Make a black-box model of a dc amplifier, including the input offset voltage in the model.
12. Give a definition of the bandwidth (BW) of an amplifier using the term *midband gain*.
13. The human ear is responsive to frequencies from about 60 Hz to 15 kHz. The bandwidth of the amplifier in a phonograph prior to 1955 was generally less than 10 kHz, with the low-frequency limit at about 100 Hz. What did this bandwidth do to the phonograph sound? (Actually, the bandwidth of the pickup and loudspeaker were also limited.)
14. (a) Generally, what sort of power supply configuration is required for a dc amplifier?
    (b) Name two desirable features of dc amplifiers that result from this power supply configuration.
    (c) Which feature is virtually necessary if several dc amplifiers are to be cascaded? Why?
15. A differential amplifier could well be marked with two leads each marked + and two more each marked −. However, in each case the two should be distinguished as to use. Explain.
16. (a) What is meant by *common-mode rejection* of an amplifier? What general type of amplifier does this term apply to?
    (b) What is the definition of CMRR?

17. (a) In electrocardiography, how can the 60-Hz interference problem be modeled electrically?
    (b) What can be done to reduce the 60-Hz interference problem?
18. (a) What is meant by white noise?
    (b) What is the origin of thermal noise or Johnson noise in a resistor?
    (c) How does flicker noise differ from white noise?
    (d) Which resistor type is superior in terms of noise: carbon composition or metal film?
19. Which amplifier type is likely to have the greater problem with flicker noise: a dc amplifier or an ac amplifier? Explain.
20. (a) Explain why an amplifier with the minimum possible bandwidth is generally preferable in terms of noise problems.
    (b) What generally limits a scientist from using an amplifier with exceedingly small bandwidth?

PROBLEMS

1. A certain amplifier is specified by its manufacturer to have a power gain of 45-dB when used with a 50-Ω load. Its input impedance is 1000 Ω. If 0.1 V_{rms} is established at its input, what voltage appears across the 50-Ω load connected to its output?
2. (a) It is not unusual to express virtually any voltage ratio in dB, according to $dB = 20 \log (V_1 / V_2)$. Argue that according to the fundamental definition of dB, the two voltages should appear across identical resistances (if not actually the same resistor). What should be the voltage-ratio-to-dB relation for nonidentical resistances?
   (b) A certain amplifier is characterized by an input impedance of 1000 Ω and an open-circuit voltage gain of 100. Argue that strictly speaking, it is improper to say that its open-circuit voltage gain is 40 dB. (Nevertheless, when voltage ratios are quoted, it is common practice to ignore the resistances involved and simply use the simple voltage-ratio-to-dB relation. Then 40 dB is correct.)
3. It is possible to purchase voltage calibrators for calibrating VOMs, DVMs, and oscilloscopes. One such device was available quite inexpensively for a number of years from a well-known "kit" company. Although it provides only a dc output with 1% precision, it remains a convenient device. The calibrator produces 1.00 V out with no load; however, when 5 kΩ is connected to the output, the output decreases to 0.9 V. Show that the output impedance for this dc signal source is 555 Ω.
4. Suppose the voltage calibrator of the above problem is set to 10 mV *on its dial* and connected into a certain dc amplifier. A 100 Ω/V_{dc} voltmeter is connected to the output of the dc amplifier and the 1.5-V scale of the voltmeter is used.

The dc amplifier has a voltage gain (open-circuit) of 50, an input impedance of 1 kΩ, and an output impedance of 20 Ω.
(a) Make a diagram similar to Figure A5–2 for this problem.
(b) Find the reading that will appear on the voltmeter.

5. A certain audio oscillator has an output impedance of 200 Ω. It is connected to a good oscilloscope, and its output is adjusted to 0.2 $V_{p\text{-}p}$. Then the oscillator is connected to an amplifier that has an input impedance of 400 Ω, an output impedance of 50 Ω, and an open-circuit voltage gain of 100. A 100-Ω resistor is connected to the output of the amplifier.
(a) Diagram the problem using a black-box model for the parts of the system.
(b) Find the peak-to-peak voltage that develops across the 100-Ω resistor.
6. Suppose we have an audio oscillator with a 50-Ω output impedance and another with a 600-Ω output impedance. They each present an open-circuit output voltage of 20 V_{rms}. What is the maximum possible power that may be output by these respective generators to some (proper) load connected to each?
7. A certain audio power amplifier has an output impedance of 2 Ω and is connected to an 8-Ω loudspeaker. Compute how much the power output to the loudspeaker would increase if the amplifier operated into a matching loudspeaker.
NOTE: Calculate the percentage increase in power.
8. Suppose 3 identical ideal ac amplifiers are cascaded, each with a voltage gain of 20. A signal source of 0.1 $V_{p\text{-}p}$ is connected to the input of the overall amplifier.
(a) What is the voltage expected from the output of the overall amplifier?
(b) What minimum power supply voltage is required for each stage of the overall amplifier?
9. Consider the 3-stage cascaded amplifier of Figure A5–5. Assume the three stages are identical, each with input impedance R_{in} and output impedance R_{out}. Solve for the open-circuit voltage gain of the amplifier; that is, find a formula for v_{out}/v_{in}.
10. Consider an amplifier constructed by cascading 3 ideal amplifiers with voltage gains A_1, A_2, and A_3. Show that if A_1, A_2 and A_3 are expressed in dB, the overall voltage gain in dB is $A = A_1 + A_2 + A_3$. Thus the voltage gains are simply additive when expressed in dB.
11. We have on the lab bench an ac amplifier and also a dc amplifier. Each has a voltage gain of 4.
(a) Suppose a 1.5-V battery is connected to the respective amplifiers. What is the voltage out of each?
(b) Suppose a 1.5-V_{rms} ac voltage is connected to the respective amplifiers. What is the voltage out of each? What frequency assumption is made?
12. Let us consider the low-frequency bandwidth limit of ac-coupled amplifiers. Suppose two identical amplifiers are capacitively coupled as shown in Diagram A5–1. For each the output resistance is considered to be ideal, the input resistance is 1 kΩ, and the voltage gain is 10. They are coupled with a 0.1-μF capacitor. An *ideal* signal generator with 0.1-V_{rms} output signal is attached to the input A of the overall amplifier.

(a) Suppose the frequency of the signal generator is very high. Find the voltages appearing at ports *B* and *C*, respectively.
(b) Suppose the frequency of the signal source is 500 Hz. Find the magnitude of the voltages appearing at ports *B* and *C*, respectively.
(c) Argue that the frequency response of the overall amplifier is determined by a high-pass filter behavior. What resistor and capacitor compose the filter?
(d) What is the low-frequency end (−3 dB point) of the amplifier?

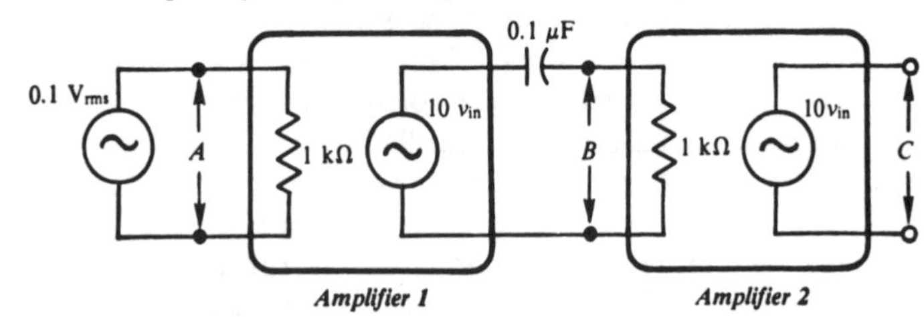

DIAGRAM A5–1.

13. Consider the high-frequency bandwidth limit of cascaded amplifiers. Suppose two identical amplifiers are cascaded as shown in Diagram A5–2. The input resistance of each is considered to be ideal, but there is an internal shunt capacitance of 200 pF. The open-circuit voltage gain is 10, and the output resistance is 10 kΩ for each amplifier. An ideal signal generator with 0.1-V_{rms} output signal is attached to the input *A* of the overall amplifier.
(a) Suppose the frequency of the signal source is very low. Find the voltages appearing at ports *B* and *C*, respectively.
(b) Suppose the frequency of the signal source is 1 MHz. Find the magnitude of the voltage appearing at ports *B* and *C*, respectively.
(c) Argue that the frequency response of the overall amplifier is determined by a low-pass filter behavior. What resistor and capacitor compose the filter?
(d) What is the bandwidth for this amplifier?

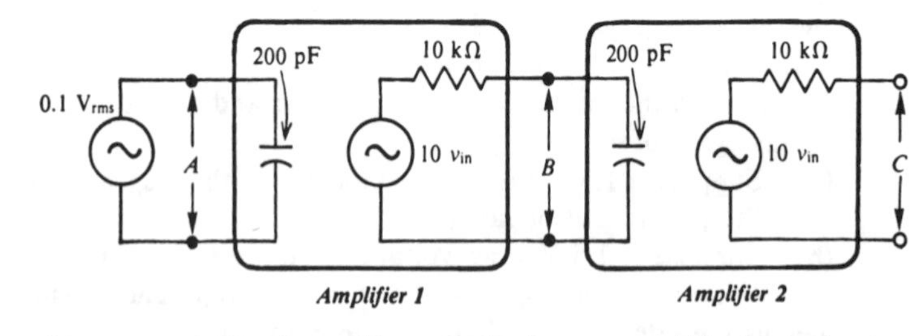

DIAGRAM A5–2.

14. A_v can be positive or negative, depending on whether an amplifier is noninverting or inverting.
(a) Justify this statement.
(b) Which is positive?

15. A 0.1-V battery is connected to a dc noninverting amplifier and also to a dc inverting amplifier, as shown in Diagram A5–3. The magnitude of A_v for each

amplifier is 5. Three DVMs are connected with the red lead placed as indicated by R in each case. When the red lead is more positive, the DVM reading shows a + sign. What is the reading, including sign, for each DVM?

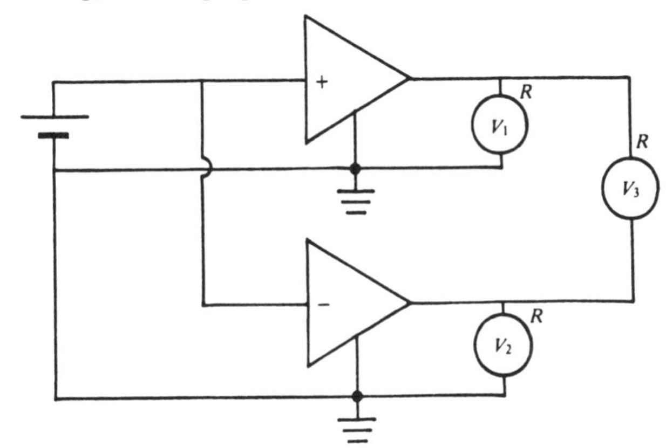

DIAGRAM A5–3.

16. (a) An ideal differential amplifier with a difference-mode gain of 20 has +50 mV applied to its + input and −100 mV applied to its − input. What is the output signal?
    (b) What assumption about common-mode gain did you make in part (a)?
17. Consider an ideal difference amplifier that is connected as pictured in Diagram A5–4. The voltage gain of the amplifier is 20. What is the voltage reading, including sign, of the DVM? (Note the "red" DVM lead R.)

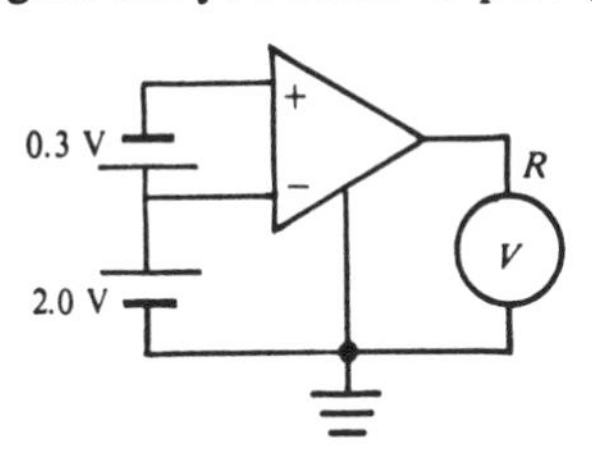

DIAGRAM A5–4.

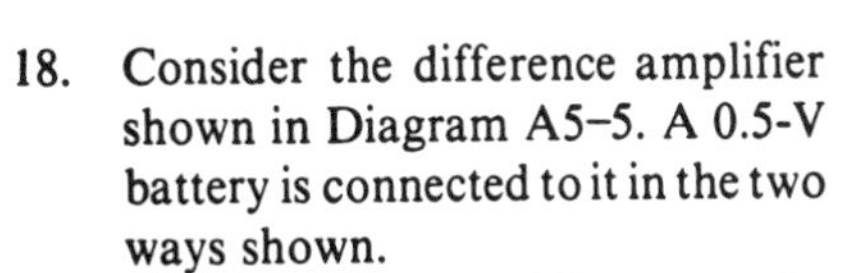

18. Consider the difference amplifier shown in Diagram A5–5. A 0.5-V battery is connected to it in the two ways shown.
    (a) What is the difference voltage gain for this amplifier?
    (b) What is the common-mode voltage gain for this amplifier?
    (c) What is the CMRR for this amplifier in dB?

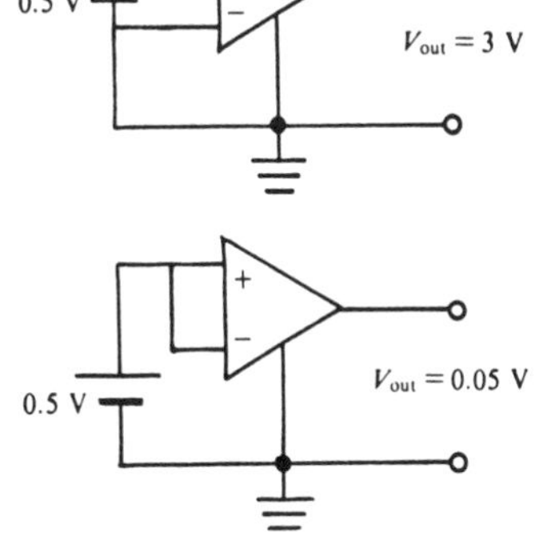

DIAGRAM A5–5.

19. A difference amplifier (or differential amplifier) has a difference gain of 100 and a common-mode rejection ratio of 60 dB. Suppose a 1-$V_{p\text{-}p}$ signal is applied to both inputs simultaneously with respect to ground. What signal will appear at the output?
20. Suppose a Wheatstone bridge is operated in out-of-balance mode in conjunction with an ideal noninverting amplifier that has a voltage gain of 160. The excitation voltage to the bridge is 4.5 V, and the arms of the bridge are each 300 Ω, except for the transducer resistance, which we write as 300 Ω + Δ. See Diagram A5–6.
    (a) The source of excitation voltage to the bridge must *not* be grounded in this circuit. Why?
    (b) Show that the output V_{out} (in V) is given by +0.6Δ, if Δ is in ohms.
    (c) Suppose the excitation voltage must be grounded at the negative terminal. We still desire the output signal V_{out} to be relative to ground. Draw a circuit illustrating how it can be accomplished using a difference amplifier.
    (d) What is the nominal common-mode voltage into the difference amplifier of part (c)?

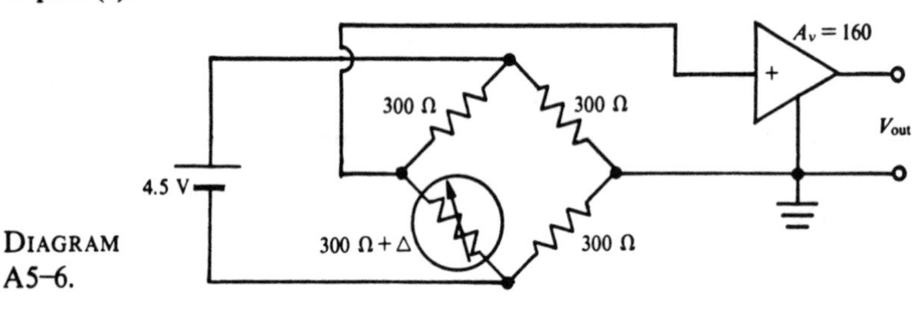

DIAGRAM A5–6.

21. (a) An ideal *bandpass filter* passes no signal from input to output for frequencies outside the bandwidth or *pass band* of the filter; however, it passes all signals unattenuated if their frequencies lie within the pass band. Let v_{in} and v_{out} be the input and output frequencies, respectively, for a bandpass filter, and let $A_v = v_{out}/v_{in}$. If f_1 and f_2 are the lower and upper frequencies of the pass band, draw a graph of A_v versus frequency f for the ideal bandpass filter.
    (b) Suppose a metal film resistor with a resistance of 100 kΩ is connected to a noise-free ideal bandpass filter for which the band-limit frequencies are 4000 Hz and 6000 Hz. What is the rms noise voltage expected at the filter output? The temperature of the resistor is 25° C.
    (c) Repeat part (b) except change the band-limit frequencies to 29 kHz and 31 kHz.
22. A 1-kHz signal is input to an amplifier that has a bandwidth of 100 Hz about the 1-kHz center frequency. The signal-to-noise ratio from the amplifier output is 20. A bandpass filter with a 10-Hz bandwidth about a 1-kHz center frequency is connected to the amplifier output. What is the signal-to-noise ratio expected for the signal from the bandpass filter output?

23. A transducer with a noise voltage output of 8 $nV_{rms}/\sqrt{Hz}$ is connected to an amplifier that has an equivalent input noise voltage of 4 $nV_{rms}/\sqrt{Hz}$. The voltage gain of the amplifier is 3000 and its bandwidth is 100 kHz. What rms noise voltage is expected from the amplifier output? What is the minimum desired voltage signal from the transducer output for a signal-to-noise ratio of 20?

CHAPTER A6

Operational Amplifier and Electronic Function Blocks

Until about 1962, amplifiers were constructed from *discrete active devices;* that is, from tubes or transistors that could be used in conjunction with passive elements and a power supply to increase the power level of a voltage signal. Since that date it has been possible to produce a number of transistors interconnected with resistors and perhaps capacitors on a single small *chip* of semiconductor. An amplifier is produced all at once by techniques used to manufacture single transistors. Such an electronic circuit is termed an *integrated circuit* (IC) because transistors and resistors are combined or integrated together on one chip.

The focus for this analog electronics book is the study of one versatile type of amplifier called the *operational amplifier* (OA or *op amp*). The operational amplifier was first constructed in the 1940s from vacuum tubes and used in analog computers to perform various mathematical operations such as addition or multiplication of voltages. This function is the reason for the term *operational.* However, the operational amplifier has great versatility and has since been put to use in all sorts of applications. It is probably fair to credit Malmstadt, Enke, and Toren with popularizing the op amp among scientists with their book *Electronics for Scientists* published in 1963. At that time, the op amp was still a vacuum tube design.

The revolution in electronic system design caused by the op amp began in earnest when a young electronic engineer named Bob Widlar (three years out of college) designed the "702" integrated circuit operational amplifier for Fairchild Semiconductor in 1963. In 1965 he designed the improved "709" that quickly became a classic IC and is manufactured by almost all important sources of solid-state devices. Fairchild introduced the "741" in 1968, and this unit is a widely used "workhorse" (and the one used in most of our discussion). It is an amplifier containing 20 transistors. By 1972 the price of these units was about $1.00—less than the single price for many transistors or tubes. Subsequently, the price declined to about $.25.

Properties of the Operational Amplifier

An *operational amplifier* is any high-voltage-gain dc amplifier with high input impedance and low output impedance, usually having differential inputs. *High* is a relative term and is qualified later on. The op amp is universally represented as in Figure A6–1a; that is, as a triangle with an output at one corner and the noninverting and inverting inputs marked plus (+) and minus (−), respectively. Most units are operated from positive and negative voltage supplies so that the amplifier output may swing either positive or negative. The power supply connections are usually omitted from the symbol.

The operational amplifier may be modeled as in Figure A6–1b. The similarity of the op-amp model in this figure to the general amplifier model given earlier in Figure A5–1 should be quite evident. The *output impedance* R_{out} of the op amp corresponds directly; however, the *input impedance* R_{in} for the op amp is between its + and − input terminals since the op amp is a difference amplifier. Notice that v_d is the *difference voltage* between the two input terminals of the actual operational amplifier. Then the size of the voltage generator in the model is given by $A_{vo}v_d$, where A_{vo} is called the *open-loop voltage gain* of the operational amplifier. A_{vo} corresponds directly to the A_v of Figure A5–1, but the subscript *o* is added to emphasize or distinguish that this is the voltage gain of the op amp *itself.* (The reason for the term *open loop* will become clear shortly as we consider amplifier circuits that incorporate op amps.) If no load is connected to the op-amp output, we see that the output voltage is the same as the voltage of the internal voltage generator, so that

$$v_{out} = A_{vo}v_d \quad \textbf{(A6–1)}$$

This relationship simply says the output voltage v_{out} of the op amp is more than its input difference voltage v_d by the factor of the open-loop voltage gain A_{vo} of the op amp.

NOTE: Arrows are used in Figure A6–1b and also in the remainder of the book to indicate points between which a given voltage is defined; notice the v_d and v_{out} arrows in Figure A6–1b. The "tail" of the arrow tells the reference point for the voltage. If the point at the arrow's "head" is the more positive of the two, the designated voltage is a positive quantity.

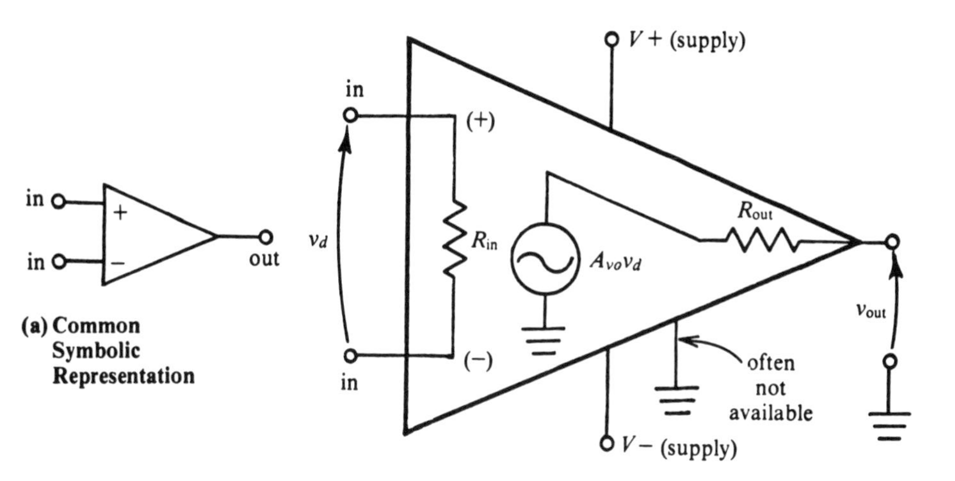

FIGURE A6–1. *Operational Amplifier*

Let us consider some values for the parameters in the model of Figure A6–1b for an actual operational amplifier. For the popular 741 IC op amp, the resistance between inputs R_{in} is typically 2 megohms, the output resistance R_{out} is 75 ohms, and the *open-loop voltage gain* A_{vo} is 200,000 (106 decibels). However, this gain begins to fall for frequencies greater than about 10 hertz. The maximum supply voltage is ±18 V (use of ±12 V to ±15 V is common). The dc power supply connections are made to the $V+$ and $V-$ leads as diagrammed earlier in Figure A5–8.

Ideal Operational Amplifier and Basic Circuits

If the operational amplifier has a very large (approaching infinity) input resistance R_{in} and also very large (approaching infinity) voltage gain A_{vo}, it has the essential ingredients of being an *ideal op amp*. Two more requirements should be added: (1) If the voltage between the inputs v_d is zero, the output voltage should be exactly zero (the extent to which this is not true is called the *input offset-voltage error* and is discussed later under real op-amp considerations; and (2) the voltage gain should remain large independent of frequency.

The operational amplifier is almost always used with a large amount of *negative feedback;* that is, some of the output signal is fed back to the input in a way that tends to negate or counteract the original signal into the amplifier. The case of *positive feedback* is usually much less desirable. Most students have encountered a case of positive feedback: When the microphone of a PA system is placed in front of the loudspeaker, positive feedback from the loudspeaker output to the microphone input causes a loud squeal.

If the two requirements hold true, the analysis of various amplifier configurations is quite straightforward and the amplifier behavior turns out not to depend on the character of the op amp itself. The behavior of the resulting amplifier in general depends only on the values of the components connected to the op amp. If these values are accurately known, the behavior of the constructed amplifier is just as accurately predictable; herein lies the beauty of the approach. Actually, this sort of situation has long been true of amplifiers that include a high gain element and a large amount of feedback. Enough said; let us see by example.

Inverting Amplifier

Certainly the basic configuration and prototype illustration of how the op amp may be put to use is the *inverting amplifier* shown in Figure A6–2a. Only two resistors are involved: a resistor R connected between the amplifier input and the op-amp inverting (or −) input, and a resistor R_f connected between the op-amp output (as well as the amplifier output) and the op-amp − input. The noninverting (or +) input is connected to ground; the "ground" (or reference) terminals of the amplifier input and output are also connected to ground.

The reader should be clear on two points here. First, although often no power supply connections are shown connected to the op-amp triangle, they must in fact be connected. They are shown only as dashed lines in the drawing. In general, both positive $V+$ and negative $V-$ supply voltages are used (equal in magnitude), and the zero potential or *ground* of the supply is connected to the operational amplifier circuit in some way. In this circuit, it is to the + input of the op amp.

NOTE: This ground might *not* be truly connected to *earth* however; the amplifier is then said to be *floating* relative to true ground.

Second, the op amp along with resistors and wires is placed in a box, and the whole now becomes the inverting amplifier—or simply amplifier—with which we subsequently work. It has a pair of input posts and a pair of output posts as shown in Figure A6–2a.

Notice that the voltage into the amplifier is v_{in}, while the voltage into the operational amplifier *itself* is v_d. The voltage out of either is the same: v_{out}. The arrows drawn by these voltages in Figure A6–2a show the sense of positive polarity (arrowhead near more positive point). If the voltage is reversed in any case, the value is negative.

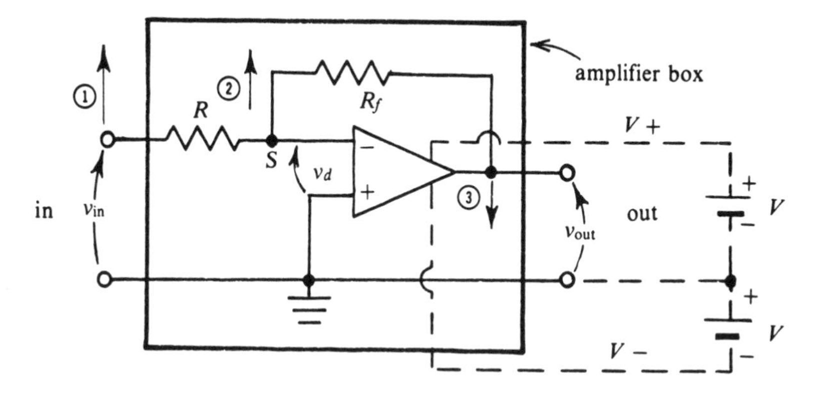

(a) Connection of Two Resistors to an Op Amp to Make an Inverting Amplifier

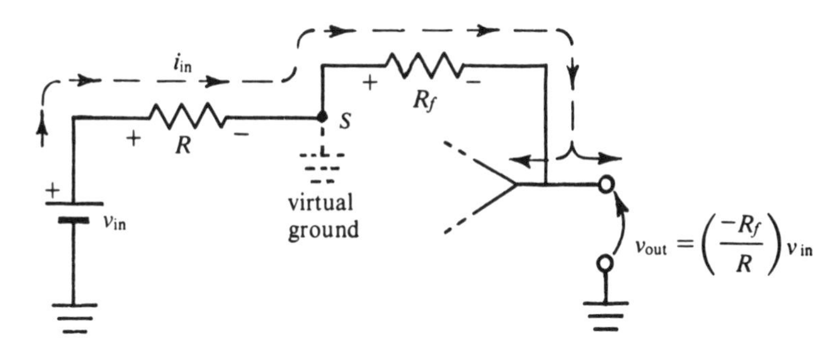

(b) Current Path and Voltages with the Ideal Op Amp

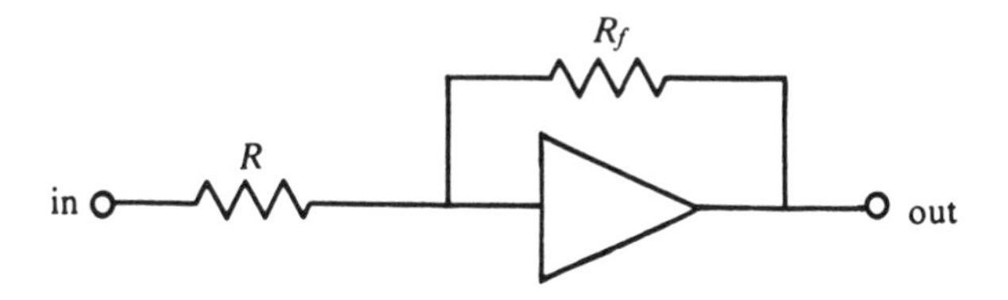

(c) Very Simplified Representation

FIGURE A6–2. *Inverting Amplifier Constructed from an Operational Amplifier Gain Block*

There are now three points that are central to a quick and easy derivation of amplifier behavior using op-amp design analysis:

1. The magnitude of the output voltage of the op amp must be less than the supply voltage value. For example, if the supply voltages used are ±15 V, there is *no way* the output voltage can be more than 15 V above or below ground. (Refer to Chapter A5.) If the op amp is *functioning properly,* the output voltage of the op amp is directly proportional to the voltage into the *op-amp input.* Then from Equation A6–1 we have

$$v_d = \frac{v_{out}}{A_{vo}}$$

But if v_{out} is limited to 15 V and A_{vo} is 150,000, for example, then v_d must be less than 15 V/150,000 = 0.1 mV = 100 μV. If A_{vo} were arbitrarily large, as for an ideal op amp, the voltage v_d would be vanishingly small; even so, v_d is a small voltage. Thus we have the rule: If a linear operational amplifier circuit is functioning, the voltage difference between the − and + inputs is virtually zero (*is* zero for an ideal op amp).

2. Due to the large impedance between the − and + inputs, virtually no current passes into an op-amp input (zero current for an ideal op amp).

3. In linear amplifier circuits—that is, not digital circuits—a connection is always made between the output and the inverting − input in some way. This is *negative feedback,* wherein the output tends to oppose the change at the input (which caused the output). (If the connection were made to the noninverting input, such *positive feedback* generally causes the output to oscillate or else to simply move to one supply voltage and stay there. Therefore, this is never done, except in the case of oscillator or digital circuits.)

With these points in mind, we next consider the circuit in Figure A6–2a. First, let's try to get a qualitative understanding of what happens. Notice that there is a connection between the output and the − op-amp input. Thus, if the voltage v_2 should increase [as shown by arrow (1) in the drawing], the voltage into the inverting input also increases [shown by arrow (2)]. The output then tends to go in the inverse sense or down a great deal as shown by arrow (3) at the output. However, the output voltage is connected back to the input through R_f. Thus, the output tends to greatly counteract the upward movement of the inverting input. This causes the voltage v_d at the − input to be much less than the voltage v_{in} that is applied to the amplifier input. Consequently, the output voltage is much less than $A_{vo} \times v_{in}$. (However, v_{out} is always equal to $A_{vo} \times v_d$.) We can see qualitatively that the negative feedback effect is less if R_f is made larger; therefore we expect a larger amplifier gain if larger R_f is used.

Next we consider a quantitative analysis that is exact if an ideal op amp is used and quite accurate for many real op-amp situations. Because the voltage v_d between the − and + inputs is so small and the + input *is* connected to ground, the − input is always "virtually at ground." Indeed, the − input is referred to as a *virtual ground* in the circuit. It is labeled as point S and is often called the *summing point.* We see why this term is useful later in the situation of the *summing amplifier.*

Regard the input voltage v_{in} as a battery as shown in Figure A6–2b. A current then flows through R toward the − input of the op amp, but it does *not* flow into the op amp because of its high input resistance (for example, for the 741 there are 2 megohms to the + terminal and thus to ground). It therefore flows through R_f toward the output, which should be at a negative voltage anyway. Now the current flowing from v_{in} through R is completely predictable since one end is at voltage v_{in} and the other is virtually at ground. We have a voltage v_{in} across the resistance R, and the current i_{in} through it (from the battery) is

$$i_{in} = \frac{v_{in}}{R} \tag{A6–2}$$

This same current flows through the feedback resistor R_f so the voltage across it is clearly

$$v_{R_f} = i_{in} R_f = v_{in}\left(\frac{R_f}{R}\right) \tag{A6–3}$$

But the end of R_f connected to the − input is virtually at ground and is the more positive as drawn in Figure A6–2b because the current is flowing *in* at that end. The negative end of R_f is connected to v_{out}. Then the output voltage is negative and just equal to the voltage across R_f as calculated in Equation A6–3. We have

$$v_{out} = -v_{R_f} = -v_{in}\left(\frac{R_f}{R}\right) \tag{A6–4}$$

The minus sign in this equation reflects the fact that v_{out} will be negative when the input is positive (or it will be positive when the input is negative). It is an inverting amplifier. Finally the ratio of the output voltage v_{out} to the input voltage v_{in} is the voltage amplification of the amplifier. Dividing both sides of Equation A6–4 by v_{in} gives

$$\frac{v_{out}}{v_{in}} = A_v = -\frac{R_f}{R} \tag{A6–5}$$

Thus the voltage gain of an inverting amplifier using an ideal op amp is simply the ratio of the feedback resistance to the value of the input resistor.

This gain is quite different from the voltage gain of the op amp itself, which is A_{vo}. A_{vo} is called the *open-loop gain* of the op amp since it is the gain without feedback or with an *open* in the feedback loop (R_f = infinity).

The *input impedance* of the inverting amplifier is examined next and is very easy to obtain. In Chapter A5 we examined the concept of amplifier input impedance; quite clearly it is that an amplifier presents some impedance (resistance) R_{inA} to a voltage v_{in} connected to its input. Thus, a current i_{in} flows into the amplifier input that must be given by $i_{in} = v_{in}/R_{inA}$. This expression can be turned around. If the current that flows into an amplifier input as a consequence of a certain applied voltage is known, we can quickly solve for R_{inA}. In our amplifier the current in was argued to be $i_{in} = v_{in}/R$. Then

$$R_{inA} = \frac{v_{in}}{i_{in}} = \frac{v_{in}}{v_{in}/R} = R \qquad \textbf{(A6–6)}$$

Therefore, the input impedance to this amplifier is simply R.

Actually, this result is obvious. Note that the end S of resistor R (connection to the − input of op amp) acts as though it is connected *virtually* to ground and the other end is connected to the voltage in. But the "bottom end" of voltage v_{in} is connected to ground, too, so it is simply connected across a resistance R, and this resistor must then be the input impedance presented to the outside world by the input of the inverting amplifier circuit.

Unfortunately, this resistance for practical amplifiers cannot be terribly large, and although the operational amplifier itself may have quite an ideal input resistance, the *constructed* inverting amplifier does *not*. The reader may ask: Why not simply use a large value for R; for example, 1 megohm? Suppose the desired gain is 50. Then the feedback resistor R_f must be $A_v R_f = 50 \times 1$ megohm. The 50-megohm resistor is larger than any standard resistor and could not be easily obtained with high accuracy. Also, high humidity conditions cause leakage of current over the resistor surface or mounting board that changes its value in an unknown and variable way. Finally, R and R_f are then of the order of R_{in} for the op amp itself, and the ideal op-amp analysis we have used for voltage gain fails. Not all of the current i_{in} in Figure A6–2b flows through R_f; some flows into the − input of the op amp.

The *output impedance* of an amplifier constructed from a high-gain amplifier through use of negative feedback can be shown to be very low. In general, the output impedance R_{outA} of a feedback amplifier based on an op amp is on the order of R_{out} (A_v/A_{vo}), where R_{out} is the corresponding impedance of the op amp itself. If the open-loop gain of the op amp is very large, the output impedance of the constructed amplifier approaches zero even though the output impedance of the op amp itself is quite finite (recall that it is 75 ohms for the 741). Thus, the op amp with feedback approaches the ideal in its output behavior. But a warning is needed: The current that small integrated circuit op amps can deliver is typically limited to a few milliamperes (10 to 20).

Finally a comment should be made about the highly simplified op-amp amplifier representation that is sometimes encountered. Figure A6–2c shows this approach where even the ground line is omitted, as well as − input designation. Of course, the ground connections must really be there, and the in and out voltages are assumed to be relative to ground.

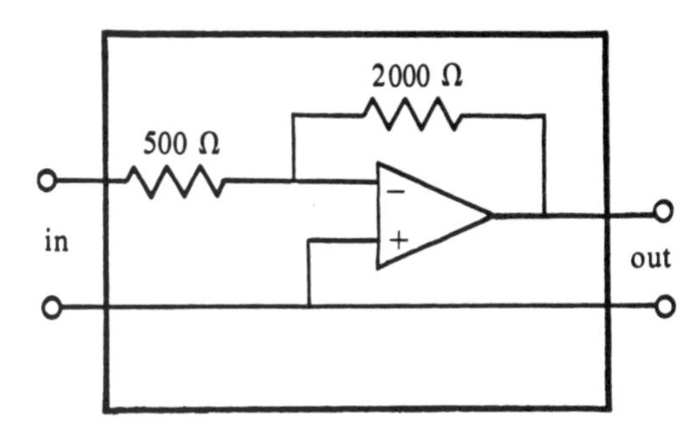

FIGURE A6–3. *Operational Amplifier Circuit*

> ***Example:*** Consider the operational amplifier circuit pictured in Figure A6–3. Find the *black-box equivalent amplifier* (general amplifier model of Chapter A5) for this circuit.
>
> *Solution:* The voltage gain A_v for this circuit is $A_v = -2000\ \Omega / 500\ \Omega = -4$. The output impedance is virtually zero. The input impedance is just the 500 Ω of that resistance *R*. We therefore obtain the black-box equivalent or model shown in Figure A6–4. Notice we have a minus sign beside the magnitude of the output generator to indicate that the polarity of v_{out} is inverted from v_{in}.

Noninverting Amplifier

The fact that the output of the inverting amplifier is reversed in polarity from the input can be inconvenient; also the low input impedance of that amplifier

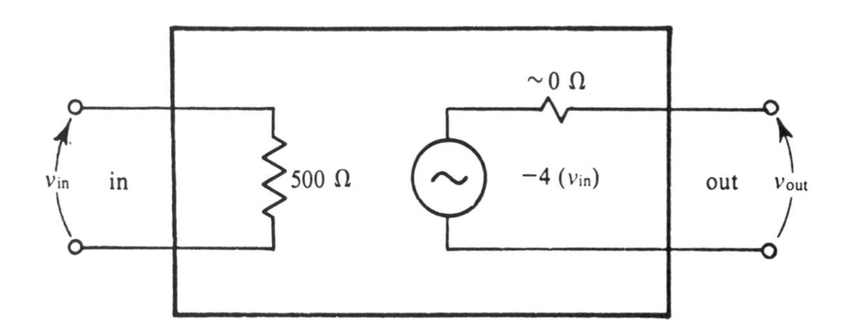

FIGURE A6–4. *Black-Box Model*

is less than ideal. Both of these features are remedied in the *noninverting amplifier;* its configuration is pictured in Figure A6–5a.

Again two resistors, R and R_f, are involved and again notice that the feedback connection is between the output and the inverting (or −) input. This must be so for negative feedback. However, the input voltage v_{in} comes directly into the noninverting (or +) input. This causes the output to change in the same voltage direction as the input voltage so that a noninverting amplifier is produced.

To derive the voltage out when an ideal (or at least "good") op amp is used, consider the important current path from the op-amp output through R_f and R to ground. (The fact that the op amp also sends current to R_L does not matter unless the current capability of the op-amp output is exceeded.) If v_{in} is positive, we expect the output to also go positive so that the current will flow in the direction shown for i in Figure A6–5b. Because of the high impedance of the op-amp input, none of the current flows into the − input. The output voltage appears across the series R and R_f combination, so that the current i is given by

$$i = \frac{v_{out}}{R + R_f} \tag{A6–7}$$

The voltage at point S relative to ground is just the current i times the resistance R. But the voltage difference between point S (the − input) and the amplifier input (the + input) is virtually zero if the amplifier is functioning properly. Therefore, the input voltage v_{in} and the voltage at S are virtually the same relative to ground. Then

$$v_{in} = v_S = iR = \left(\frac{v_{out}}{R + R_f}\right) R \tag{A6–8}$$

We solve Equation A6–8 for the ratio of v_{out}/v_{in} to obtain the voltage gain of the amplifier; that is,

$$\frac{v_{out}}{v_{in}} = A_v = \frac{R + R_f}{R} = 1 + \frac{R_f}{R} \tag{A6–9}$$

Notice the formula gives a positive number in accord with a noninverting amplifier. Also notice the ratio of R_f to R is involved, although the gain is more than that of the inverting amplifier by just 1. Again, to obtain high amplification the R_f/R ratio should be large. In this amplifier, however, there is no particular penalty for having R quite small. It is *not* the input impedance here.

To obtain an idea of the input impedance R_{inA} for the noninverting amplifier, consider where the current from v_{in} must go. It must pass through R_{in} of the operational amplifier toward S. But R_{in} itself is a large value, and we are assured the amplifier input impedance is at least this large. More subtle is the fact that S is at almost the same voltage as v_{in}. Therefore, the input current has very little inclination to travel through R_{in} to S. The value of the

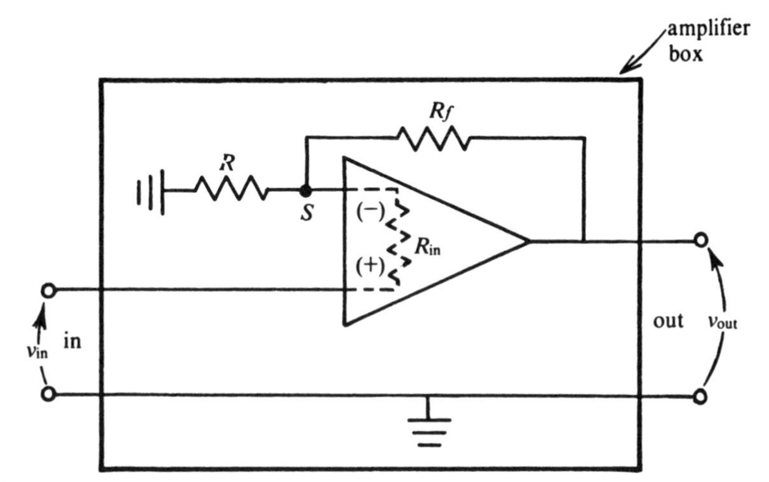

(a) Two Resistors Connected to an Op Amp to Produce a Noninverting Amplifier

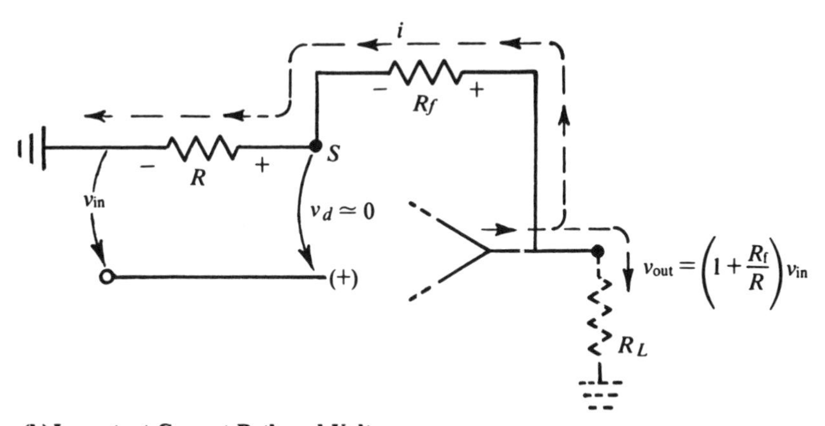

(b) Important Current Path and Voltages

FIGURE A6–5. *Noninverting Amplifier Based on an Operational Amplifier*

input current through R_{in} is obtained by realizing that the voltage across it is v_d and then solving for this input current from the equation, $v_d = i_{in}R_{in}$. Now the output voltage is *always* more than v_d by the factor of the op-amp open-loop gain. So $v_d = v_{out}/A_{vo}$. Next remember $v_{out} = A_v v_{in}$, where A_v is the gain of the amplifier, so that

$$v_d = \left(\frac{A_v}{A_{vo}}\right) v_{in} \qquad \textbf{(A6–10)}$$

The current through R_{in} is then

$$i_{in} = \frac{v_d}{R_{in}} = \left(\frac{A_v}{A_{vo}}\right)\left(\frac{v_{in}}{R_{in}}\right) \qquad \textbf{(A6–11)}$$

Finally we solve for the ratio of v_{in}/i_{in} from Equation A6–11, because *this is the effective resistance* R_{inA} presented by the overall amplifier to the input voltage. We see

$$\frac{v_{in}}{i_{in}} = R_{inA} = \left(\frac{A_{vo}}{A_v}\right) R_{in} \qquad \textbf{(A6–12)}$$

The result for R_{inA} from Equation A6–12 can be an *extremely* large number (for example, $R_{in} = 2$ megohms, $A_{vo} = 200{,}000$, and $A_v = 100$ for the 741 op amp). Actually, the resistance from the + input as well as the − input to ground in the operational amplifier itself is likely to be less. Also, the capacitance between the input and ground will present an impedance that can be lower than this at higher frequencies. Capacitance effects must always be watched at higher frequencies. There will be capacitance between structures in the op amp and between circuit structures on the circuit board. Finally, leakage of current along the surface of the circuit board or the IC socket (for example, through moisture) can easily limit the input resistance of the amplifier to something less than 500 megohms.

The resistance from the + input terminal to ground within the op amp is called the *common-mode input impedance* R_{icm+}; there is a similar resistance from the − input terminal to ground, R_{icm-}. The parallel resistance of these two common-mode input resistances can be measured (in principle) by connecting the + and − inputs together (or in common) and measuring the ac resistance to ground, R_{icm}. Typically this is on the order of 100 times larger than the *differential* input resistance R_{in}, but it is very seldom specified by manufacturers. In a noninverting amplifier, R_{icm+} appears in parallel with the resistance given by Equation A6–12 and is often the determining factor for the input resistance of the amplifier, R_{inA}. The output impedance of the amplifier is very low due to negative feedback, just as in the inverting amplifier case.

Voltage-Follower Amplifier

If the feedback resistance R_f is made equal to zero—that is, a short circuit or wire—in Figure A6–5, the *amplifier* shown in Figure A6–6a results. The voltage gain and input impedance for this amplifier are still given by Equation A6–9 and Equation A6–12, respectively. But since $R_f = 0$, the voltage gain is just unity from Equation A6–9 and the input impedance should be extremely large.

The output voltage is just the same as the input voltage; it *follows* the input wherever it goes (but between the values of the supply voltages). There

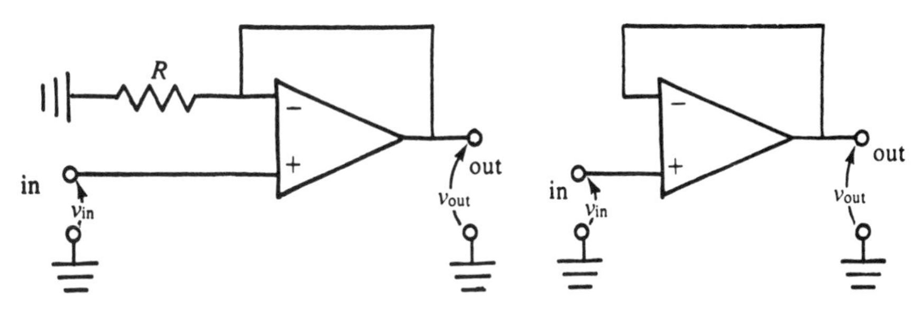

(a) Resistance R_f Equals Zero **(b) Common Voltage-Follower Circuit**

FIGURE A6–6. *Voltage-Follower Amplifier Constructed from an Operational Amplifier*

is no voltage gain, but the amplifier is *not* useless. It may amplify power a great deal, as discussed in Chapter A5. The very large input impedance and small output impedance of this amplifier are its great virtues. However, the current that can be delivered by the op amp should again be kept in mind, as was mentioned earlier.

In fact, the resistance R can be made infinite, and the voltage would again be seen from Equation A6–9 to be unity. Omitting R entirely is the same as making it infinite. Therefore, the usual op-amp voltage-follower circuit used is shown in Figure A6–6b. Simply connect a wire from the output to the − input, and use the + input as the amplifier input.

Summing Amplifier

We next consider the first circuit for which the operational amplifier can be used to perform a mathematical operation: *the summing amplifier.* We want to arrange an amplifier that will produce as output a voltage that is the sum of several input voltages. Thus, if 2.1 V, 3.6 V, and −1.6 V are put to the amplifier inputs, the output voltage will be (2.1 + 3.6 − 1.6) V = 4.1 V. We would then have an electronic adding machine, where a voltage proportional to a number (the *analog* of a number) is input, and a voltage proportional to the result is output. This is the first requirement for an analog computer, but is not the most common application in this world of cheap digital computers. The summing amplifier is useful in solving various instrumentation problems in science and industry. We may simply want a signal that is the literal sum of two signals that are available to the user.

The circuit used for a summing amplifier is a simple extension of the inverting amplifier, and its analysis is simple if the analysis of the inverting amplifier has been well understood. A circuit is illustrated in Figure A6–7. Notice that three resistors (there could be more) are shown connected to the

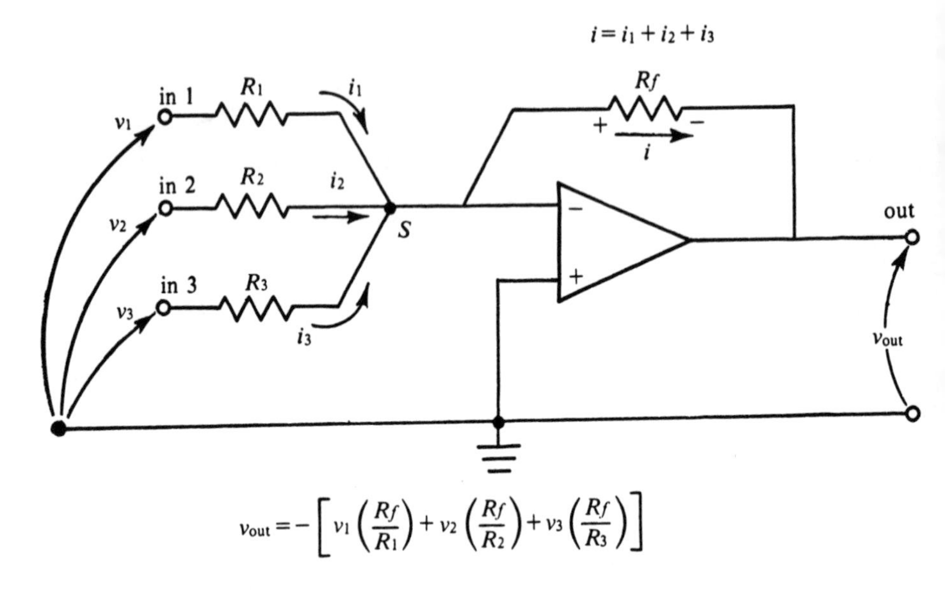

FIGURE A6–7. *Summing Amplifier Constructed from an Operational Amplifier*

summing point S, the inverting input of the op amp. These resistors are each rather like the R resistor of the simple inverting amplifier. Voltages v_1, v_2, and v_3 may be independently connected between each input and ground as illustrated. But the point S remains virtually at ground for the same reasons as argued earlier. Then, the current through each *input resistor* is determined by the input voltage effectively appearing across it. We have

$$i_1 = \frac{v_1}{R_1} \quad i_2 = \frac{v_2}{R_2} \quad i_3 = \frac{v_3}{R_3} \tag{A6–13}$$

These currents merge at the summing point S and flow through R_f to the op-amp output. (Here we see reason for the term summing point.) They do *not* flow into the op-amp − input because of its high impedance.

Now the voltage out is just the voltage across the feedback resistor R_f because the end at the S point is virtually at ground and the other end is connected to the output. We see that

$$v_{out} = v_{R_f} = -iR_f = -(i_1 + i_2 + i_3)R_f \tag{A6–14}$$

Finally inserting the currents from Equation A6–13, we have

$$v_{out} = -\left[v_1\left(\frac{R_f}{R_1}\right) + v_2\left(\frac{R_f}{R_2}\right) + v_3\left(\frac{R_f}{R_3}\right)\right] \tag{A6–15}$$

Now suppose $R_1 = R_2 = R_3 = R_f$ in Equation A6–15. Then

$$v_{out} = -(v_1 + v_2 + v_3)$$

and the output is just the sum of the input voltages (the sum is inverted, however) and we have the desired summing amplifier.

However, the more elaborate expression of Equation A6–15 shows a nice possibility. It is clearly possible to take a *weighted sum* of the input voltages by using different values for R_1, R_2, and R_3. For example, if $R_f/R_1 = 2$, $R_f/R_2 = 7$, and $R_f/R_3 = 4$, the output is $-(4v_1 + 7v_2 + 4v_3)$.

Difference Amplifier

We considered the behavior of a difference or differential amplifier in Chapter A5. It is really an *analog subtractor* because it subtracts the voltage at the inverting input from the noninverting input in determining the output. The operational amplifier itself is a difference amplifier, and we might think there is then no need for any special design work to produce a difference amplifier. Actually, the need exists because the operational amplifier cannot be used "naked": Its gain is too high.

The trick is to combine inverting and noninverting amplifier designs in one package. As the first effort, consider the design shown in Figure A6–8a. If the v_{in-} input is grounded, the output is just $+(1 + R_f/R)v_{in+}$ because we clearly have the simple noninverting amplifier. If v_{in+} is grounded, the output is $-(R_f/R)v_{in+}$ since we merely have an inverting amplifier. If voltages are present at both inputs simultaneously, it is reasonable (though not really proven by these statements) that the voltage out is the sum of these two voltages. The proof is left as an exercise. Then

$$v_{out} = +\left(1 + \frac{R_f}{R}\right)v_{in+} - \left(\frac{R_f}{R}\right)v_{in-} \qquad \textbf{(A6–16)}$$

If $R_f/R >> 1$ we have (approximately) a difference amplifier; that is,

$$v_{out} = \frac{R_f}{R}(v_{in+} - v_{in-})$$

The common-mode rejection by this approach will not be extremely good due to the approximation used. Also, the input impedances presented to v_{in+} and v_{in-} are *very* different. The problem is that the gain from the noninverting amplifier is more than that from the inverting amplifier. This situation could be remedied by putting a voltage divider between v_{in+} and the + input of the op amp (see Figure A6–8b). Simply reduce the voltage from v_{in+} by a factor so that this factor times $(1 + R_f/R)$ is the same as R_f/R. The voltage divider factor is $R_2/(R_1 + R_2)$ so we want

$$\frac{R_2}{R_1 + R_2}\left(1 + \frac{R_f}{R}\right) = \frac{R_f}{R} \qquad \textbf{(A6–17)}$$

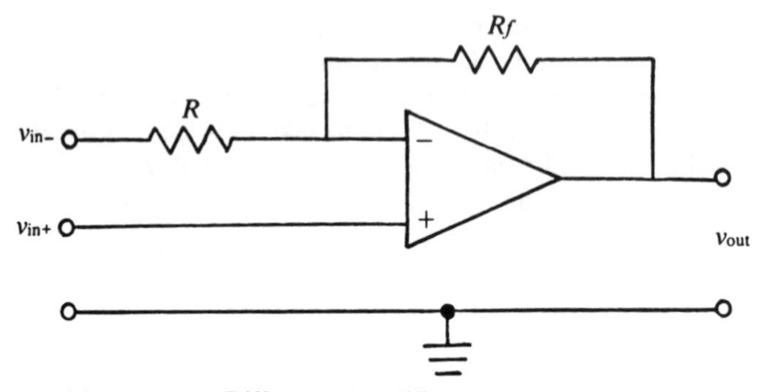

(a) Attempt at Difference Amplifier

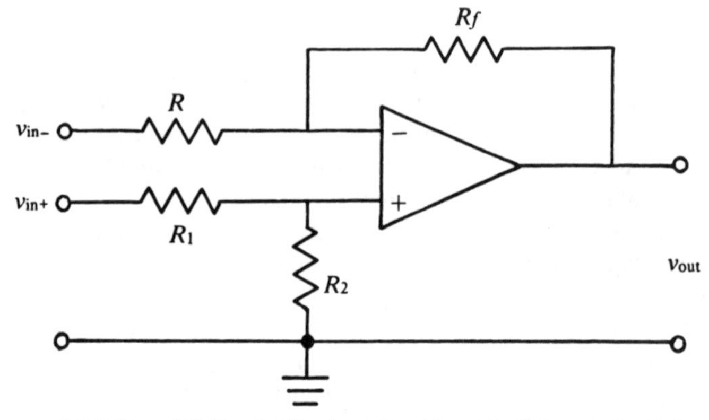

(b) Voltage Divider in Noninverting Input to Balance the Gain

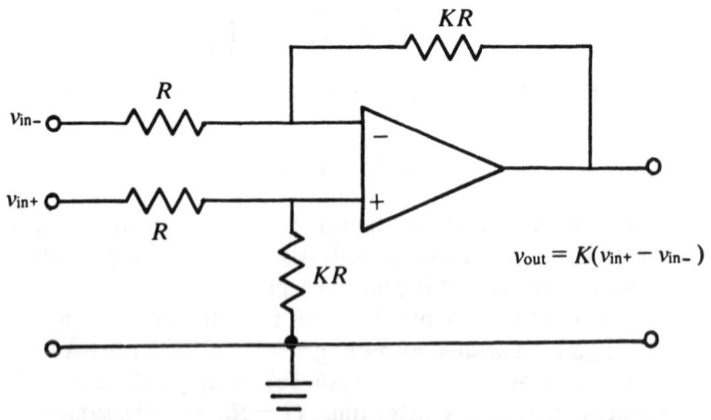

(c) Difference Amplifier with Balanced Gain K at Each Input, but Unbalanced Input Impedances

FIGURE A6–8. *Difference or Differential Amplifier*

Move this factor to the right side of the equation as follows

$$\left(1+\frac{R_f}{R}\right)=\frac{R_f}{R}\left(1+\frac{R_1}{R_2}\right)$$

We can solve for the ratio R_1/R_2 in this expression (or simply notice the solution by inspection). We find

$$\frac{R_f}{R}=\frac{R_2}{R_1} \quad \textbf{(A6–18)}$$

as the requirement on the resistance ratio of the voltage divider. In particular, one possibility is to set $R_1 = R$ and $R_2 = R_f$. Let $K = R_f/R$. From Equation A6–14 we then have

$$v_{out}=\frac{R_f}{R}\,(v_{in+}-v_{in-})=K(v_{in+}-v_{in-}) \quad \textbf{(A6–19)}$$

This situation is pictured in Figure A6–8c. We can show (with some work) that the common-mode rejection ratio of the resulting amplifier is the same as that of the operational amplifier itself, provided the resistors are exactly matched.

Rather than having to select resistors so carefully, a simple approach is to use a potentiometer for the voltage divider in the noninverting input. The proper setting of this potentiometer can be done experimentally by applying a signal to both inputs at once—that is, in common, or common mode—and adjusting the potentiometer for a *minimum* signal at the output. Figure A6–9 shows the amplifier and experimental arrangement. Since the potentiometer permits optimizing the common-mode rejection ratio (CMRR) for the differential amplifier, it is given the label *CMRR adjustment* in the figure. The amplifier is shown connected to a common-mode input signal to adjust the CMRR potentiometer.

The amplifier pictured in Figure A6–8c does have the disadvantage of different input impedances relative to ground at the two inputs. The input impedance at v_{in-} is just R (like an inverting amplifier), while that at v_{in+} is $(R + KR)$ because these two resistors are simply in series to ground and no current flows into the op-amp + input. It is possible to work with other resistances for R_1, R_2 in the voltage divider network of Figure A6–8b. Choose $R_1/R_2 = R/R_f$, and also $R_1 + R_2 = R$. These are two simultaneous equations that satisfy both the balanced-gain and equal input-impedance requirements. It is left as a problem to show that R_1 and R_2 must satisfy

$$R_1=\frac{R^2}{(R+R_f)} \qquad R_2=\frac{RR_f}{(R+R_f)}$$

The difference gain of the amplifier in Figure A6–8b with these values for R_1 and R_2 will still be R_f/R. However, the *simplest solution* is to use a potentiometer—CMRR *pot*—with resistance R, as shown in Figure A6–9. Then, the v_{in+} input impedance is clearly R, just as it is for v_{in-}.

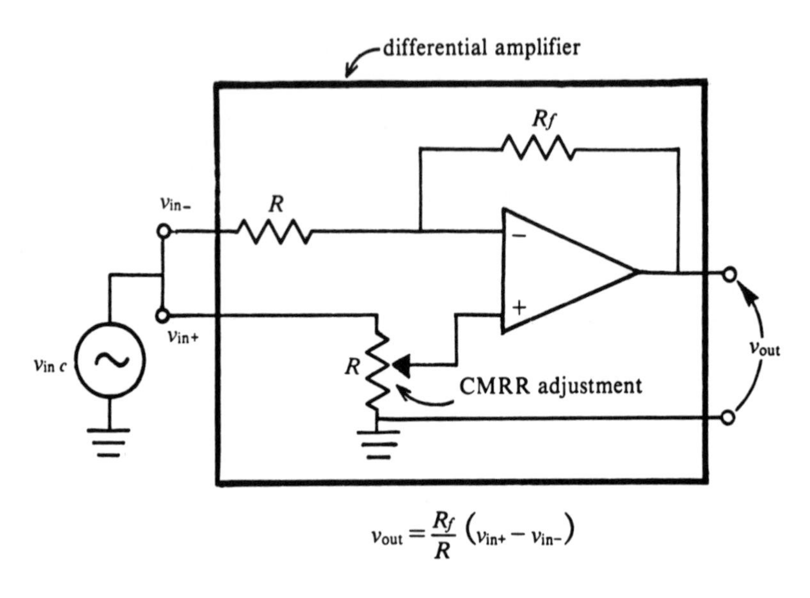

$$v_{out} = \frac{R_f}{R}\left(v_{in+} - v_{in-}\right)$$

FIGURE A6–9. *Difference Amplifier with CMRR Adjustment*

Instrumentation Amplifier

The difference amplifier of Figure A6–9 still falls short of the ideal in that it does not have a very high input impedance. This situation can be remedied by using a pair of voltage followers ahead of the two inputs as shown in Figure A6–10. These followers are often called *buffer amplifiers* since they act as an intermediary to present a desirable high impedance to the outside world. If the op amps used are all low offset voltage, low drift, very high gain, and wide bandwidth (in short, excellent) units, the resulting amplifier is often called an *instrumentation amplifier* because it is desirable for use in various instruments.

Current-to-Voltage Converter

Suppose we want to measure current very accurately using a voltmeter (for example, to construct an ammeter from a DVM). The simple approach is to place a resistance in the current-carrying line and measure the voltage across the resistance. Unfortunately, a finite voltage then clearly develops across the resistance, and hence the ammeter; the ammeter is not ideal since it has this resistance.

A nice approach can be thought of that is a simple alteration of the inverting amplifier. Simply remove the *input resistor* R in Figure A6–2a to

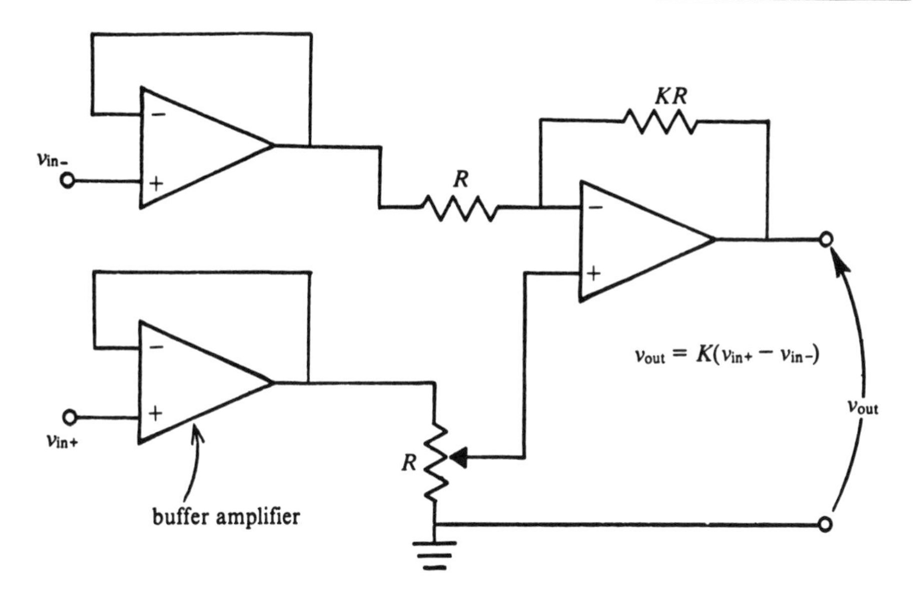

FIGURE A6–10. *High Input-Impedance Difference Amplifier*

obtain the circuit shown in Figure A6–11. Then, whatever current i that flows in toward S will flow through R_f by arguments that should now be familiar. A voltage develops across R_f that becomes the amount of voltage that the output is below ground. Thus, the output voltage is directly proportional to the current in and is given by

$$v_{out} = -iR_f \quad \textbf{(A6–20)}$$

A voltmeter connected across the output can then be calibrated in terms of the input current. If R_f is a precision resistor, the current-to-voltage conversion that is affected here is very accurate.

A nice feature of this approach is that the voltage that develops between the ammeter leads (1) and (2) in Figure A6–11 is very small. This voltage is simply the v_d between the op-amp inputs that must in fact be there to produce v_{out} at the output. But $v_{out} = A_{vo}v_d$ so this voltage is $v_d = v_{out}/A_{vo}$. Because of the large value of A_{vo}, v_d is a very small voltage and the ammeter approaches the ideal of causing a small voltage drop when inserted in a circuit.

Example: Suppose we have a voltmeter that reads 1 V full scale and we wish to construct an ammeter that reads 10 microamperes full scale. Find R_f and also the voltage that appears between the microammeter leads if a 741 op amp is used.

Solution: The voltmeter will be connected across the output terminals of the circuit in Figure A6–12. We want 1 V out when 10 microamperes flow through R_f. Then,

$$1\ \text{V} = (10^{-5}\ \text{A})R_f \text{ or } R_f = 10^{+5}\ \Omega$$

For this current in, the voltage between the microammeter leads will be

$$v_d = \frac{1\ \text{V}}{A_{vo}} = \frac{1\ \text{V}}{200{,}000} = 5\ \mu\text{V}$$

for a typical 741. The circuit is shown in Figure A6–12.

One application of this circuit involves measuring the short-circuit current of photovoltaic cells (solar cells). A silicon photovoltaic cell produces an open-circuit output voltage of about 0.6 V when in bright light. As mentioned in Chapter A3, the short-circuit current from a photovoltaic cell is accurately proportional to the light intensity falling on it and can be used to monitor light intensity. However, when an ammeter is used to measure the short-circuit current, the voltage across it must be much less than 0.6 V. The op-amp–based current-to-voltage converter fits this requirement very well, as we see from the preceding example.

NOTE: There is one practical problem and warning with respect to the use of the op-amp microammeter of Figure A6–12 in a general electrical circuit. In fact, this problem often occurs in using electronic instruments and was discussed briefly in Chapter A2. We will examine the problem in more detail as an aside here.

If the leads of the op amp are inserted into a general circuit that is at some point *truly grounded to earth,* the op-amp circuit itself must

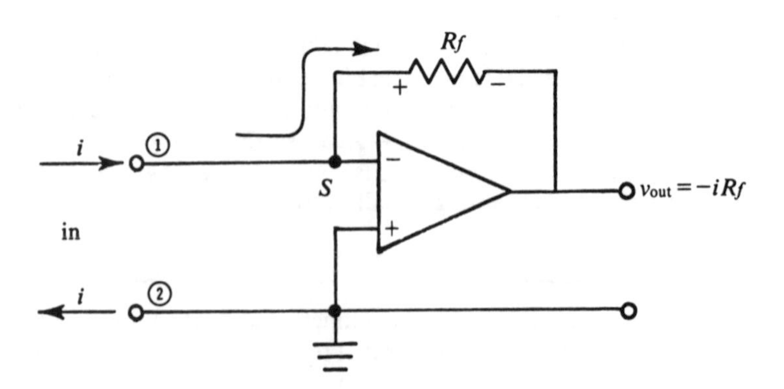

FIGURE A6–11. *Current-to-Voltage Converter*

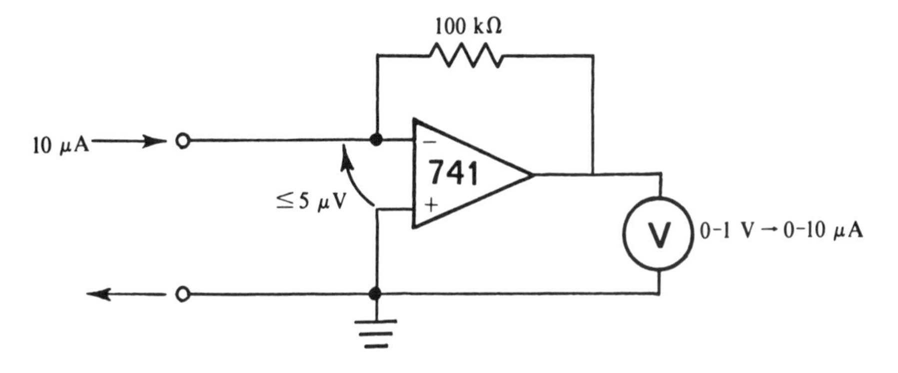

FIGURE A6–12. *Microammeter Using 1-V Voltmeter*

have a *floating ground.* Indeed (as we have mentioned earlier), the term *common* rather than ground is preferred here, since the ground of the V± supply is then not really "earthed," but merely is a common point used in the power supply and the op-amp design.

Figure A6–13 illustrates the problem. Suppose we wish to measure the current through the resistor R_3; therefore we open the circuit at that point and insert our electronic microammeter. However,

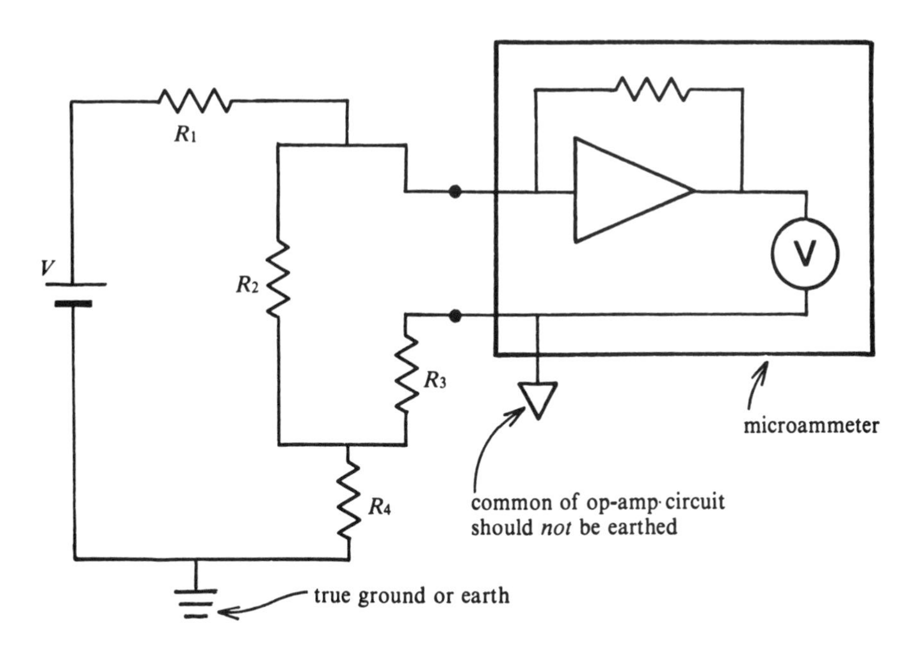

FIGURE A6–13. *Circuit Illustrating the Need for a Floating Ground or Common*

the circuit is already truly grounded or earthed at one point, and consequently both ends of R_3 are *off ground.* Connecting the microammeter—with one of its leads being true ground—would cause a short to ground at this place in the circuit. There can be *only one true ground point* (or wire) in a circuit.

Actually, no point in the transformer secondary or regulating circuit of the power supply for the op amp need be truly earthed; then the power supply is *floating relative to true ground* as required in this situation. There may be a problem of current leakage between the secondary and the primary windings of the transformer in the power supply, however, since one lead of the transformer primary is automatically connected to true ground in the United States. (The white wire of the 110-V line is connected to earth.) Sometimes the simplest solution for a true floating power supply is to use batteries for the op-amp power.

Voltage-to-Current Converter (Constant Current Source)

There are times when we want to produce a well-defined and controlled current. This current could be controlled by a voltage input that is itself constant or varying as the situation may require. The voltage-to-current converter performs this function and is quite easy to understand on the basis of how the fundamental inverting amplifier operates.

We simply put the *load* through which the current should pass in place of the feedback resistor, R_f (see Figure A6–14). Since point S is virtually ground, the current i that flows from the battery V_s is determined to be

$$i = \frac{V_s}{R} \qquad \textbf{(A6–21)}$$

This current passes through the load (virtually none goes into the − op-amp input) independent of what the load resistance is (within limits). Thus, a current i is generated that is proportional to the input voltage V_s only, and a *voltage-to-current converter* has been designed. The limit referred to above is that the voltage that appears across the load must always be less than the supply voltage since this load voltage is actually the output voltage of the op amp, and the output voltage cannot exceed the supply voltage used.

It should be noted that if only a current meter is at hand, it is possible to construct a voltmeter with this circuit; simply use the ammeter as the load in this circuit. In fact, this method is very similar to the way a voltmeter has always been constructed from a d'Arsonval meter. R becomes the *series resistance* of the voltmeter. The difference is that the resistance of the ammeter does not matter in this approach. This circuit could be used to calibrate an ammeter (as the load) with the aid of an accurate V_s and R.

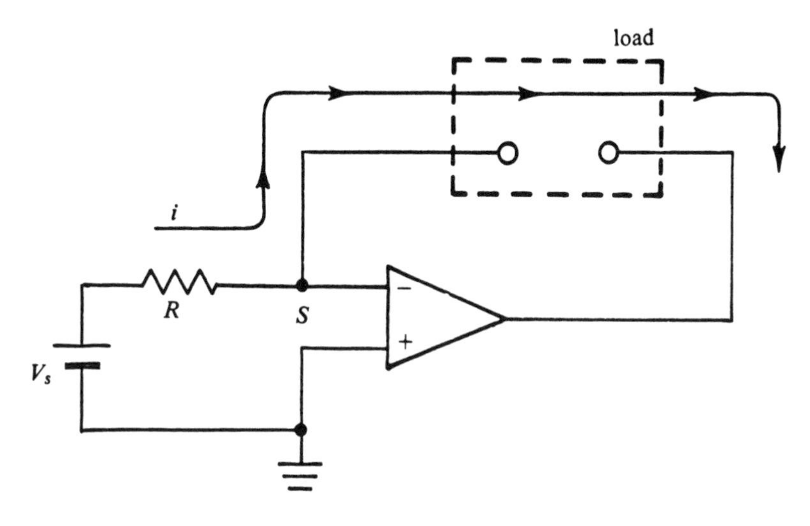

FIGURE A6–14. *Voltage-to-Current Converter*

More important is the fact that if V_s is constant in Figure A6–14 and Equation A6–21, the current i is also constant. Therefore, we have designed a circuit that sends a *constant current* through a load, no matter what its resistance is. This circuit is the opposite of an ideal battery, which supplies a *constant voltage* across a load, regardless of its resistance.

Logarithm Function Block

There are a number of situations where the laboratory signal runs over a large range of amplitude, and we want to take the logarithm of the signal voltage in order to display the relative variations at low levels as well as high. For example, the ear is responsive to a wide range of sound intensity. To obtain the *reverberation time* in a room involves measuring the time for the sound after a noise burst (for example, a pistol shot) to decay by a factor of 1,000,000 or 60 decibels. This is a large change that could not possibly be read from a meter with a single range and would be difficult to measure by switching ranges on a multirange meter.

Another application would be to linearize a voltage that is either exponential or follows a power law; this can be done by using the property of logarithms. Thus, an exponential (in time) signal sent to a logarithm converter and then to a stip-chart recorder can be expected to generate a straight line on the recorder's ordinary graph paper. Finally, the fact that $\log AB = \log A + \log B$ and also $\log A^n = n \log A$ permits us to perform multiplications, square roots, and so on, of voltage signals with op-amp–based *logarithm function blocks.* This is certainly a powerful capability.

The basic log function block or *log amplifier* makes use of the exponential characteristic of the semiconductor diode as given in Equation A4–1 of Chapter A4. The diode is simply placed in the inverting amplifier in place of the feedback resistor, as shown in Figure A6–15.

The current flowing through the diode is determined by the input voltage v_{in} and resistor R to be $i = v_{in}/R$ (think voltage-to-current converter). This current flows through the diode and a modest and decidedly nonlinear voltage develops across it. The output v_{out} is the negative of this voltage since the point S is at virtual ground.

To find the voltage across the diode, we use the theoretical diode law Equation A4–1. For diode voltage V_D greater than about 0.12 V for germanium and 0.25 V for silicon, the exponential term dominates. We then have

$$i \simeq i_R e^{|e| V_D / nkT} \qquad \textbf{(A6–22)}$$

Here a genuine "fudge factor," n, has been introduced to more closely approximate the experimental situation for silicon as opposed to germanium. The n is approximately 1 for germanium diodes and 2 for silicon. Reverse diode leakage current is represented by i_R.

We solve for the voltage across the diode V_D (and thus get at the amplifier output voltage) by taking the logarithm of Equation A6–20). This gives

$$\log \frac{i}{i_R} = \left(\frac{|e| V_D}{nkT}\right) \log e \qquad \textbf{(A6–23)}$$

Solving for V_D (at room temperature and inserting $\log_{10} e$, Boltzmann's constant k, and the charge of the electron e) gives

$$V_D = 0.059 \text{ n} \left(\log \frac{i}{i_R}\right) \text{V} \qquad \textbf{(A6–24)}$$

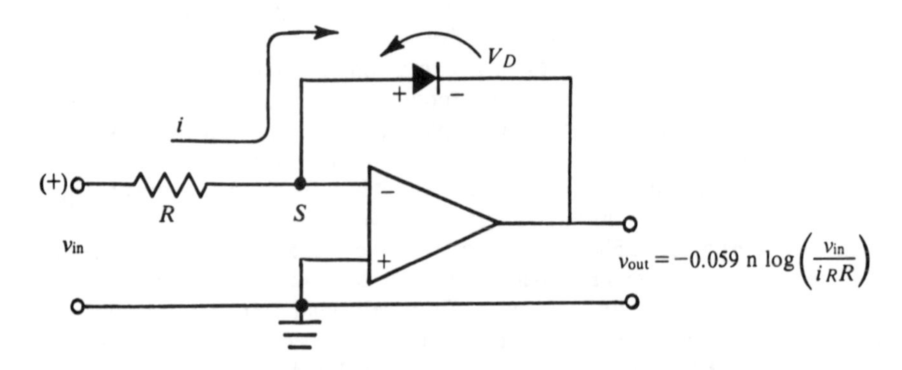

FIGURE A6–15. *Logarithm Function Block*

Now i_R is a constant, (of order 10^{-8} amperes for silicon) and $i = v_{in}/R$. Also $v_{out} = -V_D$ so that

$$
\begin{aligned}
v_{out} &= -0.059 \text{ n} \log \frac{v_{in}}{Ri_R} \text{V} \qquad \text{(in volts)} \\
&= -0.059 \text{ n} \log v_{in} + 0.059 \text{ n} \log i_R R \\
&= -0.059 \text{ n} \log v_{in} + \text{constant}
\end{aligned}
\qquad \textbf{(A6–25)}
$$

Thus, the output is proportional to log v_{in} and of the order of a few tenths of a volt. The voltage can be increased by use of an amplifier following this circuit.

This circuit can operate over 3 orders of magnitude (factor of 10^3) change in the input voltage. Notice it will only work for positive input voltage (true of the logarithm function anyway). However, it is sensitive to temperature due to T in Equation A6–22 and is sensitive to offset errors for small input voltages. Commercial units are available that involve more elaborate design to give more input range and temperature tolerance.

Antilogarithm Function Block

To do the inverse of the logarithm, or the antilogarithm, simply interchange the diode and the resistor in Figure A6–15 (see Figure A6–16). Then the current that flows through the diode is obtained from the fact that its cathode is virtually at ground. The current is given (for V_D greater than 0.25 V) by Equation A6–22, where the voltage across the diode is the input voltage v_{in}. To the base 10 we have

$$\frac{i}{i_R} = 10^{(|e| \log e/nkT)v_{in}} = 10^{Kv_{in}} \qquad \textbf{(A6–26)}$$

where K is a constant. This current flows through the feedback resistor R_f so that

$$v_{out} = -i_R R_f 10^{Kv_{in}} \qquad \textbf{(A6–27)}$$

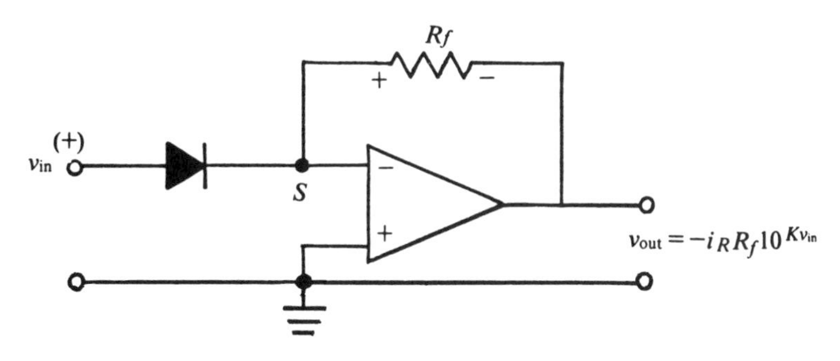

FIGURE A6–16. *Antilogarithm Function Block*

Multipliers and Square Root Functions

We now briefly examine how two voltages may be multiplied together and how the square root of a voltage may be taken using the op-amp–based function blocks. These are basic operations for an analog computer, but may be desirable in a laboratory situation as well.

Let us represent the summing amplifier of Figure A6–7 and the log and antilog amplifiers by boxes so marked. Now using the properties of logarithms

$$\log AB = \log A + \log B \tag{A6–28}$$

and

$$\text{antilog}\,(\log AB) = AB = \text{antilog}\,(\log A + \log B) \tag{A6–29}$$

Thus, we may take the antilog of the sum of the logs of A and of B to obtain the product of A and B (see Figure A6–17).

As a second example of what can be done, consider taking the square root of a voltage. We use

$$\log A^{1/2} = 1/2 \log A \tag{A6–30}$$

$$A^{1/2} = \text{antilog}\,(1/2 \log A) \tag{A6–31}$$

Multiplication by $1/2$ can be performed by an amplifier with voltage gain of $1/2$ (for example, $R_f/R = 1/2$ in Figure A6–2). The block diagram of the circuit is shown in Figure A6–18.

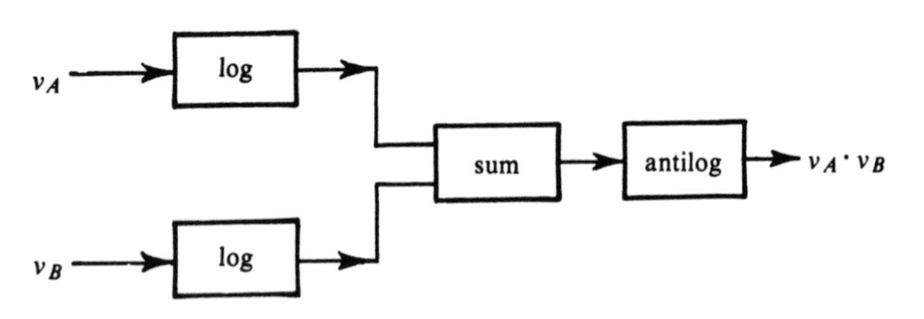

FIGURE A6–17. *Construction of a Voltage Multiplication Function from Log, Sum, and Antilog Function Blocks*

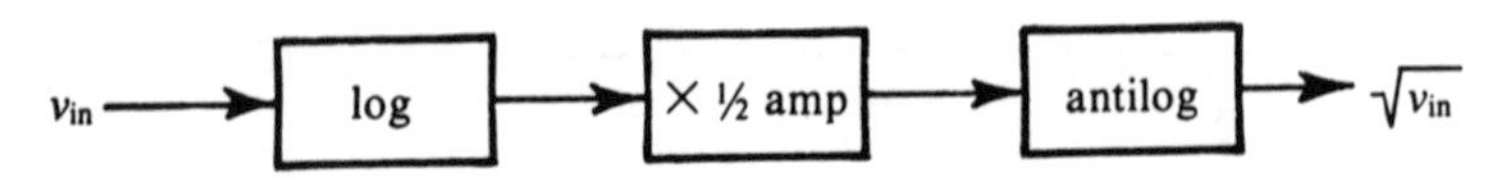

FIGURE A6–18. *Construction of the Square Root Function*

Integrating Function Block

It is quite easy to construct an op-amp–based circuit that produces an output in time that is an integral with respect to time of the input voltage; that is, an *integrating function block.*

We simply take the inverting amplifier and replace the feedback resistor with a capacitor (see Figure A6–19). The current that flows from v_{in} through R is determined again only by v_{in} and R. This current *accumulates* on the capacitor and the capacitor charges, so the voltage across the capacitor changes. The voltage across the capacitor (which becomes the output voltage) is proportional to the charge q_C that has accumulated on the capacitor, and this is simply the integral of the input current. By Ohm's Law:

$$i = \frac{v_{in}(t)}{R} \qquad \text{(A6–32)}$$

Also

$$i = \frac{dq}{dt} = \frac{dq_c}{dt}$$

so that

$$\frac{dq_c}{dt} = \frac{v_{in}(t)}{R} \qquad \text{(A6–33)}$$

Then we solve for q_c by doing the integral of Equation A6–33.

$$\int_0^t \left(\frac{dq_c}{dt}\right) dt = \int_0^t \left(\frac{v_{in}(t)}{R}\right) dt \qquad \text{(A6–34)}$$

Thus

$$q_c(t) - \overset{0}{\cancel{q_c(0)}} = \frac{1}{R} \int_0^t v_{in}(t)\, dt \qquad \text{(A6–35)}$$

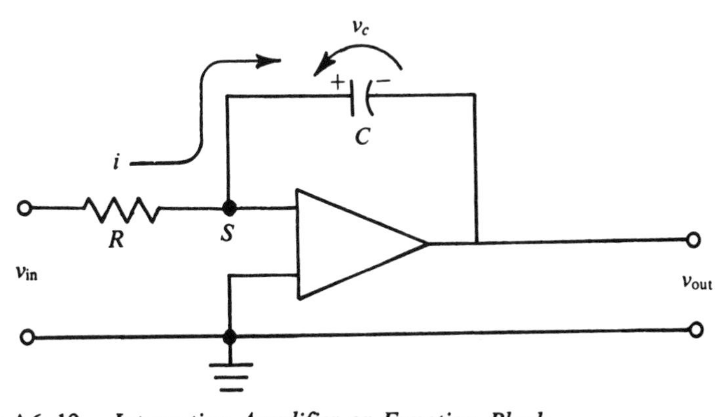

FIGURE A6–19. *Integrating Amplifier or Function Block*

To simplify matters we *assume* the capacitor was discharged at $t = 0$. Next

$$v_{\text{out}} = -v_c = -\frac{q_c}{C}$$

so that

$$v_{\text{out}} = -\frac{1}{RC}\int_0^t (v_{\text{in}}t)\,dt \tag{A6–36}$$

Thus, the output voltage is proportional to the integral of the input voltage, with a proportionality constant of $(1/RC)$. Also, the output voltage is inverted.

Example: We wish to obtain a voltage that increases linearly in time with a high accuracy.

Solution: Simply connect a battery V to the input of an integrating amplifier. Then

$$v_{\text{out}} = \frac{-1}{RC}\int_0^t V dt = \left(\frac{-V}{RC}\right)t \tag{A6–37}$$

For example, if $V = 0.5$ V, $R = 100$ kilohms, and $C = 10$ microfarads, we have

$$v_{\text{out}} = \left(\frac{-0.5 \text{ V}}{10^5 \times 10^{-5}\text{s}}\right)t = (-0.5 \text{ V/s})t$$

The circuit is shown in Figure A6–20a, and the output voltage behavior is shown in Figure A6–20b. The switch S is assumed to be held closed until it is opened at time $t = 0$; consequently the output voltage is held to virtual ground (or zero volts) until the switch is opened. After the switch is opened, the output voltage grows steadily more negative.

Notice in the example just discussed that if the switch connected across the capacitor is closed every T seconds, a sawtooth generator with period T is formed. (The switch can be a field-effect transistor to make a completely electronic circuit.) Thus, we have a sweep generator for an oscilloscope.

If a square wave that is symmetric in voltage and time such as in Figure A1–12b is applied to the input, the reader should be able to reason that a "triangle wave form" is produced at the output of the integration circuit.

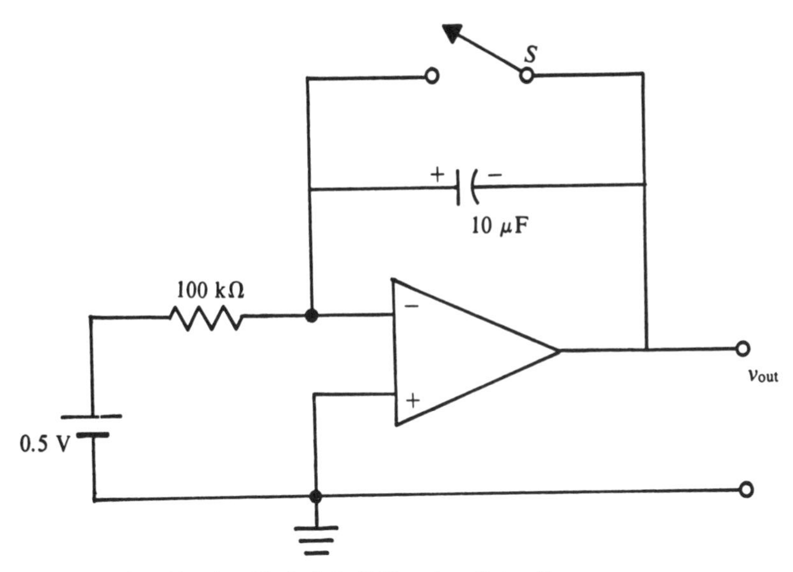

(a) Integrating Circuit with Switch S Closed until $t = 0$

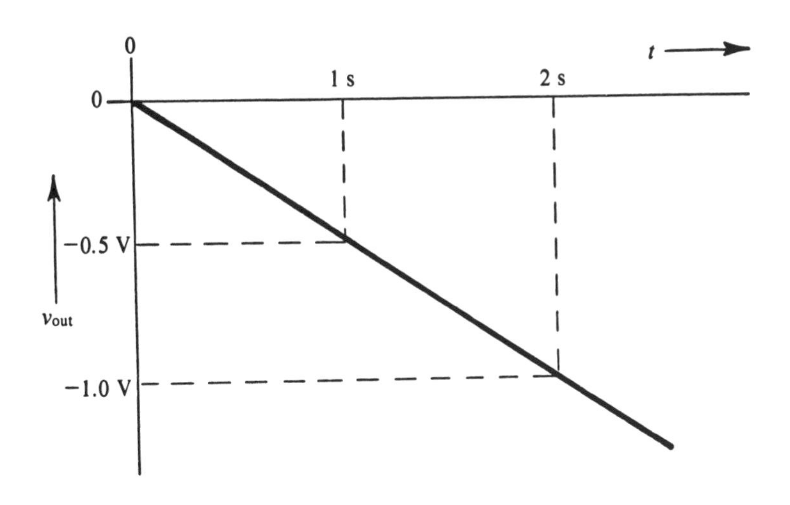

(b) Voltage Out of Integrating Circuit

FIGURE A6–20. *Circuit to Produce Voltage That Changes Linearly in Time*

Limitations and Operation of Real Operational Amplifiers

Up to this point in the chapter, the circuit behavior and the formulas to describe it have assumed ideal operational amplifiers. If the op amp were ideal, the formulas would be generally quite simple. They are often referred to as *classical op-amp formulas.* Although modern operational amplifiers are remarkably good in most respects, they only approximate the ideal op amp. In critical applications, it is important to understand the limitations of real operational amplifiers and to attempt to minimize their effects as much as possible. The remainder of the chapter speaks to these rather practical matters, including actual op-amp connections and circuit construction.

Finite Voltage Gain

The classical op-amp formulas that have been derived thus far assume an ideal op amp with infinite open-loop gain. In fact, IC operational amplifiers typically have an open-loop gain of perhaps 200,000. To what extent are the closed-loop—that is, op amp with feedback—formulas in error?

For the basic inverting amplifier of Figure A6–2a it is possible to derive the closed-loop voltage gain A_v in terms of the op-amp open-loop gain A_{vo}, R_f, and R. It is

$$A_v = \frac{-R_f/R}{1 + (1/A_{vo})\,[1 + (R_f/R)]} \tag{A6–38}$$

Now $R_f/R = A_{vI}$, the ideal op-amp gain formula. Since the term $(1/A_{vo})$ $[1 + (R_f/R)]$ is a small quantity, we may use the relation $(1 + x)^{-1} \cong (1 - x)$ for small x. Then

$$A_v \simeq -A_{vI}\left[1 - \frac{1}{A_{vo}}(1 + A_{vI})\right] \tag{A6–39}$$

Suppose $A_{vI} = 100$ and $A_{vo} = 200{,}000$. Then

$$A_v = -A_{vI}\left(1 - \frac{1}{2000}\right)$$

and the classical formula is in error by one part in 2000 or 0.05%.

Finite Bandwidth

The open-loop gain of the op amp does not remain at the high value of 200,000 to particularly high frequencies. For the 741 op amp, at about 10

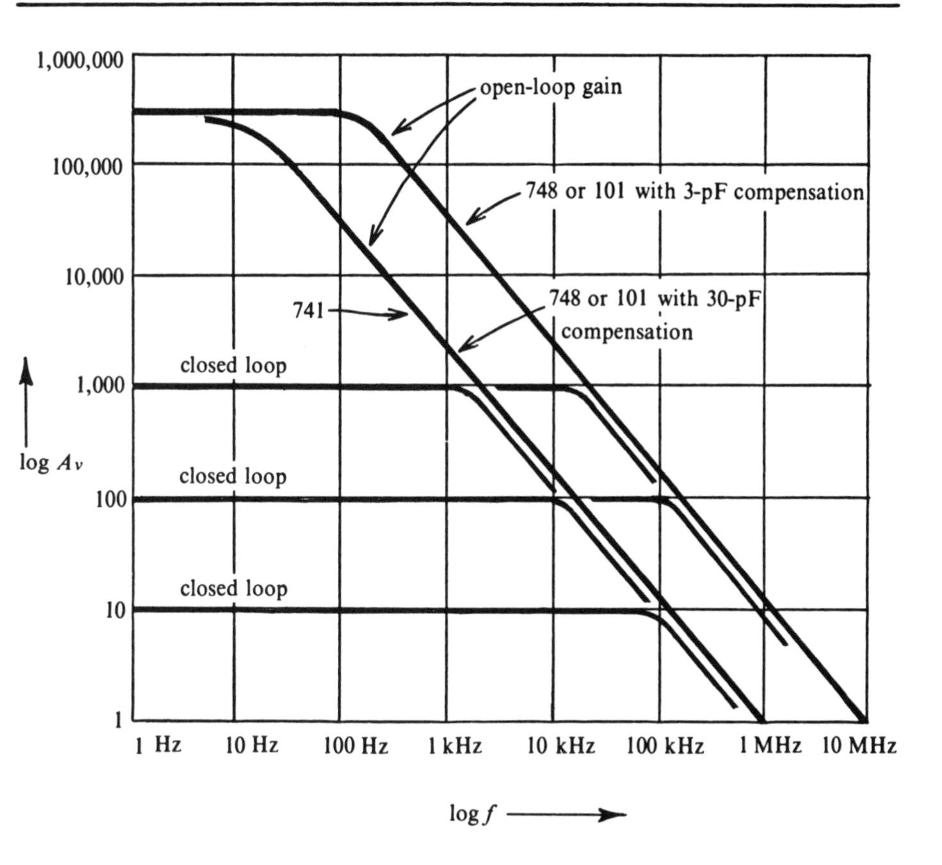

FIGURE A6–21. *Frequency Response for Op-Amps Models 741, 748, and 101*

hertz the gain begins falling at a rate of a factor of 10 for every factor of 10 increase in frequency. Figure A6–21 illustrates gain versus frequency or *frequency response* for the 741 and 748 operational amplifiers. The 101 model op amp behaves similarly to the 748.

The closed-loop gain for the op amp can only hold constant for frequencies up to where the open-loop gain of the op amp falls to the closed-loop value. Then the gain of the op amp with feedback falls, too. Notice the curves labeled *closed loop* in Figure A6–21. Thus, for the 741 op amp the gain is *flat* for a gain of 1,000 up to about 1 kilohertz; flat for a gain of 100 up to about 10 kilohertz; flat for a gain of 10 up to about 100 kilohertz; although not shown, it is flat for unit gain to almost 1 megahertz. There is a moral about amplifiers here that holds for any type of amplifier design, not only with op amps: The product of the gain times the bandwidth is constant for a given amplifier. Notice that this *gain-bandwidth product* (GBP) is about 10^6 for the 741 op amp. Also notice that the 748 (and the 101 op amp) has a larger gain-bandwidth product of about 10^7. However an external *compensation capacitor* must be connected between the appropriate pins of this type of op amp to prevent oscillations. Three picofarads should be used for a gain of 100 or more; increasing capacity must be used for amplifiers with lower closed-loop gain. Thus, 30 picofarads must be used for a unity-gain amplifier; then its performance is about the same as for the 741.

Unwanted op-amp instability and oscillations occur because of positive feedback. Although we have been careful to connect the feedback in a negative fashion for all of the designs presented thus far, the phase of the signal passing through the op amp becomes increasingly shifted at higher frequencies (due to internal capacitances). For example, significant phase shifts begin at about 100 kilohertz for the 748 and 101 devices. This means that the signal fed back to the input can reinforce the signal already there at high frequencies, and this reinforcement can produce an oscillator. The compensation capacitor reduces the gain at high frequencies so that the level of the signal fed back is insufficient to sustain oscillations.

The 741 is called an *internally compensated* operational amplifier, because no fuss with an external capacitor is required. The more recent (introduced in 1976) CA3140 op amp from RCA is also internally compensated and has a bandwidth about the same as the 748 (the GBP is quoted as 4.5 megahertz). The CA3140 is a combination MOSFET-bipolar op amp that has an ultra low input bias current; this device and property are in the section on input bias current. Other op amps with *very* wide bandwidth have been introduced. Thus, in 1979 Signetics announced the NE5539 op amp with a GBP of 1200 megahertz (1.2 gigahertz). (They have advertised the NE5539 with the catchy phrase, "dc to daylight"; what does this imply?)

An output (pulse-response) limitation related to the gain-bandwidth product is the *slew rate* of an operational amplifier. Figure A6–22 tells the story. Suppose a *square pulse* is applied to a 741 op amp connected as a voltage follower. Although the voltage of the input pulse rises and falls very rapidly, the voltage of the output pulse from the amplifier responds to the pulse *edges* at a very definite rate. The output voltage moves along a slant line at the rate of 0.5 volts/microsecond; this rate is called the *slew rate* of the op amp. Generally, an op amp with a larger GBP also has a faster slew rate. The fact that the 741 is of limited application in digital and computer circuits becomes evident. Computer circuits often use pulses on the order of 5 V in amplitude and 1 microsecond in duration. The 741 cannot be used to follow or produce a pulse with these *high-speed* characteristics. Other op amps have been designed to be high-speed devices. An example is the LM118/218/318 from National Semiconductor; its GBP is 15 megahertz, and its slew rate is 70 volts/microseconds. The slew rate of the NE5539 mentioned above is 600 volts/microsecond, a thousand times faster than the 741.

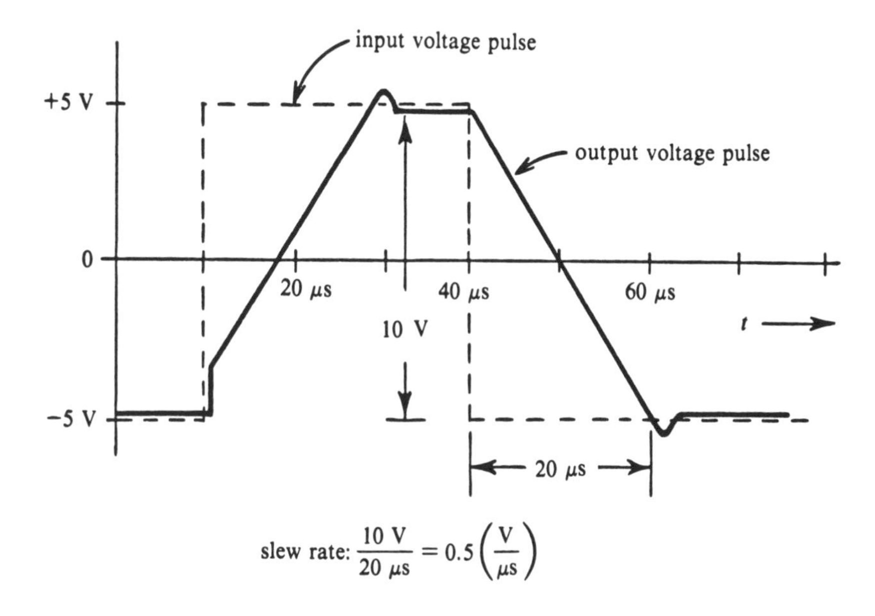

FIGURE A6–22. *Pulse-Response Limitation (Slew Rate) for a 741 Op Amp Connected as a Voltage Follower*

Input Offset Voltage

As mentioned in Chapter A5, real dc amplifiers have a voltage at the output even though none is present at the input. The amplifier behaves as though a small battery—called the *input offset voltage* v_{io}—is inserted at the amplifier input. The voltage that appears at the output (when the input leads are shorted) is the *closed-loop* amplifier gain times the input offset voltage.

For the 741, 748, and 101 bipolar transistor op amps, the input offset voltage is of the order of 1 millivolt. Field-effect–transistor (FET) op amps usually have a larger op-amp offset, although the CA3140 has only 2-millivolts typical v_{io}.

Example: An inverting amplifier with a gain of −50 has been constructed from a 741 op amp. When the input to the amplifier is shorted to ground, the output reads +0.2 V.

1. What is the input offset voltage?
2. Is it necessary to live with this rather large error?

Solution:
1.
$$v_{io} = \frac{v_{out}}{A_v} = \frac{+0.2\ \text{V}}{-50} = -0.004\ \text{V} = -4\ \text{mV}$$

This voltage is somewhat more than the voltage the 741 typically has, although it is within the specification sheet maximum of ±6 millivolts.
2. No. Read on.

Several op-amp types provide special pins to *null the voltage offset* to zero. For example, a 10-kilohm potentiometer (10K pot) should be connected between the null pins for the 741 or the CA3140 op amp, with the sliding tap of the pot connected to the V− supply voltage. The actual pin connections are illustrated in Figure A6–23 and discussed further in the next section. To correct for the output offset voltage error, the input leads to the constructed amplifier (*not* the input leads to the op amp itself) should be shorted together and the *offset null pot* adjusted until the output from the amplifier is very small.

Although the offset can be nulled in this way, the *input offset voltage drift* causes the output to drift away from zero when the temperature changes. The 741, 748, and 101 op amps typically drift about 10 microvolts per centigrade degree. A special precision op amp, the 725, has a v_{io} drift of only 0.5 microvolts/centigrade degrees. (It costs about 10 times the 741 price.) The OP–7 op amp from Precision Monolithics, Inc. has a similar offset drift specification, but also has an *input bias current* that is an order of magnitude smaller than the 725 and 741. Later we find that this current can be an important parameter.

Op-Amp Connections

Figure A6–23a illustrates pin connections for the 741, 748, 101, and 3140 op amps. The numbers shown are the pin numbers to the 8-lead metal can as well as the 8-lead mini-DIP IC packages (these packages are pictured later in Figure A6–28). Note that this group of widely used op amps may all be connected in the same fashion; they are basically pin interchangeable. Pins 2 and 3 are the inverting and noninverting inputs, respectively, while pin 6 is the op amp output.

As we have mentioned, the 101 and 748 op amps each require a phase compensation capacitor. Notice the placement of the compensation capacitor C_1 between pins 1 and 8 in the circuit of Figure A6–23a. Since the 741 and 3140 are internally compensated, Figure A6–23b reveals that pin 8 is not used on these op amps.

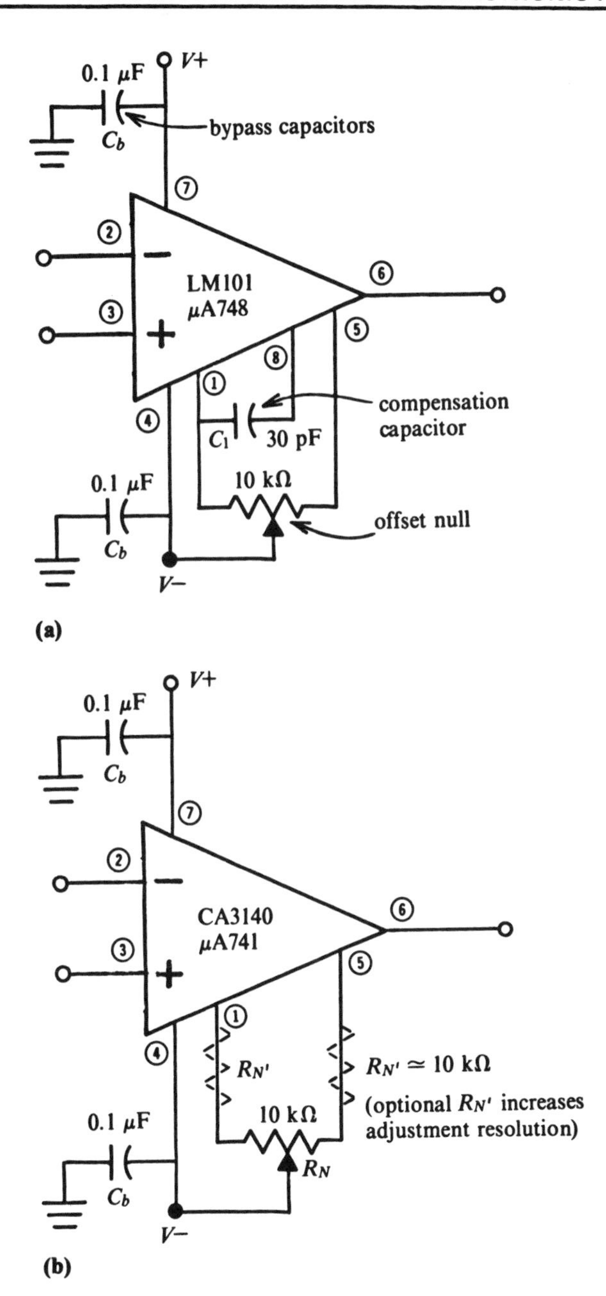

FIGURE A6–23. *Pin Connections, Including Compensation and Offset Null, for Popular Op-Amp ICs*

The null pot connections for any of the four op amps may be seen to be the same from Figures A6–23a and b. Optional additional resistors R_N' are shown in series with the null pot R_N in Figure A6–23b; these permit increased resolution in setting the null.

Notice the *bypass* capacitors C_b shown connected to the V+ and V− power-supply terminals (pins 7 and 4, respectively) in Figures A6–23a and b. These capacitors are connected in order to suppress unwanted oscillations, especially for op amps with a high gain-bandwidth product and in circuits with high closed-loop gain. Rather like the oscillation problem requiring a compensation capacitor, the root cause of the problem is positive feedback of high-frequency signals from output to input. This time the problem comes about because of feedback from the op-amp output stages through the power-supply connections and back to the input stages through the supply connections. At least one pair of bypass capacitors should be used on each printed circuit board. In troublesome cases they should be connected directly between the supply pins for the IC and ground. Low inductance capacitors should be used—for example, disc-ceramic or tantalum types—because inductive impedance increases at higher frequencies. The purpose of the bypass capacitors is to provide a *short* or bypass to ground for high frequencies. Long wires to the capacitors should be avoided since even a straight length of wire possesses inductance.

While on the topic of power-supply connections, let us examine briefly the voltage capabilities of these popular op amps. As we have indicated earlier, the output voltage cannot exceed the power-supply voltage from which an amplifier operates. But if the power-supply voltages V+ and V− should exceed certain limits for a particular op amp, the op amp may be destroyed due to voltage breakdown within it. On the otherhand, if the supply voltages are too small, the op amp will not function properly. Generally the supply voltages V+ and V− should be nearly equal in magnitude and well regulated. Table A6–1 gives power-supply limitations for the popular op amps that we have already mentioned.

It will be noted that the μA741C has a smaller upper-voltage limit than the μA741. The μA741C differs from the μA741 in that it has generally *relaxed specifications* compared to the μA741, and μA741C is designated the *commercial* temperature range device for the op-amp type. Similarly, the

TABLE A6–1. *Power-Supply Limitations for Popular Op Amp IC's*

Op Amp	*Power-Supply Limitation*
μA741, μA748, LM101	$3\text{ V} \leq \lvert V\pm \rvert \leq 22\text{ V}$
CA3140B	$2\text{ V} \leq \lvert V\pm \rvert \leq 22\text{ V}$
CA3140	$2\text{ V} \leq \lvert V\pm \rvert \leq 18\text{ V}$
μA741C, μA748C, LM301	$3\text{ V} \leq \lvert V\pm \rvert \leq 18\text{ V}$

TABLE A6–2. *Operating Temperatures for Popular Op Amp IC's*

Op Amp	*Operating Temperatures*
μA741, μA748, LM101, CA3140, CA3140B	−55°C to +125°C (military range)
μA741C, μA748C, LM301	0°C to +70°C (commercial range)

LM301 is the commercial version of the LM101; that is, the LM301 has relaxed specifications (and a lower price) relative to the LM101. The temperature ranges for satisfactory operation of these op amps is given in Table A6–2. The μA741 and LM101 are called *military temperature range* devices, while the μA741C and LM301 are called *commercial temperature range* devices.

Input Bias Current

Any of the electronic active discrete devices—bipolar transistor, field-effect transistor (FET), vacuum tube, or metal-oxide-silicon field-effect transistor (MOSFET)—require some current (that is, *bias current*) flowing into the controlling input of the device (base, gate, or grid) in normal operation. Relatively speaking, the bipolar transistor requires much more bias current than the others; this is a shortcoming of the ordinary (bipolar) transistor. The FET and especially the MOSFET (transistors) are the most recent and have a high input impedance rather like the venerable vacuum tube. The MOSFET significantly exceeds most vacuum tubes. The principles of operation of these devices are described in Chapter A8.

The 741, 748, and 101 are examples of bipolar-transistor integrated-circuit op amps. For operation they require a bias current on the order of 0.1 microamperes = 100 nanoamperes = 10^{-7} amperes *into each input.* There must be a dc path into the + input and the − input that supplies this current. Thus, Figure A6–24a illustrates an amplifier that will *not* work because the capacitor blocks any dc path into the + input. The input voltage connected to the noninverting amplifier must supply this bias current to the + input (see Figure A6–2b). However, if the voltage source connected to the op amp has a high output impedance, there can be a severe and literally upsetting voltage drop across its output impedance due to the bias current.

A case in point is the pH meter that is a common fixture in a chemistry department. The pH electrode, which determines acidity of a solution, consists of glass and reference electrodes that can produce up to 1-V emf, but has an output impedance (because the current must flow through glass) of about

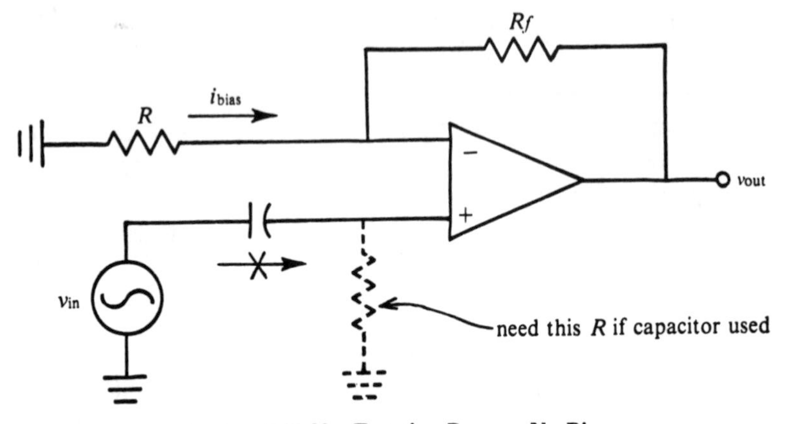

(a) Op Amp Circuit That Will *Not* Function Because No Bias Current Can Flow into the + Input

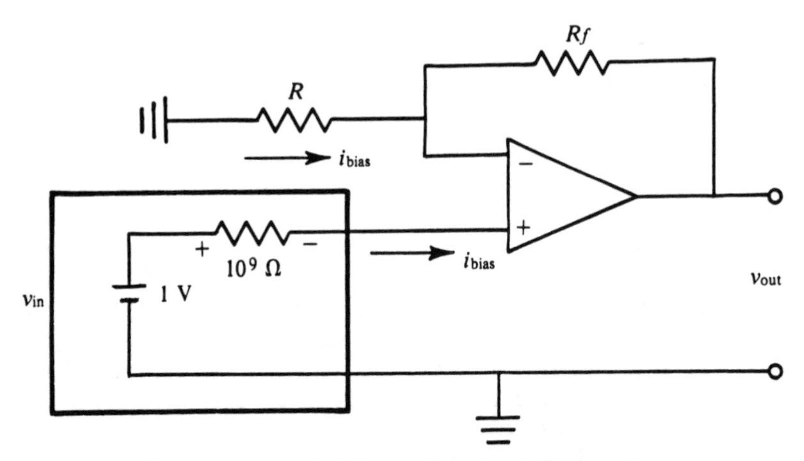

(b) Effect of Bias Current in a High Output Impedance Voltage Source

FIGURE A6–24. *Bias-Current Problems*

10^9 ohms. See the situation illustrated in Figure A6–24b. If the op amp is a 741, the bias current is about 10^{-7} amperes so that a voltage drop of 10^{-7} amperes $\times$ 10^9 ohms = 100 V would occur across the output impedance of the source. The op amp would *not* function. Even if the input current were reduced to 1 nanoampere, the voltage drop would be 1 V across the output impedance, and a 100% error would occur in the voltage actually reaching the + input of the op amp.

In the recent past, special *electrometer* vacuum tube triodes were required to work at the extremely low bias currents required by such items as pH electrodes. However, MOSFETS and the CA3140 MOSFET-input op amp have opened a new era. The bias current required by the CA3140 is about 10 picoamperes = 10^{-11} amperes. The voltage drop across the 10^9-Ω impedance of the pH electrode is then 10^9 ohms $\times$ 10^{-11} amperes = 0.01 V—much better. Furthermore, the input impedance of the CA3140 is about 10^{12} ohms (compare to 10^6 ohms for the 741).

There is one worry about MOSFET devices however: They may be ruined by static electricity. RCA claims the CA3140 has protected inputs that require no special handling procedures. However, in seasons or regions of low humidity, it is generally recommended that a person be grounded to discharge static electricity before handling MOSFET devices. These devices are generally quite safe after connection in a circuit. Other manufacturers are now also producing MOSFET-bipolar–transistor op amps, as well as FET-bipolar–transistor op amps. We will shortly look at the Op–15/16/17, which is an FET-bipolar type. Standard FET transistors are not subject to static damage; therefore, this op-amp type is not either. However, FET-bipolar op amps often do not have such extremely low input bias currents as the MOSFET-input type.

When working with op amps in situations where small voltages—millivolts or less—are being measured, it is important that the same impedance to ground be present from each op-amp input. This is so that the voltage drop from the input bias currents through these resistances is the same at each input and also so that temperature problems are minimized. The bias-current changes significantly with temperature and if the currents flow through different resistors to the + and − inputs, they will experience a temperature-dependent dc offset voltage from this effect.

Figure A6–25 illustrates how a recommended resistance R_b should be placed in the two important op-amp amplifier configurations in order to minimize voltage offset from the bias currents, i_b. The resistor R_b should be placed between the + op-amp input and ground for the inverting amplifier (see Figure A6–25a), while it should be placed between the + op-amp input and the v_{in} terminal in the noninverting amplifier (see Figure A6–25b). The resistor R_b is simply the parallel equivalent resistance of R and R_f. This is indicated in the figure with the common notation $R \| R_f$. The resistance R_b does *not* affect the classical gain formula for either amplifier, provided the op amp has high open-loop gain.

The fact that R_b should equal $R \| R_f$ can be argued as follows. In either amplifier, we desire the output to be precisely zero ($v_{out} = 0$) when the input is grounded ($v_{in} = 0$). In fact, the two amplifier circuits are then identical, with identical bias-current paths for each (see Figures A6–25a and b). Now, the output voltage v_{out} can only be zero if the voltages at the − and + op-amp inputs are precisely equal (assuming no input offset voltage, of course). But the − op-amp signal input will be *below* ground due to the

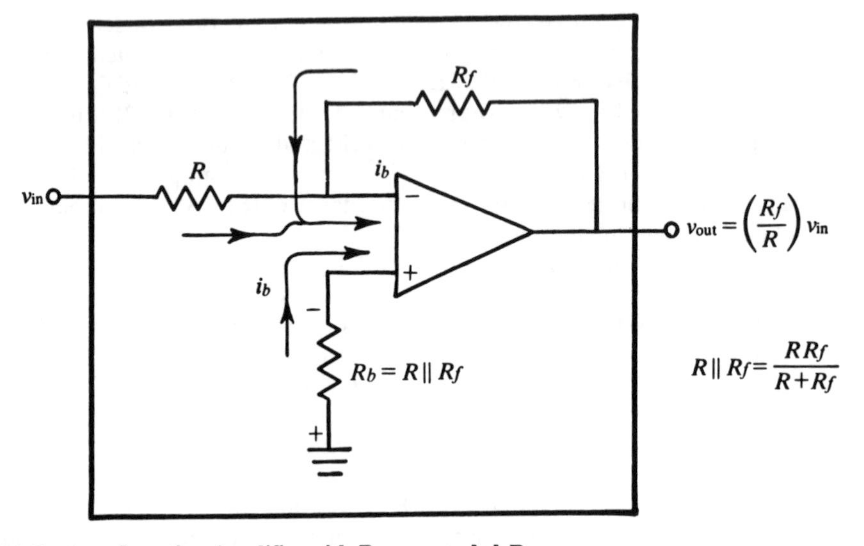

(a) Op-Amp Inverting Amplifier with Recommended R_b

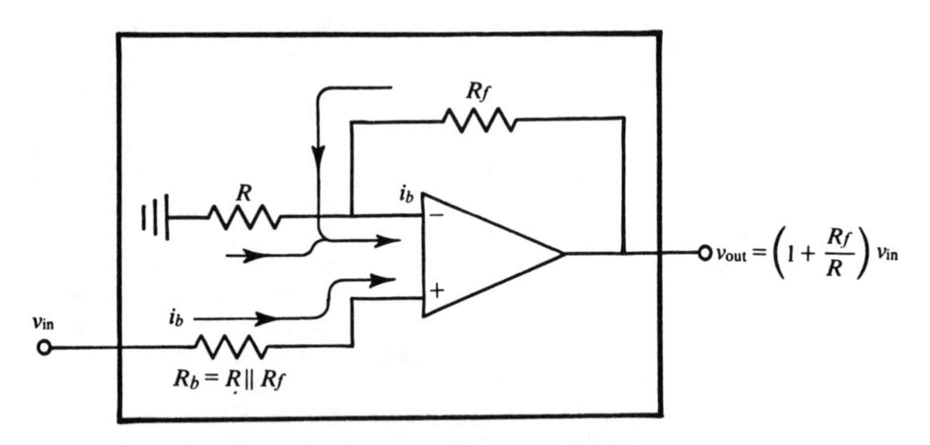

(b) Op-Amp Noninverting Amplifier with Recommended R_b

FIGURE A6–25. *Op-Amp Amplifiers, Showing Recommended Bias-Current–Drop Correcting Resistors (R_b), and Bias-Current Flow Paths*

voltage drop of i_b flowing from zero volts (from ground and from $v_{out} = 0$) through the *parallel* resistors R and R_f to the − op-amp input. We desire the + op-amp input to be below ground by just the same amount, and this will be true *if* the bias current into it flows through a resistor $R_b = R \| R_f$.

It is true that this argument assumes the bias currents into the two op-amp inputs to be identical. Unfortunately, there generally is a difference, called the *bias-current offset*. For the 741 it is on the order of 20 nanoamperes, which is 20% of the bias current. This could be corrected for by adjusting R_b; then better temperature stability could be expected. However, generally R_b is simply selected as we have discussed, and the voltage offset adjustment is used to balance the amplifier output to null.

The input bias current does change with the temperature of the op amp. Particularly if bias-current correction resistor R_b is *not* used, this bias-current drift causes an offset voltage drift that is as troublesome as the op-amp (inherent) offset voltage drift. Therefore, the *input bias-current drift* specification is normally listed on the manufacturer's specification sheets. Of course, a device that simply has a very small bias current generally causes fewer problems in regard to this effect. The bias voltages across the resistors to the + and − op-amp inputs are necessarily smaller in the first place and, therefore, can change by at most a smaller amount in the second place. The input bias-current drift for the 741 is about −1 nanoampere/centigrade degrees.

Example: Assume a 741 op amp has been especially selected so that its input offset voltage at 20°C is extremely small. However, its input bias current is typical (it must exist for op-amp operation). This op amp is connected to make an inverting amplifier as shown in Figure A6–26; note that the recommended R_b of Figure A6–25a is absent.

1. What output voltage is expected when the input is shorted to ground? What is the equivalent input offset voltage?
2. What output voltage may be expected if the ambient temperature changes to 30°C?

Assume the following typical specifications for the 741:

input bias current = 100 nanoamperes

input bias-current drift = −1 nanoampere/centrigrade degrees

input offset-voltage drift = ±10 microvolts/centigrade degrees

Solution:

1. This problem is easy provided the proper viewpoint is taken. A bias current must flow into the − input of the op amp. Now if the op amp has very high gain A_{vo}, negative feedback requires that the summing point S is virtually at zero volts (virtually at ground). This means that the potential at the two ends of the 10-kilohm resistor is the same, and no current flows through it. (This is like the case of the balanced bridge.) The bias current therefore *must* flow through the R_f resistor.

Then

$$v_{out} = +i_b R_f = 100 \text{ nA} \times 100 \text{ k}\Omega$$
$$= 10^{-7} \text{ A} \times 10^{+5} \ \Omega = 0.01 \text{ V}$$

This is equivalent to an input offset voltage of

$$\frac{v_{out}}{A_v} = \frac{0.01 \text{ V}}{10} = 1 \text{ mV}$$

Note that this is comparable to the *inherent* input offset-voltage value for a typical 741 (that is, for a 741 that is not offset nulled).

2. When the temperature changes from 20°C to 30°C, the bias current i_b changes by

$$\Delta i_b = (-1 \text{ nA/C}^\circ)(10 \text{ C}^\circ) = -10 \text{ nA (or } -10\%)$$

Then

$$\Delta v_{out} = \Delta i_b R_f = (-10 \text{ nA})(100 \text{ k}\Omega) = -1 \text{ mV (or } -10\%)$$

The output voltage also changes due to the inherent input offset-voltage drift; that is,

$$\Delta v_{out} = A_v \Delta v_{io} = (10)(10 \ \mu\text{V/C}^\circ)(10 \text{ C}^\circ) = \pm 1.0 \text{ mV}$$

We use (±) here because the drift may be upward for one 741, and downward for another. In the worst case (Murphy's Law!), the output drifts -1 mV $-$ 1 mV $= -2$ mV. This voltage change adds to the voltage output value from part 1, leaving the output voltage at -1.0 mV.

Comments: In part 1, note that the output offset voltage could be calculated by assuming that the bias current actually flows through the R (10 kΩ) resistor and multiplying the voltage drop across R by the voltage gain of the *amplifier* to get the output voltage. This somewhat intuitive approach will be discussed as a problem in the problem/exercise set for this chapter.

In part 2, notice that the influence of input current drift is equal to that of offset voltage drift. However, placement of the proper R_b in the circuit would greatly reduce the bias-current drift problem, so that the remaining drift would represent 100 μV referred to the input (10 μV/C° × 10 C°). Suppose for example that a thermocouple voltage were being amplified. Then 100 μV drift would correspond to an actual signal change for about 2 to 3 C° change in sensor temperature. It should be clear that a 741 would not permit high precision thermocouple amplification. An op amp with better offset-voltage drift specifications should be selected (for example, a 725 or even better, an OP–07).

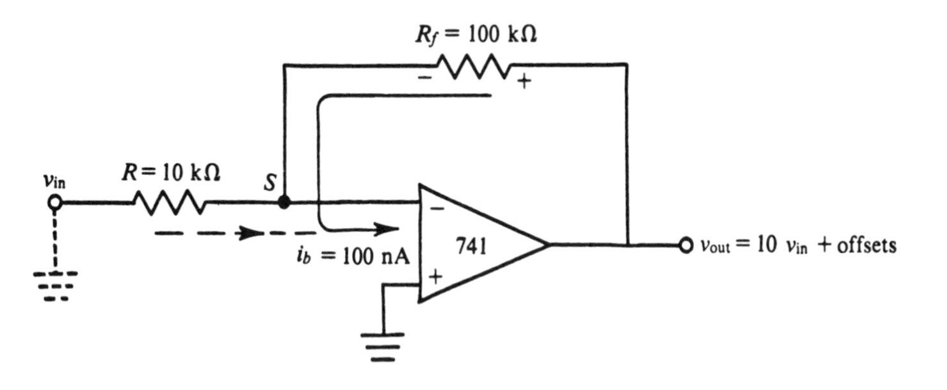

FIGURE A6–26. *Example Inverting Amplifier Circuit without Recommended R_b*

Output Capabilities of an Op Amp

Most common IC operational amplifiers cannot dissipate much heat and are not capable of supplying more than 5 to 20 milliamperes. If the voltage of the op amp reaches a value such that the limiting current is being delivered to the load resistor R_L, it will simply go no higher. The output voltage will therefore *clip* at a value

$$v_{out,max} = i_{out,max} R_L \tag{A6–40}$$

Example: For the 741 op amp, the maximum output voltage for a load resistance of 200 ohms is listed as 5 V. What maximum current can be supplied by it to a load?

Solution: The maximum current is 5 V/0.2 kΩ = 25 mA. In fact, this current is listed as the output short-circuit current for the amplifier.

NOTE: To obtain accurate amplifier operation with some margin of safety, a smaller maximum output current should be assumed.

An operational amplifier can be readily helped along to supply more current by means of a transistor connected to its output. A transistor is basically a current-amplifying device, where the amplified current, I_C—that is, *collector current*—is more than the base current I_B by some factor beta or h_{fe}, which typically is 30 to 200. Negative feedback must be connected from the *overall* output to − input (see Figure A6–27). In this way the load current

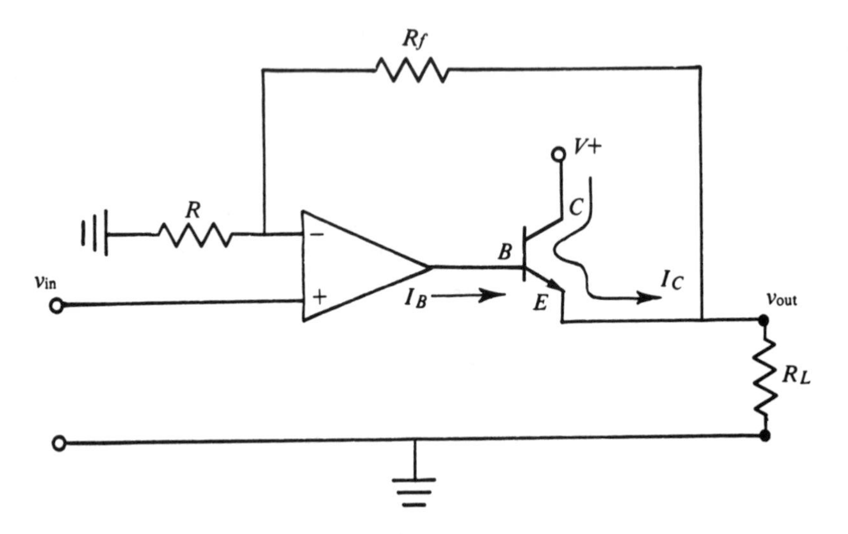

FIGURE A6–27. *"Outboard" Transistor to Increase Current Capabilities of the Op Amp by the Beta Factor of the Transistor*

can be β times as much as the current capabilities of the op amp. The op amp supplies the base current and the transistor multiplies it by the factor beta.

Selecting an Op Amp

In many rather standard applications without especially stringent requirements, the old standby—the general-purpose 741—is sufficient. However, the preceding discussion of op-amp realities shows that the 741 is only an approximation of the ideal op amp. A better approximation may be required if the 741 does not meet certain requirements. Perhaps a wider bandwidth, faster slew rate, higher input impedance, lower bias current, lower offset-voltage drift, lower power-supply drain, or higher output-current capability are needed.

In general, the op-amp designer can improve certain specifications, but only at the expense of others. Consequently, each manufacturer typically offers a spectrum of op-amp devices; there are also interesting variations among manufacturers. Table A6–3 presents a dozen specifications for five categories of op amps. (Even so, some specifications of possible interest were omitted, for example, bias-current offset and its drift, power-supply rejection ratio, large-signal bandwidth, and short-circuit output current.) The categories presented are:

1. *General purpose*—generally a balance of good specifications.
2. *Precision*—particularly outstanding *input specifications,* such as high input impedance, low offset voltage and offset-voltage drift, low bias current and bias-current drift, high common-mode rejection ratio (CMRR). These specifications typically come at the expense of bandwidth and slew rate. Therefore, these op amps are particularly appropriate for very low voltage input signals at dc to low frequencies.
3. *High speed*—especially wide bandwidth and fast slew rate, but often at the expense of the input characteristics mentioned under the "Precision" category; particularly appropriate for high-frequency (and pulse) input signals at higher voltage levels.
4. *Micropower*—very low power-supply current and/or power voltage are required. Some operate well from a single voltage supply; well suited to battery operation, although other specifications may be only fair.
5. *Power out*—capable of rather high output currents or high output voltages. Several of these devices have 741-like characteristics in other respects since they use a 741 to drive more "robust" output transistors on the chip.

The devices presented in Table A6–3 are not necessarily the best available within each category, but they are representative of devices that were available in 1980. In general, the devices are from the commercial-temperature-range offerings; for example, μAXXX*C*, LM*3*XX, LF*3*XX, LHXXXX*C*. Premium devices—literally handpicked—are often available with better specifications, but the prices are often premium also ($30 to $70 each). These devices often have additional suffix designations and are not presented in the table. The LF356A, LF357A (which are selected LF356 and LF357 parts) and the OP–07 are modest exceptions. The OP–07 is available with both tighter specifications (OP–07A) and more relaxed specifications (OP–07E, OP–07C, and OP–07D).

It is interesting to contrast properties of the devices, both between categories and within them. Note also the *comment* column at the far right of the table. For example, the FET and MOSFET input-stage devices have very outstanding input impedance (10^6 megohms = 10^{12} ohms) and very low bias current, but do not compare with the precision devices in terms of offset-voltage drift. However, the input bias current for an FET-input device does increase rapidly with temperature: a factor of two for every 10 C° temperature increase. This bias-current drift represents an exponential behavior, and the bias-current values in the table are for temperatures near room temperature.

Although the precision devices indeed have very low drift of input bias current and offset voltage, notice that they are "turtles" relative to the "jackrabbit" high-speed devices. Actually, as all-around high-performance devices, the rather recent (about 1977) LF357A and OP–17 are quite impressive. They are both placed in the high-speed category. The OP–17C was priced at about $10 when it was introduced.

TABLE A6–3. *Typical (Not Guaranteed) Specifications for Operational Amplifiers in Five Categories*

	Minimum-Maximum Supply Voltage (V)	*Compensation Required*	*Open-Loop Gain (dB)*	*Supply Current ($v_{out} = 0$) (mA)*	*$A_{vo} = 1$ Bandwidth (MHz)*
General Purpose					
μA741C	±3 to ±18	no	110	2	1
μA748C	±3 to ±22	yes	110	2	4
LF356A	±5 to ±18	no	105	2	4.5
LM301	±3 to ±18	yes	110	3	1
CA3140	±2 to ±22	no	100	4	4.5
Precision					
μA725	±3 to ±22	yes	130	3	2
LH0044C	±2 to ±20	yes	140	1	0.4
LM308	±3 to ±18	yes	110	0.3	1
OP-07	±3 to ±22	no	115	4	0.6
High Speed					
LF357A	±5 to ±18	no	105	2	20
LH0032C	±5 to ±18	yes	70	18	70
LM318	±5 to ±20	yes	110	5	15
OP-17C	to ±18	no	110	7	26
Micropower					
LH0001AC	±5 to ±20	yes	95	0.06	1
OP-20	±1.5 to ±15	no	110	0.03	0.1
Power Out					
μA759C	to ±18	no	95	12	1
μA791C	to ±18	yes	100	25	1
LH0021C	±5 to ±18	yes	100	3	1
LH0061C	to ±18	yes	95	10	5

The manufacturer of the original part may be recognized by the prefix used:
μA = Fairchild Semiconductor; OP- = Precision Monolithics, Inc. (PMI);
LM, LF, LH = National Semiconductor; CA = RCA

Op-Amp Packages (Circuit Construction Comments)

Many of the monolithic integrated circuits first available (circa 1962) were in the standard TO–5, 8-lead, round *metal can* package shown in Figure

Slew Rate (V/μs)	*Bias Current (nA)*	*Bias-Current Drift (nA/C°)*	*Input Offset Voltage (mV)*	*Input Offset-Voltage Drift (μV/C°)*	*Diff. Input Resistance (MΩ)*	*CMRR (dB)*	*Comment*
0.5	80	1	2	15	2	90	
0.5	120	1	1	6	1	90	
12	0.03	0.003	1	1	10^6	100	FET input
0.5	70	0.4	2	15	2	90	
9	0.008	0.001	5	8	10^6	90	MOSFET input
0.3	40	0.25	0.2	0.6	1.5	115	
0.06	10	0.1	12	0.2	8	140	
0.3	1.5	—	0.3	1	40	110	
0.2	±2	0.008	0.03	0.3	60	120	
50	0.03	0.003	1	1	10^6	100	FET input
500	0.025	0.002	5	25	10^6	60	FET input
70	150	0.5	4	—	3	100	
65	0.1	0.01	0.7	4	10^6	96	FET input
0.25	20	0.3	2	3	—	90	
0.02	13	0.02	0.08	1	—	110	
							Maximum I_{out}
0.5	50	—	1	—	1.5	100	0.5 A
5	80	1	2	—	1	>70	1.0 A
3	150	1	3	5	1	90	1.0 A
70	200	—	3	5	1	80	0.5 A

A6–28a. Notice the *tab* identifies pin number 8. Digital integrated circuits required more input and output pins so that the 14-pin and also 16-pin *dual-in-line-pin* (DIP) package soon appeared as shown in Figure A6–28b. The pins are spaced 0.1 inch apart in each row. An identifying U-shaped indentation and usually a dot to identify pin number 1 are found at one end. The widespread use of the DIP pin spacing soon gave rise to the *Mini-DIP* 8-pin package suitable for op amps (see Figure A6–28c).

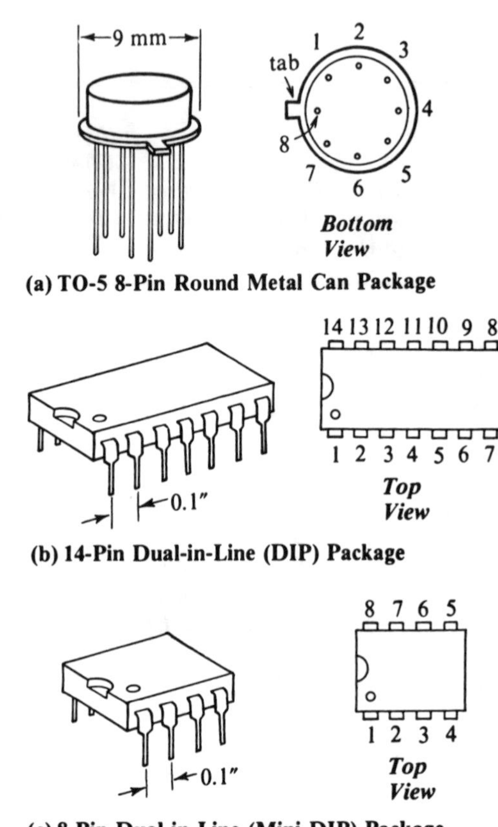

FIGURE A6–28. *Common Integrated-Circuit Packages*

NOTE: In a convention that dates back to vacuum tubes, the pin order for the package goes in the clockwise direction when the package is viewed from the *bottom*.

These devices may be soldered into printed circuit boards using appropriate soldering irons (for example, 15-watt irons), but it is a project to remove them for testing. Generally, the solder must be almost completely removed by some means; for example, a suction bulb or *solder wick*. Therefore, sockets are convenient for troubleshooting, although often they are as expensive as the IC. Perforated epoxy/fiberglass boards—perf-boards—with holes prepunched on a rectangular grid with 0.1-inch spacing are commonly available and are convenient for mounting DIP sockets and components for prototype or small-quantity designs. *Wire wrapping* has become popular in small-quantity digital circuitry and is also useful for analog circuitry. In this technique, socket pins and posts for holding components are nominally 2-centimeters long with square cross section. Number 28-30—

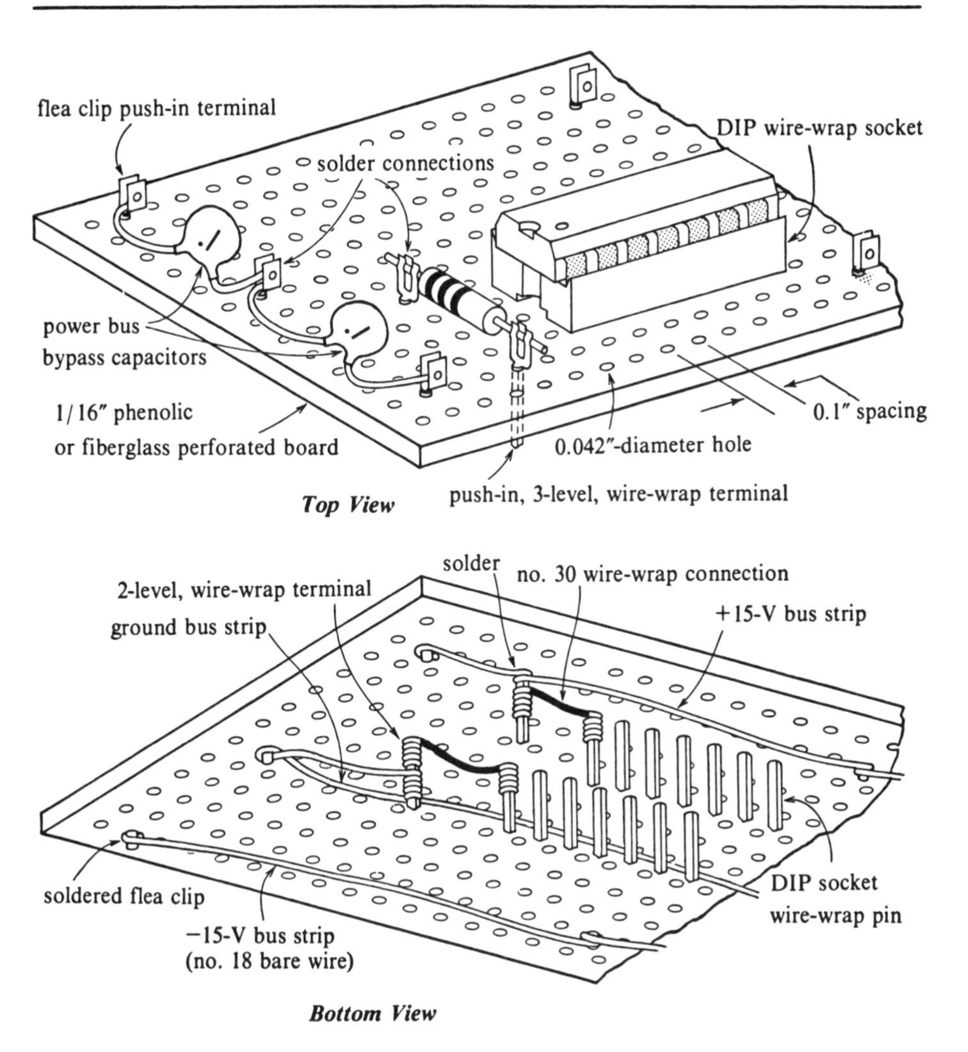

FIGURE A6–29. *Circuit Board Construction Using Perforated Board, Push-Through Terminals for Mounting Components, and Solder as well as Wire-Wrap Connection Techniques*

gauge, silver-plated, wire-wrapping wire with Kynar insulation is wrapped perhaps 7 times around a post using a wire-wrapping tool. This gives a surprisingly secure, gas-tight, electrical contact. However, this very small wire is only suitable for currents less than about 50 milliamperes. Number 22-gauge wire is suitable for electronic work where greater currents are involved (up to 500 milliamperes) but is generally not wire wrapped; it is soldered to *solder-tail sockets* or to *flea clips* that hold various components on one side of the board. Flea clips and also wire-wrap posts push through the *perf-board* holes with a friction fit.

Power supply and ground wires should be hefty (20-gauge or heavier) wires or bars to carry supply current to circuit components with a minimum of resistance and inductance. They are frequently referred to as power-supply *bus strips.* It is good practice to place low-inductance *bypass* capacitors between V+/V− and ground right on the board to reduce the possibility of circuit oscillations. Also, the ground lines should not form any closed loops because such a *ground loop* provides an inductive loop for ground-loop oscillations. Further, the output of high-gain circuits should be kept physically separated (perhaps shielded) from the input circuit to avoid oscillations. Figure A6–29 illustrates a number of these construction components and techniques.

Summary

The interior circuit of a modern operational amplifier represents very sophisticated electronic design; yet modern processes permit its fabrication at low cost. The scientist can with comparative ease use op-amp chips to design the fundamental amplifier types: inverting, noninverting, and difference amplifiers. High-quality dc amplifiers with accurately known voltage gains are quite straightforward. (Precision dc amplifiers were difficult to design a generation ago; the actual design of an op-amp chip still is.) In addition, a veritable arsenal of signal processing functions can be accomplished: summing of voltages, current-to-voltage or voltage-to-current conversion, logarithm and antilogarithm functions of voltages, multiplication or division of voltages, integration, and so on.

It is true that for precision work we must pay attention to such traits of real op amps as offset voltage, bias current, and temperature drift of both of these quantities. In some cases, precision op amps must be selected that minimize these problems. Also, work at high frequencies or with very short pulses requires selection of particular op amps. Finally, we must either select a special op amp or help an op amp with external transistors if a significant current level or power is required from the op amp. Virtually, the only area of scientific analog electronics that is difficult to solve with the op amp at the present time is one that requires high voltages (perhaps hundreds of volts) at high frequencies (more than 10 megahertz). However, it is probable that in the near future op amps that incorporate VMOS FETs (discussed briefly in Chapter A8) as their output stages will move into even this last stronghold of discrete transistor and vacuum tube electronics.

This chapter has been designed to give the scientist a *working knowledge* of how to solve a variety of laboratory electronics problems through use of the operational amplifier, but it cannot claim to be comprehensive. For more op amp details and applications, the reader is referred to the *IC OP-AMP Cookbook* by Walter Jung, or the *Transducer Interfacing Handbook* by Sheingold from Analog Devices, as well as other texts given in the Reference section.

REVIEW EXERCISES

1. (a) What is an IC?
   (b) What is an op amp?
2. Give typical values for the following aspects of the 741 op amp:
   (a) input resistance, between + and − terminals
   (b) open-loop voltage gain
   (c) output resistance
   (d) maximum power supply voltage
3. Name 4 qualities for an ideal operational amplifier.
4. Make a drawing of the circuit for each of the following using an operational amplifier and other circuit elements are needed (no values need be included):
   (a) voltage follower
   (b) logarithmic amplifier
   (c) inverting amplifier
   (d) integrating function block
   (e) noninverting amplifier
5. Suppose you construct 10 amplifiers from op amps: Each amplifier is to have an accurate voltage gain of 20. However, the open-loop gain of the op amps used varies by ±20%. Does this cause a problem? Explain briefly.
6. The circuit drawn in Diagram A6–1 is intended to be a noninverting amplifier with a gain of 10. Find two circuit errors.

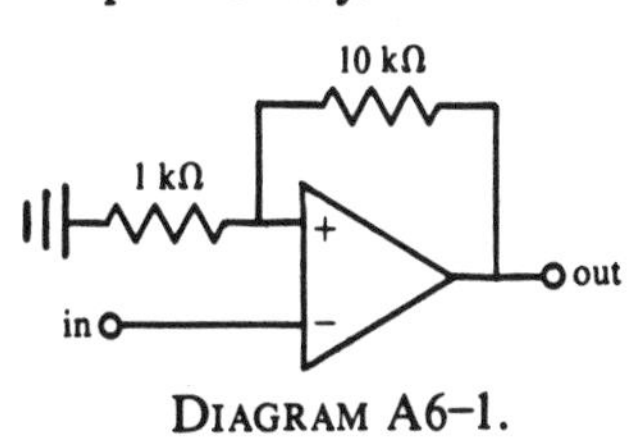

DIAGRAM A6–1.

7. The circuit shown in Diagram A6–2 can be changed to a voltage follower simply by (choose one): A. shorting R_2, B. removing R_2, C. shorting R_1, D. removing R_1.
   NOTE: More than one answer is correct.

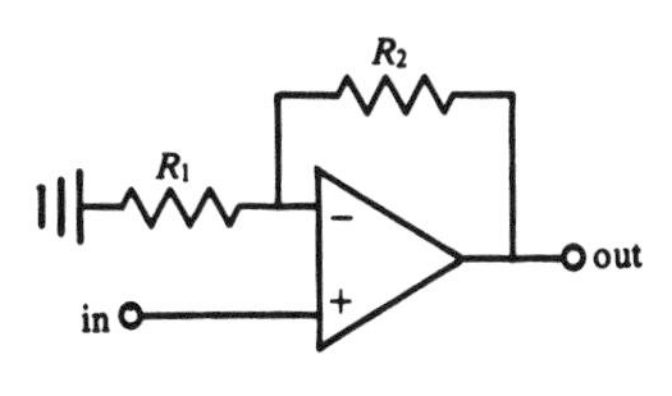

DIAGRAM A6–2.

8. Rank the following amplifiers from largest to smallest in terms of input impedance and briefly justify your ranking:
   (a) inverting amplifier, gain of 100
   (b) voltage follower
   (c) noninverting amplifier, gain of 100

9. Name 3 factors that may limit the input impedance of the voltage follower in practice.
10. Both the operational amplifier (generally) and the difference amplifier have inverting and noninverting inputs. What then distinguishes the two amplifier types?
11. (a) How many op amps are commonly used in an instrumentation amplifier? What is the principle purpose of each?
    (b) What is the important performance difference between an instrumentation amplifier and a difference amplifier constructed from one op amp?
12. What is the origin of the term *operational* in operational amplifier? Comment on its double meaning in the valid sentence: An operational amplifier is used in an operational amplifier to integrate a function.
    NOTE: Because of this problem, the term *function block* is substituted for the latter meaning in this book.
13. (a) An integrating function block can perform the integral of a function $F(x)$. For the actual circuit, what must the quantity x be?
    NOTE: x has dimensions.
    (b) What is the nature of the function F itself for the circuit; for example, what are its dimensions?
    (c) What is the nature of the output from the function block; for example, what are its dimensions?
14. (a) What is the principal problem of a simple log or antilog circuit constructed from only a diode, a resistor, and an op amp?
    (b) What is likely to be the cost-effective solution if a log or antilog circuit is desired?
15. Which of the following statements is *not* true about real operational amplifiers?
    (a) They can amplify an input signal of zero frequency, but not infinite frequency.
    (b) They require an input bias current.
    (c) They produce zero output voltage when the input is shorted.
    (d) They require connections to some power supply.
16. (a) What is typically required to be connected to an uncompensated op amp that is not required for an internally compensated op amp?
    (b) What advantage can an uncompensated op amp have, however?
17. (a) What is meant by nulling an op-amp amplifier circuit?
    (b) How is it done?
    (c) Why is this of little concern if the circuit is to amplify ac signals?
18. (a) Why is there a need for bypass capacitors on an op-amp circuit board? Where are they placed?
    (b) What type of capacitor should be selected?
19. What is the particular problem when using a bipolar op-amp circuit with a transducer that has a very high output impedance?
20. (a) What is the operating temperature range for military-range op amps as well as for commercial-range units?

(b) Select the military-range op amps from the following list of popular types: μA741, LM301, μA741C, LM101.

21. Name the two input properties of an op amp that can present significant drift problems when the op-amp temperature changes.
22. (a) What is the principle advantage of an op amp that uses MOSFETs or FETS for its input stage?
    (b) However, very few FET-input op amps are classified as precision op amps. What is their principal failing?
23. When is the slew rate of an op amp of particular concern?
24. A 741 op-amp circuit is used to amplify the output from a thermocouple before reading with a DVM. The circuit is placed in a room to monitor room temperature. Explain why the circuit will not perform satisfactorily (unless the op-amp circuit is nulled before each reading).
25. (a) An op amp and a voltmeter can be used to make an excellent ammeter. Draw the op-amp circuit involved.
    (b) What feature of the op amp permits the resulting ammeter to approach the ideal?
26. (a) Diagram a circuit constructed from an op amp that becomes a constant current source. Describe what determines the current.
    (b) What is the upper limit for the current value if typical low-power operational amplifiers are used?
27. A 741 op amp is to be used in a circuit that should output up to 5 V to a 50-Ω load.
    (a) Why will the 741 need assistance?
    (b) What can be done to provide this assistance?
28. Name 5 principal categories of operational amplifiers. Give an example application, in general terms, for each category.
29. (a) What does DIP stand for; to what does the phrase refer?
    (b) What is the direction of increasing pin numbers for TO–5 and DIP IC packages when viewed from the *bottom:* clockwise or counterclockwise?
    (c) Why is 0.1-inch-spacing perforated board convenient for contemporary circuit boards?
    (d) What wiring technique avoids soldering and is suitable for permanent circuits?
    (e) What are bus strips (also called rails) on a circuit board?

PROBLEMS

1. Draw the op-amp design for an inverting amplifier with an input impedance of 5 kΩ and a voltage gain of 20.
2. Design a noninverting amplifier with a voltage gain of 10. The smallest resistor used is to be 1 kΩ.

3. Design a noninverting amplifier with a voltage gain of 20. With no exterior load on the amplifier, the output current from the op amp is to be 1 mA at 10-V output.
4. Assume an ideal op amp is used in the circuit pictured in Diagram A6–3.
   (a) What is the reading of the voltmeter V? Justify your answer.
   (b) What is the reading of ammeter I? Justify your answer.

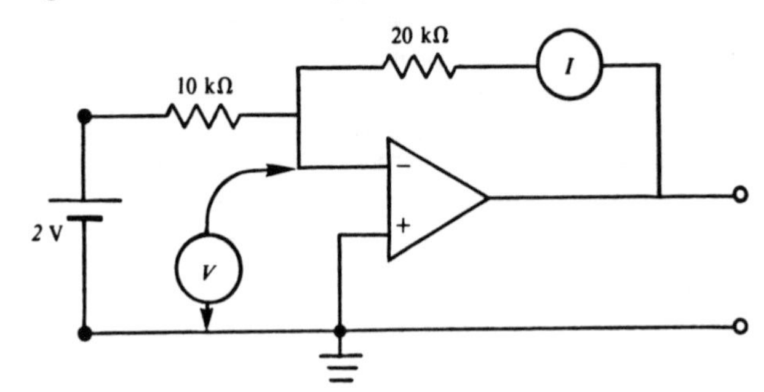

DIAGRAM A6–3.

5. Consider the op-amp circuit shown in Diagram A6–4.
   (a) If B is grounded and A is used as the amplifier input, what is the voltage gain (including sign) and input impedance of the amplifier?
   (b) If A is grounded and B is used as the amplifier input, what is the voltage gain (including sign) of the amplifier? Comment on the order of magnitude of its input impedance.

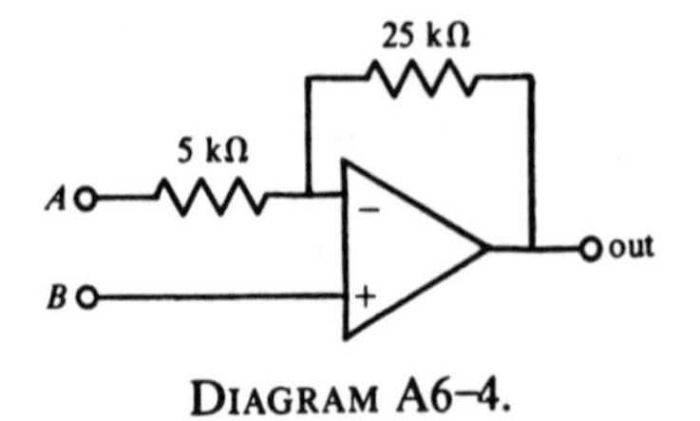

DIAGRAM A6–4.

6. (a) Can an amplifier (noninverting) be constructed with gain +1/2 from a single op amp? Explain.
   (b) Design a circuit that provides a gain of +1/2 constructed from two op amps.
   (c) Discuss what factors limit how small the resistors may be in this circuit.
   (d) Design a circuit that has high input resistance and provides a gain of +1/2. 3 op amps may be used.
7. Consider the expanded interior model for only the op-amp inputs, as shown in Diagram A6–5.
   (a) Suppose the two inputs are connected together, and the resistance between the inputs and ground is measured. What is obtained from the

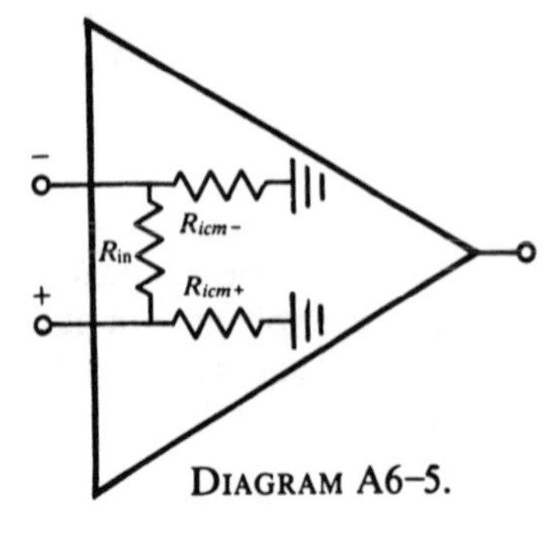

DIAGRAM A6–5.

model for this common-mode input resistance R_{icm}?

(b) Diagram a voltage follower constructed with this op-amp model. Show that R_{icm+} represents a direct shunt to ground at the follower input. (The effective resistance presented by R_{in} remains as argued in the chapter, and R_{icm-} is of no consequence here.)

(c) Assume that a voltage follower is constructed with a 741. Assume R_{in} and A_{vo} are typical and the $R_{icm+} = 100\ R_{in}$. Find the expected input impedance of the voltage-follower amplifier.

NOTE: Actually, this should be called the ac input impedance because the input bias current affects the dc value.

8. Consider the derivation of the voltage gain for the noninverting amplifier assuming finite open-loop gain, A_{vo}.

(a) Justify that the following expression holds for the quantities shown in Diagram A6–6:

$$v_d = v_{in} - v_{out}\left(\frac{R_2}{R_1 + R_2}\right) = \frac{v_{out}}{A_{vo}}$$

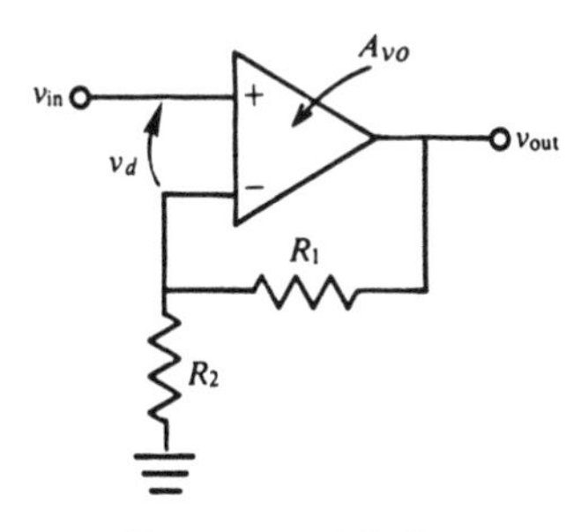

DIAGRAM A6–6.

(b) First let $A_{vo} \rightarrow \infty$ in this expression and quickly deduce the classical expression for the voltage gain of the noninverting amplifier.

(c) Let A_{vo} remain finite. Solve for v_{out}/v_{in}.

NOTE: This is the expression corresponding to Equation A6–38, but for the noninverting amplifier.

9. Consider the amplifier shown in Diagram A6–7.

(a) Is this an inverting or a noninverting amplifier?

(b) Show that the voltage gain of the amplifier is approximately 1000.

NOTE: Consider R_4/R_3 and R_2/R_1 to be successive voltage dividers.

(c) What is the advantage of this connection over the 2-resistor connection in obtaining an amplifier with very high gain?

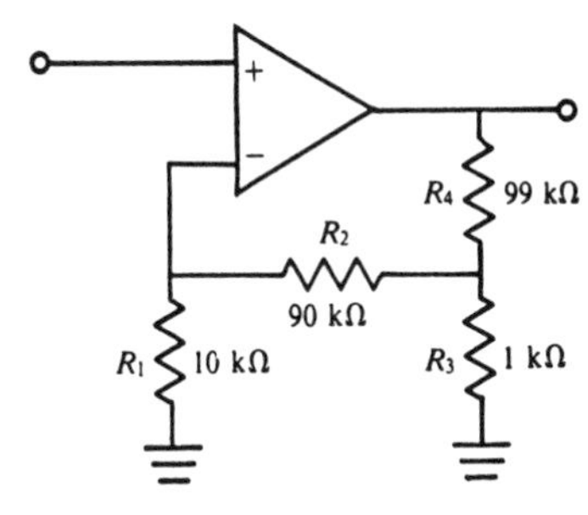

DIAGRAM A6–7.

10. The circuit shown in Diagram A6–8 represents a precision-adjustable voltage source of high stability.
    (a) Why is the voltage across the load R_L adjustable? What are the minimum and maximum values?
    (b) Why is the voltage across R_L expected to be insensitive to changes in this load resistance? However, roughly, what maximum current may be drawn by the load?
    (c) Why is a mercury (Hg) battery desirable?
    (d) Why is a multiturn pot desirable for the potentiometer in the circuit?
11. Consider Diagram A6–9. Show that even though R_x varies, this circuit maintains a constant current $I = V/R$ through R_x.

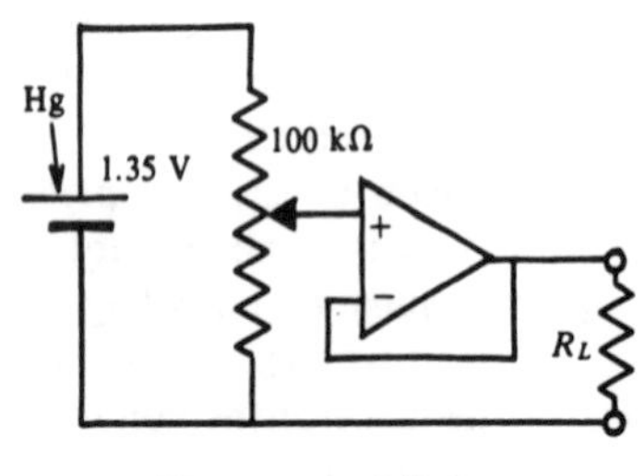

DIAGRAM A6–8.

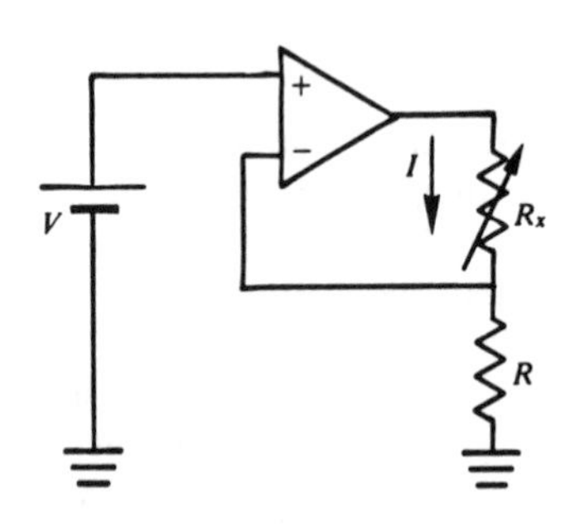

DIAGRAM A6–9.

12. Consider Diagram A6–10. We wish to design a box that receives a variable dc voltage V_1 at one input, and a 0.1 $V_{p\text{-}p}$ ac voltage at the second input. The box is to output the ac voltage at 4 $V_{p\text{-}p}$, and this ac signal is to be superposed on a dc voltage that is 5 times V_1. See the diagram. The input impedance of the ac input should be 1 kΩ. Draw the amplifier design, including component values required. Justify your design briefly.

    NOTE: The output behavior here is similar to the dc offset adjustment available on many function generators.

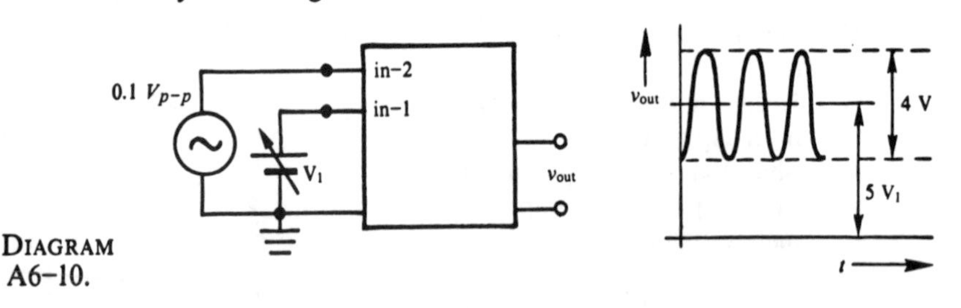

DIAGRAM A6–10.

13. Draw an amplifier design for which the output voltage v_{out} is related to three input voltages v_1, v_2, v_3 by

$$v_{out} = -(v_1 + 2v_2 + 3v_3)$$

The smallest input impedance (for any of the 3 input voltages) is to be 1 kΩ.

14. We want to justify the input voltage relation given by Equation A6–16 for the "attempt at a difference amplifier" in Figure A6–8a. Our objective is to find v_{out} in terms of v_{in-} and v_{in+}. Use the notation of the drawing as shown in Diagram A6–11.
    (a) What can be said about the voltage of S compared to v_{in+}?
    (b) Solve for the current i in terms of v_{out} and v_{in-}.
    (c) Also find a relationship between current i and v_{in+}, v_{in-}.
    (d) Solve now for v_{out} in terms of the two input voltages and compare with Equation A6–16.

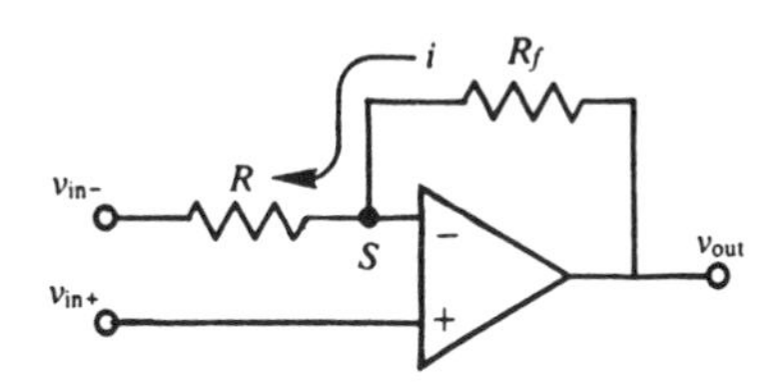

DIAGRAM A6–11.

15. We wish to use an op amp and a 5-V voltmeter to construct a milliammeter with a 10-mA full-scale reading. Draw the circuit design.

16. (a) Consider the circuit shown in Diagram A6–12, where V_{in} and R are fixed. Show that the circuit may be considered to be a resistance-to-voltage converter. What is the proportionality constant between V_{out} and R_x?
    (b) What could be the utility of this circuit in connection with a resistance-type transducer?

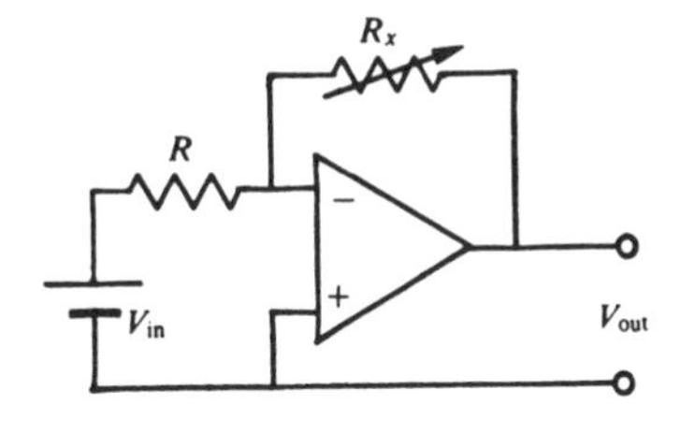

DIAGRAM A6–12.

17. What is the voltage reading of the voltmeter V in the op-amp circuit pictured in Diagram A6–13?

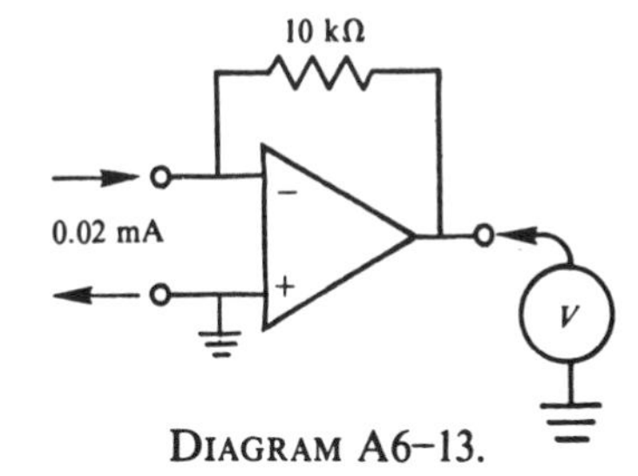

DIAGRAM A6–13.

18. (a) Consider the circuit shown in Diagram A6–14. The output voltage can be expected to be not very far from (choose

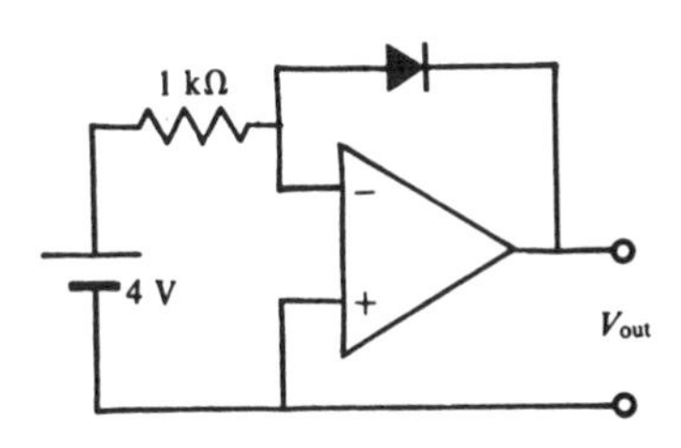

DIAGRAM A6–14.

one): A. +0.6 V, B. −0.6 V, C. +4.6 V, D. −3.4 V

(b) Give an argument to support your answer in (a) using only simple circuit and component properties.

19. The logarithm function block of Figure A6–15 operates only with positive input voltage and produces a negative voltage out. How could it be changed quite simply to accept negative voltage in and produce positive voltage out? Justify briefly.

20. (a) Draw a block diagram of a circuit to divide a voltage v_A by a voltage v_B. Briefly justify its action.

(b) Draw a block diagram of a circuit to perform the square of an input voltage. Briefly justify its action.

21. Consider the circuit pictured in Diagram A6–15. Write an expression for the output voltage, v_{out}. Justify your result.

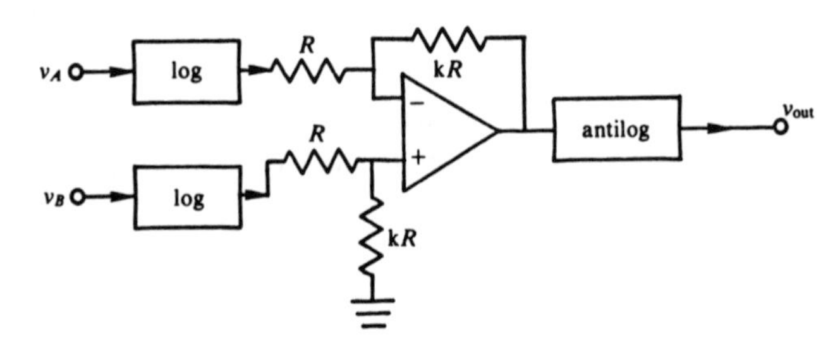

DIAGRAM A6–15.

22. (a) Suppose a short is connected across the capacitor in an integrating function block constructed from an op amp. Is the voltage source connected to the integration input liable to damage? Explain.

(b) Find the input impedance of an op-amp integrating function block.

23. (a) Suppose a unipolar square wave (see Figure A7–1a) is input to an integrating circuit. Draw the output from the integrator, justifying its character briefly.

(b) Why will the circuit cease functioning at some time after the circuit is put into operation?

24. Consider the integrating circuit shown in Diagram A6–16. Suppose the negative terminal of a 2-V battery is connected to the input of the amplifier (and the other terminal to ground). Draw a graph of the output voltage as a function of time and justify your graph briefly.

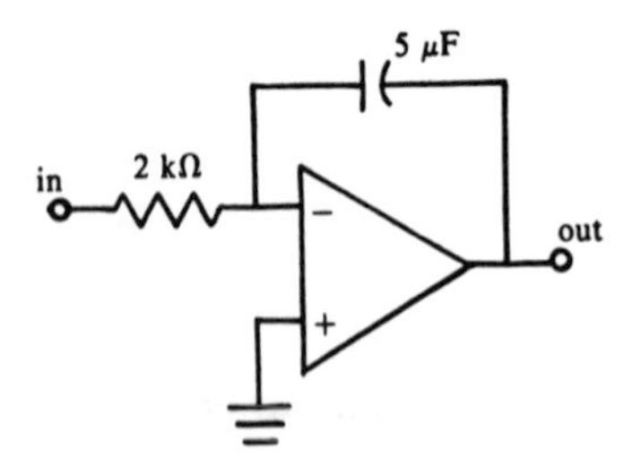

DIAGRAM A6–16.

25. (a) Consider the above integrating circuit. Suppose the input signal is given by $v_{in} = 2\text{ V} \sin(\omega t)$, where the frequency is 1 kHz. What expression for

v_{out} is expected (theoretically)? What is the phase of the output compared to the input? What is the amplitude of the output? Can a typical real op amp accommodate this amplitude?

(b) Suppose the resistor and capacitor are interchanged in the circuit. Answer the questions of part (a) for the new circuit.

26. Suppose a 741 op amp is operated in the circuit shown in Diagram A6–17. Over what frequency range can the amplifier be expected to operate with flat response?

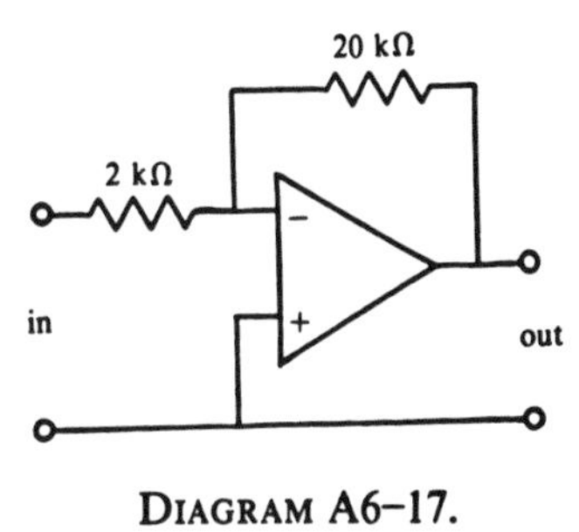

DIAGRAM A6–17.

27. (a) Suppose an amplifier with a voltage gain of 2000 is constructed from a 741 op amp. What bandwidth may be expected for the amplifier? Obtain the answer two ways: working from Figure A6-21, and using the quoted GBP for the 741.

(b) Suppose the amplifier is constructed instead with a 748 op amp using a 3-pF compensation capacitor. What is the approximate upper-frequency limit for satisfactory operation of this amplifier?

28. (a) A student constructs an op-amp amplifier with a gain of 60. With the input shorted the output voltage reads 0.015 V. What is the input offset voltage for the op amp used?

(b) The op amp has a drift of the input offset of 7 μV/C°. If the amplifier temperature is increased by 10 C°, what output would be expected with the input shorted? Give two possible values.

29. Consider the amplifier shown in Diagram A6–18.

(a) Is this an ac or a dc amplifier? Why?

(b) Give a possible reason for putting C_1 in the amplifier.

(c) Suppose $R_2/R_1 = 10$. There is probably no point in arranging an offset null for the amplifier. Why?

(d) Suppose $R_2/R_1 = 1000$. In this case there may be good reason to arrange an offset null for the amplifier. Why?

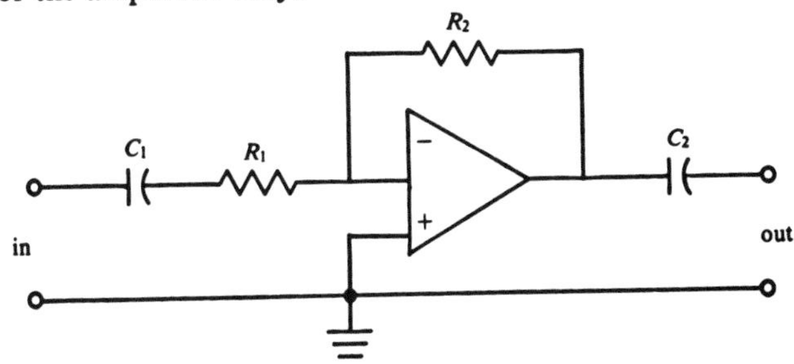

DIAGRAM A6–18.

30. The 741, 308, 318, and 3140 are four op-amp types that are widely available and inexpensive. Give the typical values of the following parameters for the respective devices. Circle the "winner" in each case.
    (a) differential input resistance
    (b) input bias current
    (c) input offset-voltage drift
    (d) slew rate
31. Suggest an op-amp type that would be appropriate for use as an amplifier in each of the following tasks. Justify your choice briefly in each case.
    (a) Voltage pulses of 5-μsec duration and 0.1-V amplitude need to be amplified to 2-V amplitude.
    (b) Essentially dc signals on the order of 100 μV are to be amplified to 10 mV. The ambient temperature changes by ± 5 C° during its use.
    (c) A 10-mV_{p-p} 500-Hz signal from a source with 1-kΩ output impedance is to be amplified to 1.0 V_{p-p}.
    (d) A 10-mV_{p-p} 200-kHz signal from a source with 1-kΩ output impedance is to be amplified to 1.0 V_{p-p}.
    (e) A 10-mV_{dc} signal from a voltage source with 50-MΩ output impedance needs to be amplified to 1.0 V.
    (f) A 100-mV_{dc} signal from a voltage source with 1-kΩ output impedance needs to be amplified to 2.0 V. The circuit is to operate from 4 penlight batteries to permit portable field operation.
32. (a) A 741 op amp is to be operated from a pair of small 8.4-V nickel-cadmium transistor radio batteries. They have a rated capacity of 0.1 ampere-hour. Approximately how long can the circuit be expected to operate before the batteries require recharging?
    NOTE: Assume the load to the circuit is high impedance.
    (b) Suppose an OP–20 is substituted for the 741 in the circuit. Find the operational duration.
33. Consider Diagram A6–19. This exercise deduces the output voltage error due to input bias current, if there is no compensating resistor connection between the + op-amp input and ground. The approach differs from that followed in the text example. Again it is assumed that A_{vo} is large, and that no input offset voltage exists, but the bias current i_b into each op-amp input is finite.

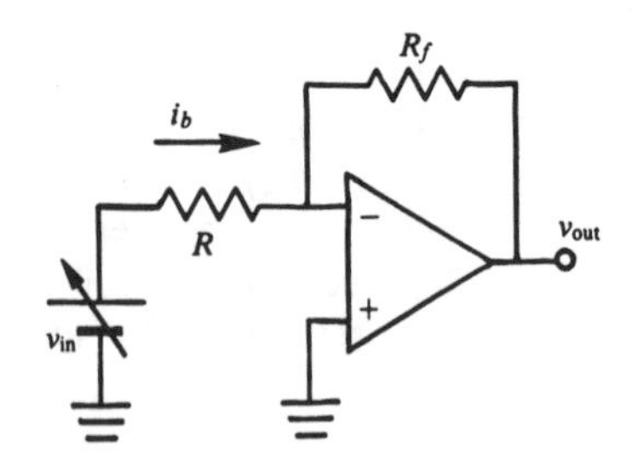

DIAGRAM A6–19.

    (a) Assume a small voltage v_{in} at the amplifier input is adjusted until v_{out} is precisely zero. Justify that the only current flowing is in R, and that it must consist of the bias current i_b.
    (b) Justify that $v_{in} = i_bR$ (the voltage drop across R).

(c) Suppose v_{in} is reduced to zero; then the amplifier input is shorted to ground. Argue that the amplifier output should be $v_{out} = +(R_f/R)v_{in}$, where v_{in} is the value obtained above. This is an output voltage error. What is its value in reference to the input?

(d) Show that $v_{out} = i_b R_f$, in agreement with the text example.

CHAPTER A7

Waveform-Shaping Circuits

The square-wave voltage waveform is that rather curious situation where the voltage changes rapidly from a low voltage level to a higher voltage, remains there for a time $T/2$, and then suddenly switches back again where it remains for a time $T/2$ before repeating the process. This situation is the symmetrical square wave (see Figure A7–1a). In the asymmetrical case, the time intervals for low and high are different (see Figure A7–1b). If the time that the voltage is low is much longer than the time that it is high, we think more of a sequence of *pulses* (see Figure A7–1c).

For *ideal* square waves and pulses the transitions between the low and high levels would be instantaneous; this feature is indicated in Figure A7–1 by the dashed vertical lines. However, real square waves and pulses have a finite *rise time* and *fall time* for these transitions. Furthermore, Figure A7–1a illustrates that a square wave may essentially be a changing dc voltage, or *unipolar* in terms of voltage, or the square wave may be true ac and therefore *symmetrical* in terms of voltage. In general, when a square wave or other waveform is not symmetrical in voltage, it is said to have an *offset.* A function generator often possesses an offset control for the waveform output.

There are a number of situations in analog electronics where these waveforms that are either high or low are of interest, although today the *"high" or "low"*—or *ON or OFF* or 1 or 0— situation is generally considered to be digital electronics. This latter field has expanded tremendously in the past twenty years and indeed, we devote *Digital and Microprocessor Electronics for Scientific Application* to the topic. Nevertheless, pulse waveforms are also useful in instruments that are primarily analog and therefore require some consideration.

Comparator

The *comparator* is a circuit that compares two input voltages V and V_r. We think of V_r as a *reference voltage;* hence the *r*. If $V > V_r$, the output voltage goes to a high voltage, while if $V < V_r$, the output voltage goes to a low voltage.

One use for a comparator is as a null detector in a Wheatstone bridge, replacing a galvanometer that has long been used to detect equality of voltage on the two sides of the bridge; that is, the comparator can be connected between points A and B of Figure A2–4 and the comparator output connected to a light bulb as a balance indicator. Actually, the comparator would not detect balance as does the galvanometer but rather the *state* of being on one side or the other of balance. The light bulb would indicate this state for small changes in R_3 near balance, and the balance point is thus accurately selected.

The reader may have realized already that a high-gain difference amplifier performs the job of a comparator. Because an operational amplifier is a high-gain difference amplifier, it is often used for the task. Figure A7–2a illustrates an op-amp comparator circuit in which the input voltage is compared to a fixed voltage V_r. In Figure A7–2b we see the output swings from V_- to V_+ when the input signal exceeds the reference voltage by a few microvolts. The output swings back to V_- when the input signal again falls below the reference level. Thus, we refer to the circuit as a *voltage level detector.* The term *threshold* is often used for the input level (V_r) at which the output switches. If we want the output to swing in the opposite sense from that shown in Figure A7–2b, we merely interchange the V_{in} and V_r connections to the op amp.

The simple circuit of Figure A7–2a has essentially the two power-supply voltages as the two possible output voltage levels (for example, +12 V and −12 V). Different levels might be desirable, such as +5 V for the high level and 0 V for the low level. A slightly clever use of diodes does the job, as shown in Figure A7–3a. Here the diode D_1 conducts when the op-amp output swings negative; this diode will not permit the voltage at its cathode to become more negative than about 0.6 V (for silicon) relative to its anode, which is at ground (see Figure A7–3b). (Almost −12 V then develops across the 1-kilohm resistor.) However, diode D_1 does *not* object to v_{out} being positive. We say diode D_1 *clips* the output near ground.

It is diode D_2 that conducts when the output swings above the voltage of the D_2 cathode, which is connected to a +5-V supply. The output thus cannot move more than about 0.6 V above the 5-V level (diode D_2 clips the output near the 5-V level). The 1-kilohm resistor protects the op amp by limit-

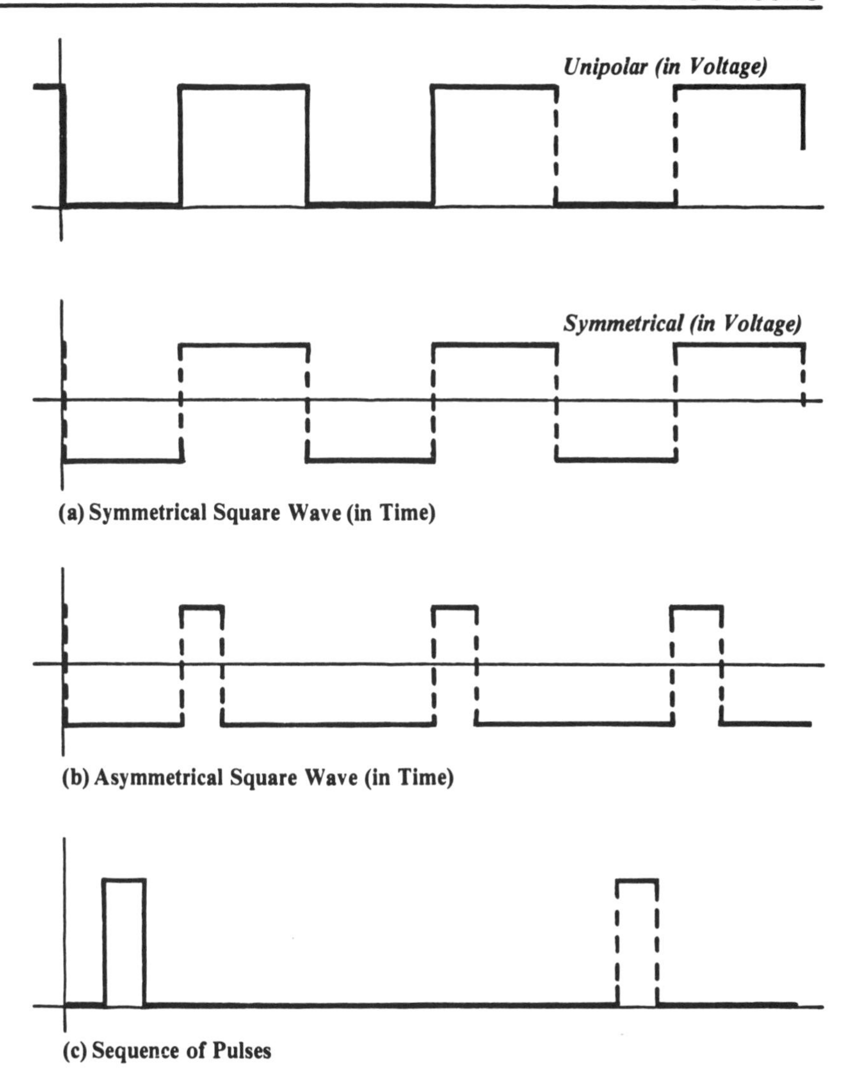

(a) Symmetrical Square Wave (in Time)

(b) Asymmetrical Square Wave (in Time)

(c) Sequence of Pulses

FIGURE A7–1. *Various Square-Wave and Pulse Forms*

ing the current flowing from the op-amp output in these two situations; the actual op-amp output itself *still* swings between V_- and V_+, and the difference between the output voltage and V_- or V_+ is absorbed across the 1-kilohm resistor. (Actually, most modern op amps self-limit the output current to protect themselves, but this resistor plays safe.) Also notice in Figure A7–3a that an adjustable reference level is made possible by the *level set potentiometer*. This is simply a voltage divider that can send an adjustable V_r to the − input.

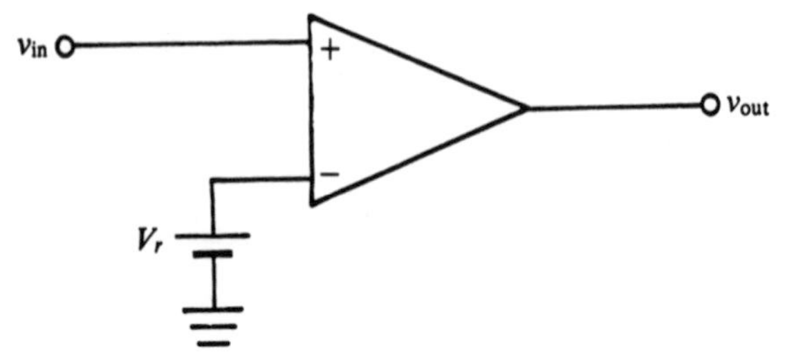

(a) Comparator Circuit

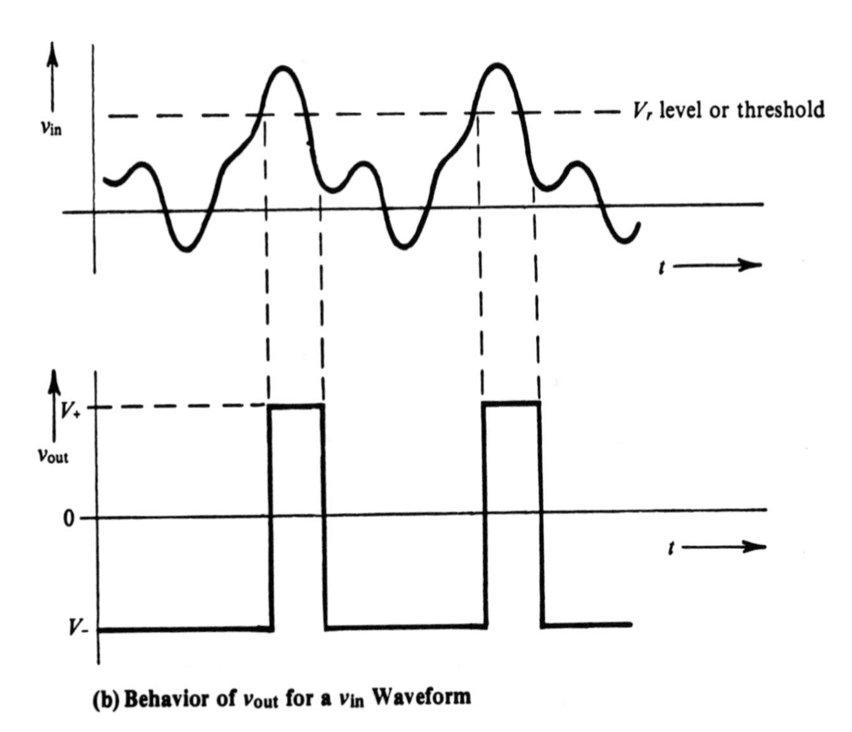

(b) Behavior of v_{out} for a v_{in} Waveform

FIGURE A7–2. *Operations of a Comparator*

Schmitt Trigger

A close relative of the comparator is the *Schmitt trigger* circuit, named after the biophysicist who invented it in the 1930s. In fact, the only difference is the presence of *hysteresis* in the switching level or threshold; that is, the level at which the output switches from low to high is different from the level at which

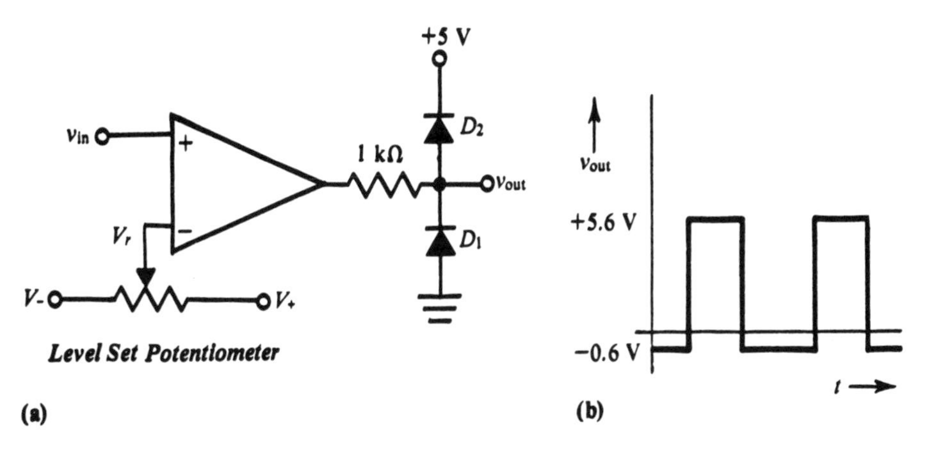

FIGURE A7–3. *Diodes Used with a Comparator to Clip Op-Amp Output to Desired Low and High Voltages*

the output switches from high to low. This little quirk is very useful in working with noisy signals, as illustrated by Figure A7–4.

Figure A7–4a shows a situation where a noisy signal is input to a single-threshold comparator; Figure A7–4b shows the same situation for a Schmitt trigger. This signal situation is common in the life sciences. For example, the pulse may be an EKG voltage generated in the body during a heartbeat. The noise represents static or high-frequency fluctuations superposed on the more slowly varying true or pure signal. It is the intention of our level-detecting circuit to detect only when the main peak of the pure signal occurs (for example, for one heartbeat) and output one pulse for this. Unfortunately, the little noise spikes generate spurious additional short pulses before and after the main pulse (see Figure A7–4a). By having separate thresholds for the upward and downward transitions, the spurious pulses are eliminated in Figure A7–4b. Notice the trick is to have these two detecting levels spaced more widely than the peak-peak noise signal. In addition, the signal may have a secondary true peak that should *not* be counted (Figure A7–4a). As shown in Figure A7–4b, the thresholds must be set above this peak (for example, so that single true heartbeats are detected).

To obtain the hysteresis behavior, two resistors should be added to the comparator or op amp in a sense that gives positive feedback; that is, between the op-amp output and the + op-amp input (see Figure A7–5a). The − input should be used for the signal input. Notice the output goes negative rather than positive when the level is exceeded in this case, as mentioned earlier. The clever scheme uses a voltage divider connected between the output voltage and the reference voltage. The divider output is connected to the noninverting or + input. Thus, the example in Figure A7–5 has the junction of the 1-kilohm and 9-kilohm resistors connected to the + input.

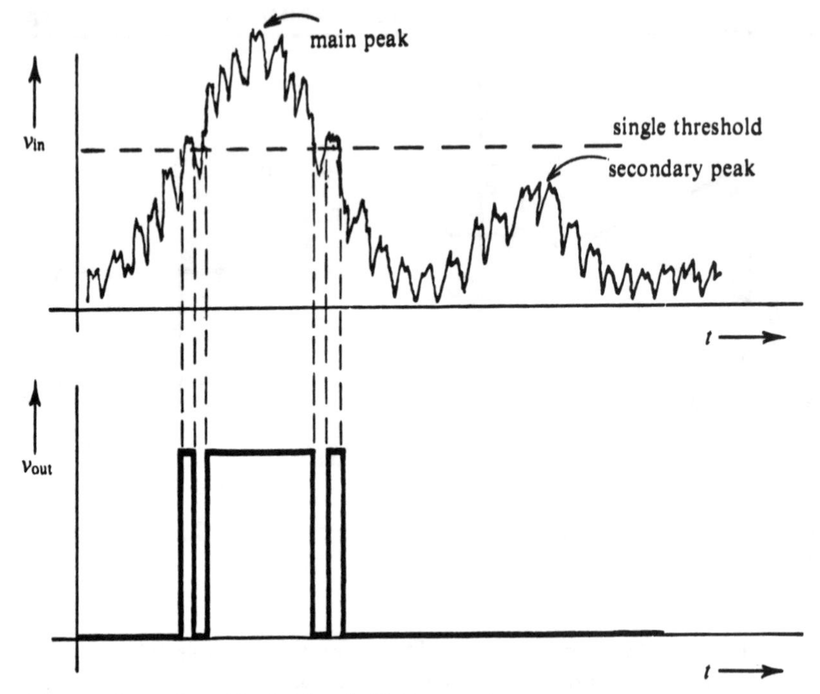

(a) v_{in} and v_{out} for a Single-Threshold Comparator

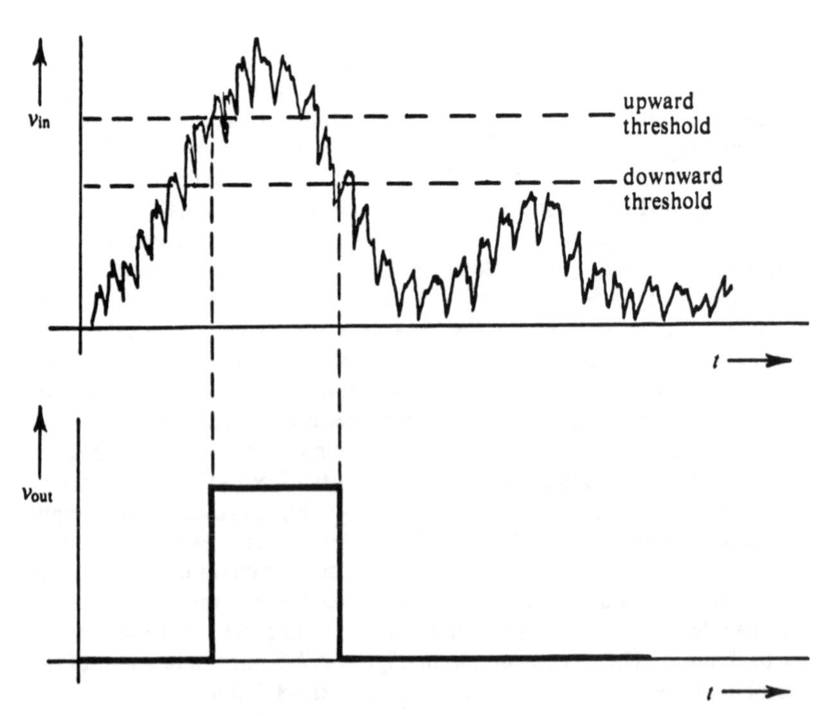

(b) v_{in} and v_{out} for a Schmitt Trigger or a Comparator with Hysteresis

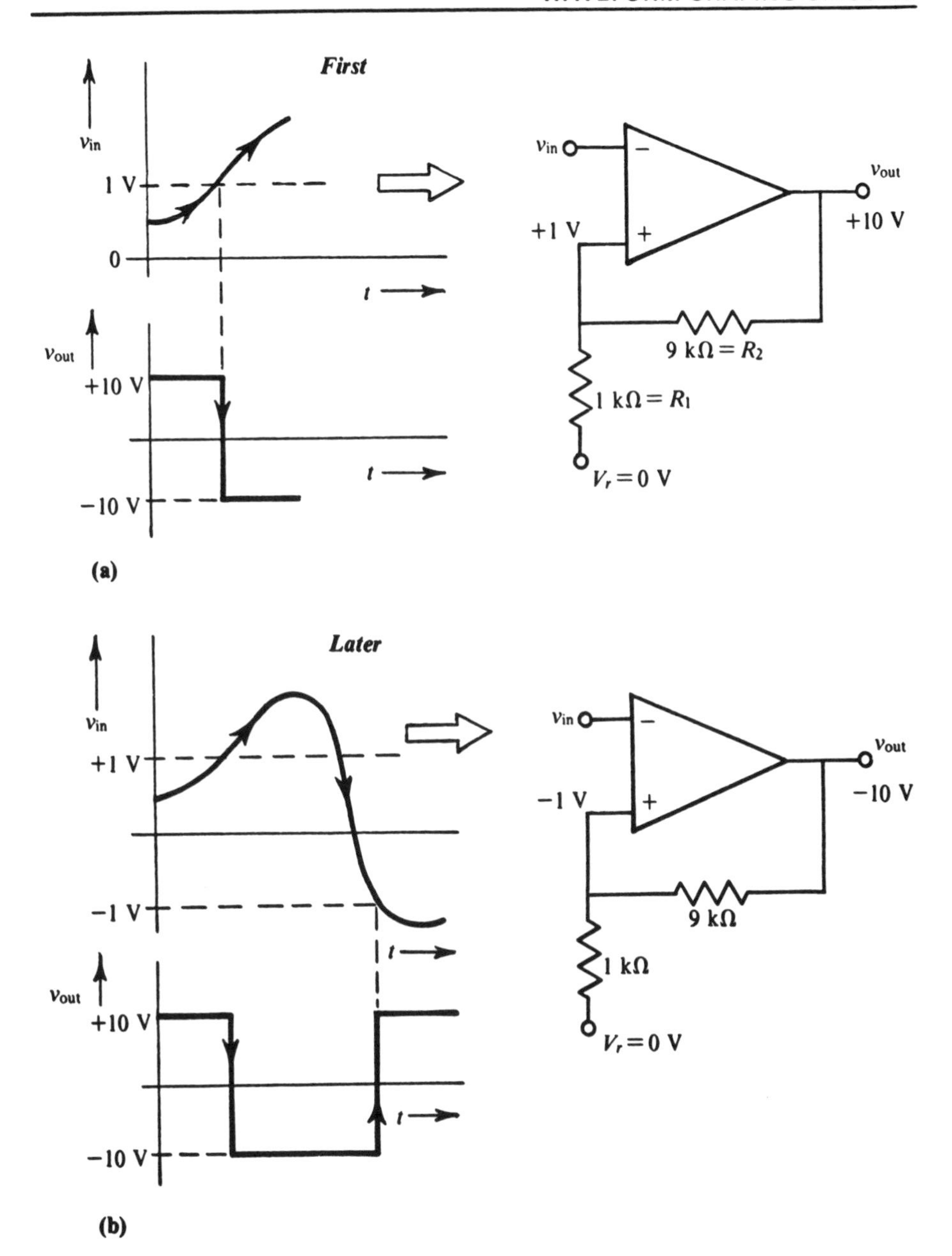

FIGURE A7–5. *A Voltage Divider from the Output That Produces Two Thresholds Symmetrically Placed about a Reference Voltage of Zero, or Ground*

◄ FIGURE A7–4 *Response of a Comparator as well as a Schmitt Trigger to a Noisy Input Signal*

Now suppose the input voltage had recently been quite negative so that the output voltage is positive with value V_+ (for example, at +10 V).* Let us assume that $V_r = 0$ V or at ground for simplicity. The voltage at the voltage divider junction is quite clearly one tenth of +10 V or 1 V (for the values of R_1 and R_2 used here). This +1-V level (and not zero volts) is presented to the + input, and only when the input voltage to the − input exceeds the +1-V level does the output switch (rapidly) from +10 V to −10 V. This situation is illustrated in Figure A7–5a.

Once the output has switched to −10 V, the reference voltage presented to the − op-amp input changes to one tenth of −10 V or −1 V (see Figure A7–5b). Only when the input drops below *this* level will the output switch back to the +10 V state again.

Notice that the hysteresis width—that is, the difference between the two trigger levels—is 2 V or twice the voltage out of the voltage divider. Quite clearly, for a general R_1, R_2, and V_+ we can write

$$\text{hysteresis width} = \frac{2R_1 V_+}{R_1 + R_2} \quad \textbf{(A7–1)}$$

Although this derivation was for the simple situation where the reference voltage V_r was ground, it is not hard to show that we get the same hysteresis width for nonzero *center reference voltages* V_r. See the illustration of v_{in} and v_{out} in Figure A7–6b.

We can obtain an adjustable Schmitt trigger center-level V_r by using a potentiometer connected between $V+$ and $V-$ (the op-amp power supplies); thus an adjustable reference voltage V_r is obtained as shown in Figure A7–6a. There is a precaution. The resistance of the potentiometer R should be small compared to R_1 and R_2 since the reference voltage presents a variable output impedance of the order R that adds to R_1 and changes the hysteresis width. Also the center of the hysteresis interval is not precisely at the voltage V_r established by the V_r-level set potentiometer. The fact that the center of the hysteresis width differs slightly from V_r is left for investigation in the Problems of this chapter. The circuit of Figure A7–6a includes clipping diodes at the op-amp output, and this is reflected in the output levels shown in Figure A7–6b.

Multivibrators

Three types of multivibrators (MVs) have been in use for many years in electronics—monostable, bistable, and astable MVs. They were first designed

*In general, the upper limit V_+ for the op-amp output can be expected to be near the voltage $V+$ that is used for the + power connection to the op amp. Similarly, the lower limit V_- can be expected to be close to the $V-$ power-supply value. We will assume $|V-| = |V+|$ here.

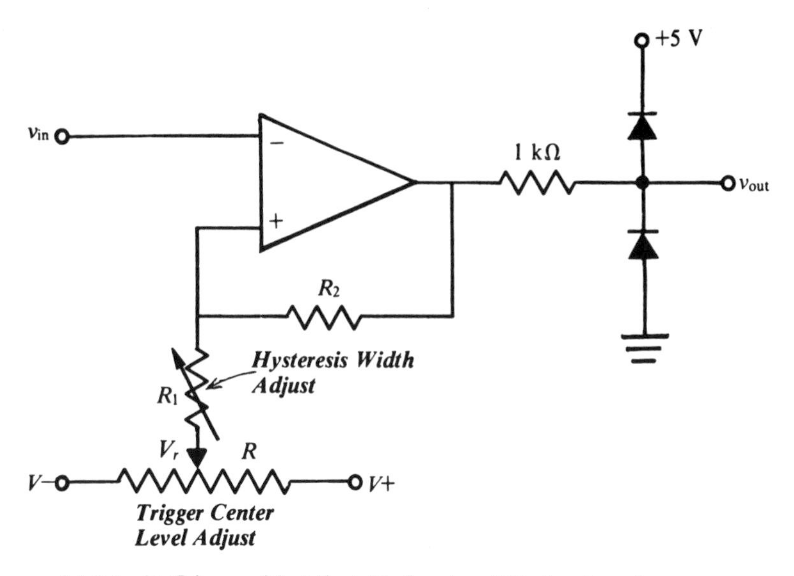

(a) Schmitt Trigger with Adjustable Level and Hysteresis and Output Clipped to Near 0 V, +5 V

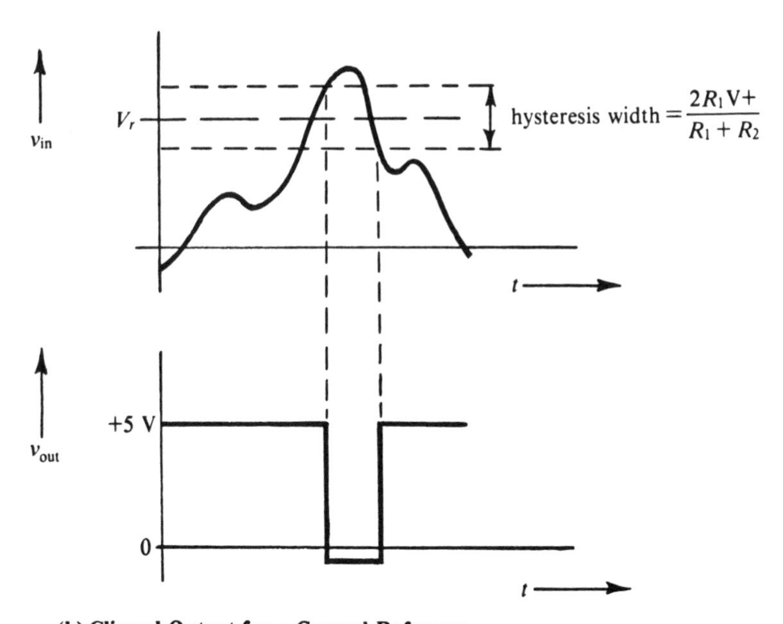

(b) Clipped Output for a General Reference Level and Hysteresis Width

FIGURE A7–6. *General-Purpose Schmitt Trigger from an Op Amp*

with vacuum tubes, later with transistors, and now are available in integrated circuits. All multivibrators deal with pulses or rectangular waveforms.

Monostable (One-Shot) Multivibrator

The *monostable* (one-shot) *multivibrator* (MMV) receives a pulse as input and, as a consequence, immediately generates an output pulse of a certain time duration that is *independent* of the length of the input pulse. Only one output pulse is generated for each input *trigger pulse*. Typically, the trigger pulse is a negative-going pulse; the output pulse begins at the time of the negative slope of the trigger square wave. The output pulse duration is controlled by the RC (charging) time constant of a resistance and a capacitance connected somewhere in the circuit. Generally, the output pulse length is of the order of RC. Figures A7–7a and b illustrate the typical situation.

There is a similarity between the Schmitt trigger and the monostable multivibrator in that one pulse out is produced for each pulse in. However, the pulse duration of the Schmitt trigger changes with the duration of the incoming pulse, while that for the MMV does *not*. Thus, one common use of the MMV is to take pulses of various lengths and "change" them to only one length. The MMV is generally operated between some positive supply voltage and ground. The low of the output pulse is also near ground and the high near the positive supply voltage.

Bistable Multivibrator or Flip-Flop

While the monostable had only one stable or "resting" state, the bistable multivibrator (BMV) has two. Its output remains low until a trigger pulse is received, at which time the output abruptly switches high; the output then remains high until another trigger pulse is received, and the output quickly switches to a low voltage. The output "flips" high and later "flops" back low—hence its alternate name *flip-flop*. Typically (but not always) the output changes on the negative-going edge of the trigger pulse. Figures A7–8a and b illustrate this situation for the BMV.

The flip-flop action is often also called *toggle* action because it reminds us of a toggle switch that can be flipped or toggled back and forth. This type of bistable flip-flop with a single input has been used in digital counting circuits since the 1920s and is often called a *toggle flip-flop* or *T-type FF;* its input is conventionally labeled T, while its output is labeled Q.

The reader should be able to justify that if a sequence of pulses with period T is fed to a bistable input, the output is a symmetrical (in time) square

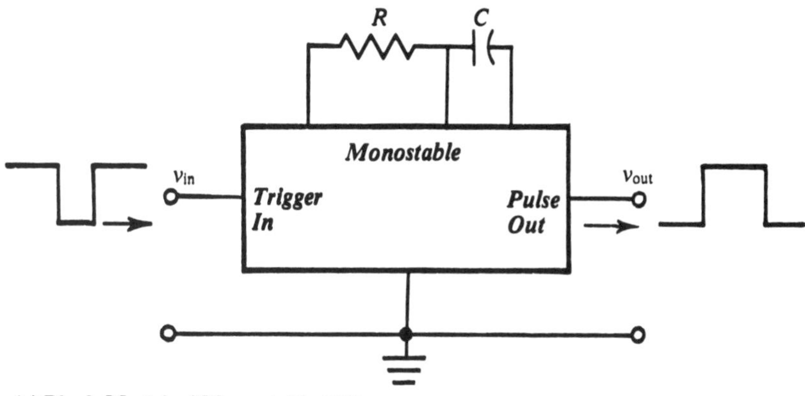

(a) Block Model of Monostable MV

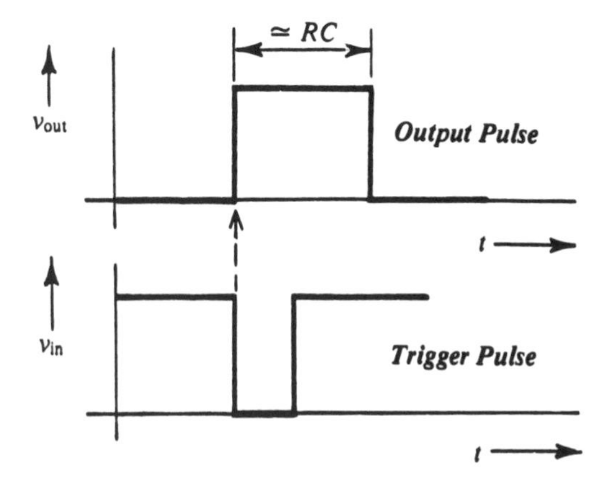

(b) Pulse Timing Relationship

FIGURE A7–7. *Monostable (One-Shot) Multivibrator*

wave with period $2T$. This bistable output could be fed to a second bistable input, and its output would be a still lower frequency or longer period square wave of $4T = 2^2T$. Feeding this to a third bistable gives a square wave of $8T = 2^3T$. In general, we can easily produce a square wave with period precisely 2^nT if a pulse train of period T is available. Figure A7–9 illustrates the bistable chain and pulses at each stage. Indeed, the system counts in binary the number of pulses that have occurred at the input. Much more can be said of counting circuits, and *Digital and Microprocessor Electronics for Scientific Application* devotes an entire chapter to the subject. Here, only a brief exposure is given.

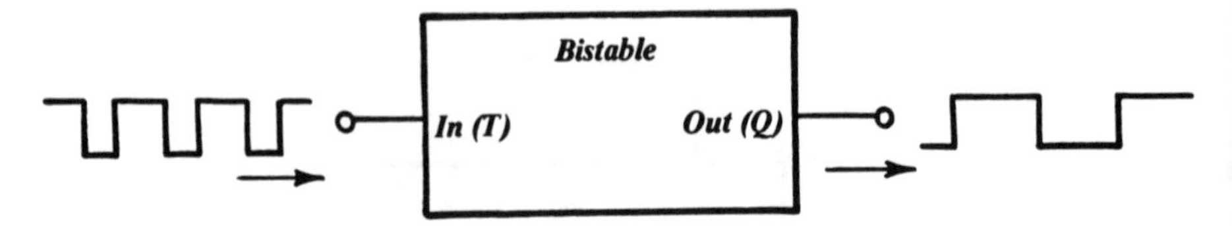

(a) Block Model of Bistable MV

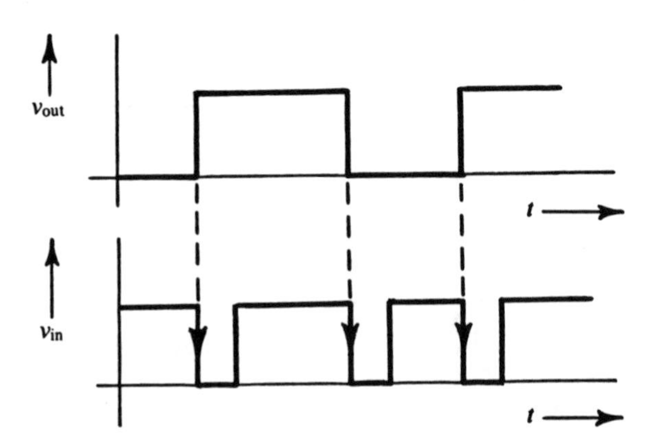

(b) Pulse and State Timing Relationship

FIGURE A7–8. *Bistable Multivibrator or T-Type Flip-Flop*

Astable (Free-Running) Multivibrator

The *astable* (free-running) *multivibrator* (AMV) is simply a square-wave generator that generates an output square wave similar to those in Figure A7–1 but uses no input. Since it runs by itself, the AMV is an oscillator.

Suppose we take two monostables and connect the output of the first to the input of the second and then connect the output of the second back into the first. The result is a self-contained pulse scheme that will perpetuate itself. (Rather like two children poking each other. The first pokes the second, but then the second pokes the first, but then....) Figure A7–10 illustrates an astable multivibrator constructed from two monostable vibrators.

555 Timer Integrated Circuit

In 1973 Signetics introduced the *555 timer integrated circuit* that has proven to be an extremely versatile device. It can be used for everything from washing dishes (timing the washing cycle) to making a siren. Basically, the 555

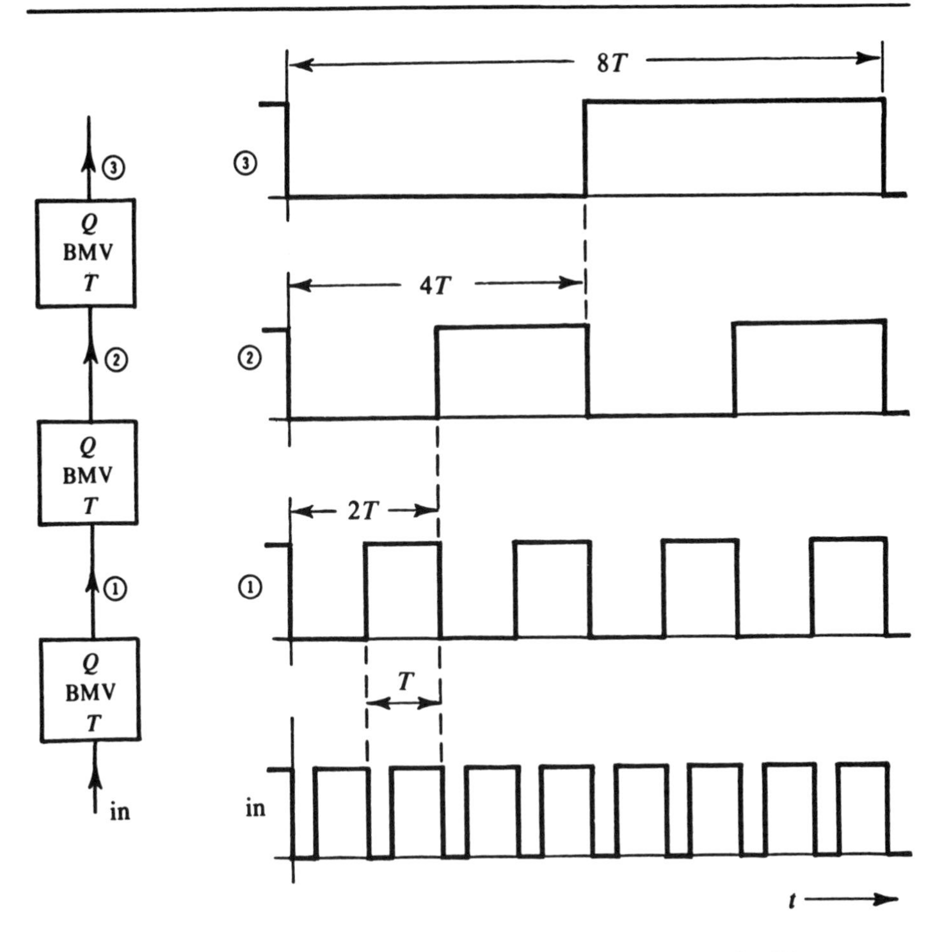

FIGURE A7–9. *Pulses in a String of Bistable Multivibrators or Flip-Flops*

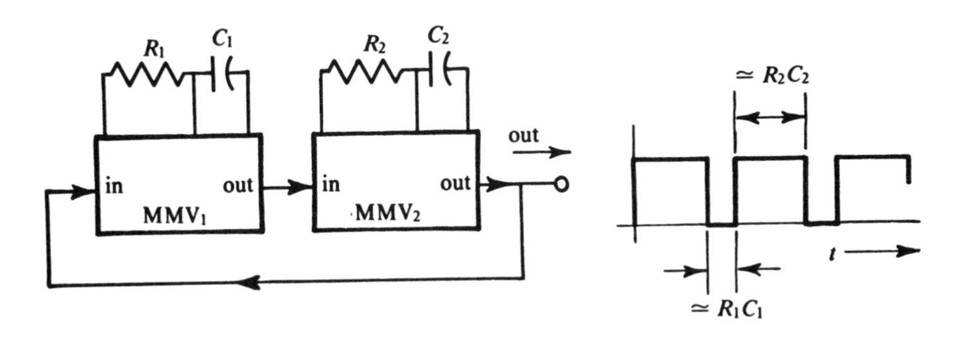

FIGURE A7–10. *Astable Multivibrator Constructed from 2 Monostable Multivibrators*

timer consists of a pair of comparators, a flip-flop, and a (transistor) switch. The pins of the IC may be connected in various ways to achieve various functions. However, the fundamental connection is the one that produces a monostable MV; as such the 555 produces a pulse of duration determined by a capacitor and a resistor connected to the device pins. The pulse duration or time interval is changeable by using an adjustable R (or C), and thus the 555 becomes a timer in the sense of an egg timer.

Figure A7–11 illustrates the important function blocks inside the 555 timer IC. The two comparators should be readily noted, as well as the switch in the lower left corner. However, the flip-flop block near the center is *not* the toggle-type flip-flop noted earlier. The bistable flip-flop within the 555 IC is called a set-reset or *RS flip-flop.* This flip-flop type has two input leads; one is labeled R for *reset* while the other is labeled S for *set.* The action of an RS flip-flop is really quite simple. If a sufficiently high voltage pulse is applied to the S input, the output Q is *set* to the high voltage state. Another pulse to the S input will *not* cause the output to toggle back to a low voltage in this flip-flop type; rather a positive pulse to the R input *resets* the output to a low voltage state. The second flip-flop output $\overline{Q}$ simply does the opposite of Q (the bar over the Q indicates inverted behavior).

Note that 8 pin connections extend from the timer. The 555 timer is available in the 8-pin mini-DIP package as well as the 8-pin round TO–5 metal can. The pin numbers for either of these 8-pin packages are given in Figure A7–11. Moreover, a dual timer is available in the 14-pin DIP package as the 556 timer; this IC simply contains two 555 timers.

Let us now examine the function of the circuit within the 555 package as it is diagrammed in Figure A7–11. Notice that three 5-kilohm resistors are connected between $V+$ and ground, and the two comparators are connected at the 1/3 $V+$ and 2/3 $V+$ voltages along this 3-resistance divider. The *lower comparator* can set the flip-flop output Q to a high voltage; the *upper comparator* can reset the flip-flop output Q to ground. Voltage transitions to initiate these actions are shown in Figure A7–11 as arrows; thus the figure shows that a low-to-high arrow to the + input of the upper comparator, called the *threshold input* (pin 6), resets the RS flip-flop. Actually, this reset happens as soon as the voltage into the + input of the upper comparator exceeds the 2/3 $V+$ that enters the − input of this comparator. Similarly, when the voltage into the *trigger input* (pin 2) or − input of the lower comparator moves below 1/3 $V+$, the RS flip-flop is set to high and the output moves to $V+$.

A negative pulse on the reset input (pin 4) can pull Q low, also resetting the flip-flop and output to low. When the output is low, the $\overline{Q}$ output is high and the switch S connects the *discharge* (pin 7) to ground. Note the rule given in the figure that a low voltage opens switch S. When a 555 timer reset is desired, the voltage into the reset pin should move momentarily from near $V+$ to near ground; for example, momentarily shorting pin 4 to ground produces a reset.

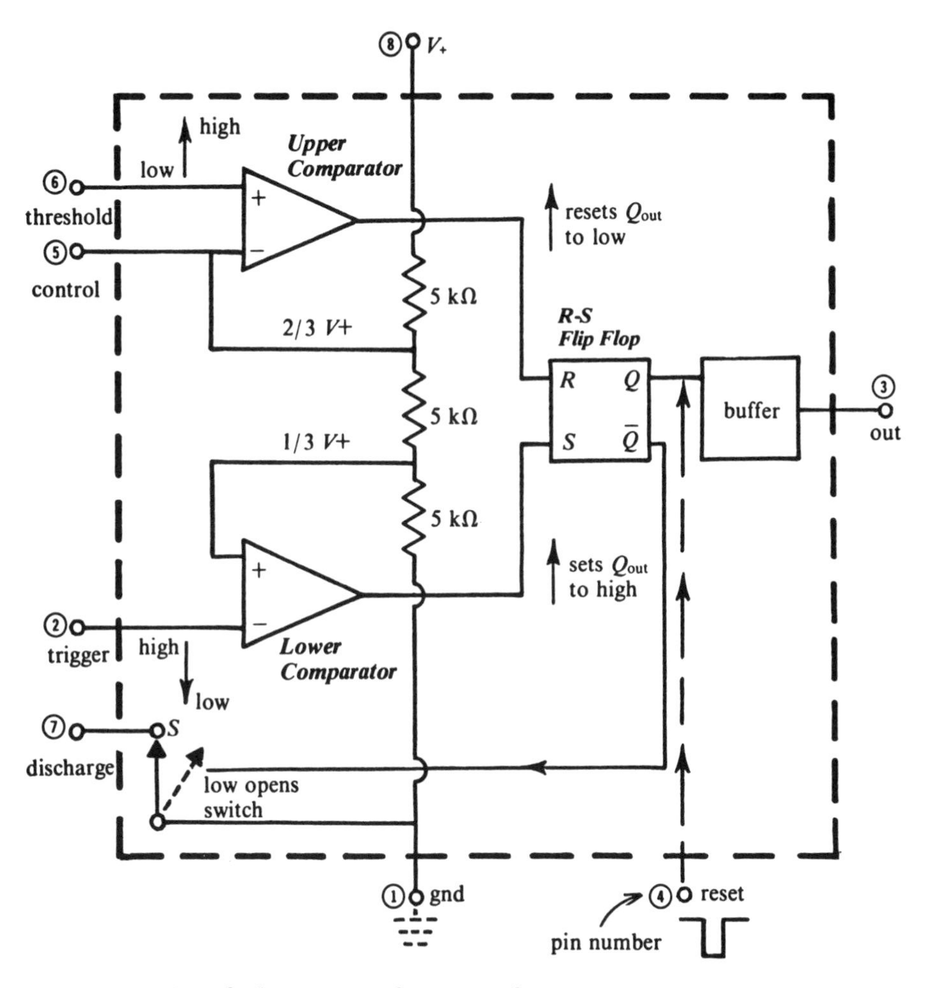

FIGURE A7–11. *Block Diagram of Interior of 555 Timer IC*

555 Multivibrators

In operation as a *monostable* MV, a capacitor *C* and resistor *R* are connected between ground and *V*+ (see Figure A7–12). The threshold of the upper comparator (pin 6) and the discharge (pin 7) are both connected to the junction between *R* and *C*. The resting state has the output low and discharge switch closed. The capacitor *C* is then shorted to ground through the switch. When an input pulse on the trigger input (pin 2) goes below 1/3 *V*+, the lower comparator goes high and sets the flip-flop. The switch *S* opens, and the 555

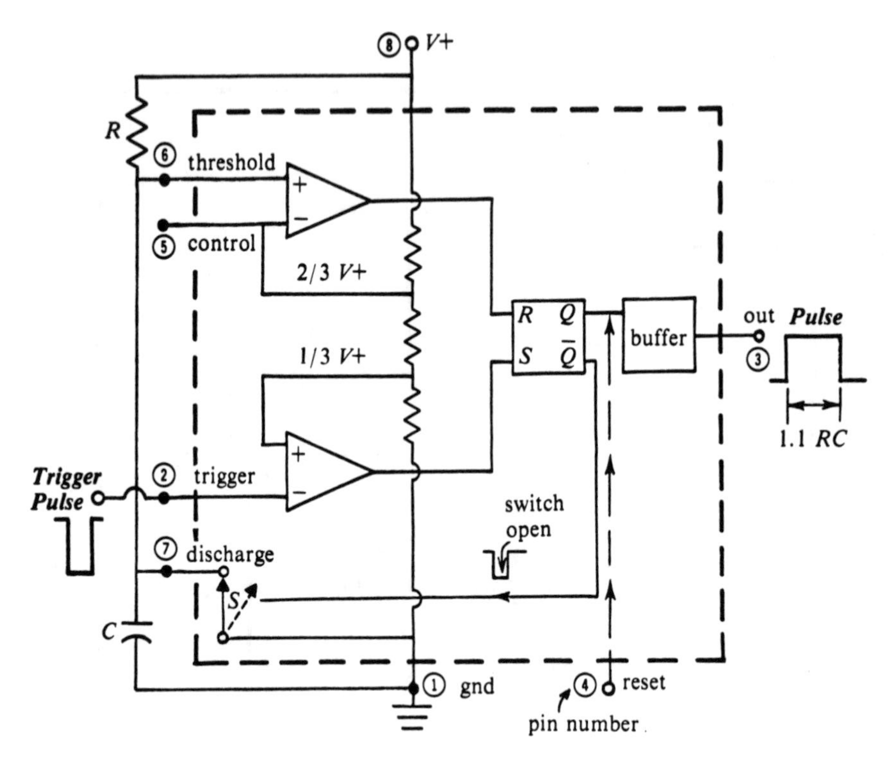

FIGURE A7–12. *Multivibrator Circuit with Monostable Connection*

out (pin 3) also goes high. Thus, the pulse begins. Current flows through R, capacitor C charges, and its voltage grows toward $V+$. However, when the voltage on C reaches $2/3\ V+$, the threshold lead to the upper comparator rises above the threshold level for this comparator and the comparator resets the flip-flop to low output. The switch S closes, discharging the capacitor, and the 555 timer output simultaneously goes low, thus ending the pulse. The timer is now back to its resting state.

The pulse duration is determined by the timer required for the capacitor C to charge through R to 2/3 of the supply voltage. Thus

$$V_C = (2/3)V+ = (V+)(1 - e^{-t_p/RC})$$

identifies the time duration of the pulse, t_p. Solving for t_p we have

$$t_p = RC \ln 3 = 1.1\ RC \qquad \textbf{(A7–2)}$$

Thus, the formula for the duration of the pulse in the monostable mode is:

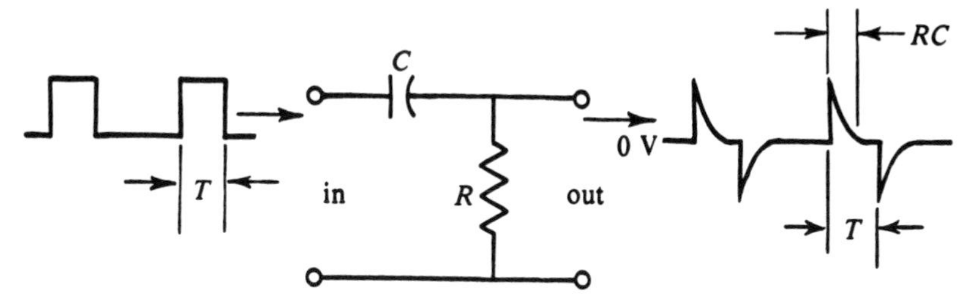

(a) Simple (Approximate) Differentiator (*RC* << *T*)

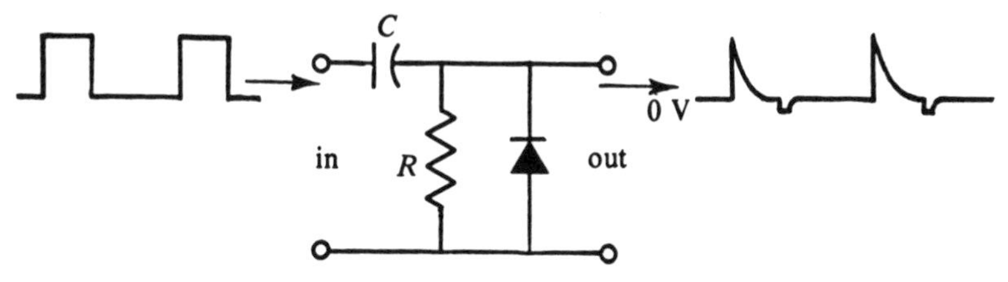

(b) Diode Placed to Clip Negative Spikes

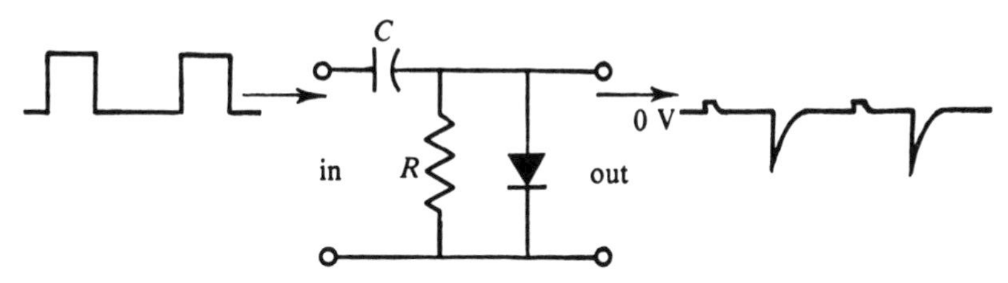

(c) Diode Placed to Clip Positive Spikes

FIGURE A7–13. *High-Pass Filter or Simple Differentiator Used to Produce Spike Pulses from Long Pulses*

Pulse length = 1.1 *RC* (*R* is in ohms and *C* is in farads for t_p in seconds). However, there is one unfortunate behavior characteristic of the 555 monostable: The input pulse must be shorter than the output pulse; otherwise the output stays high until the input returns high.

Especially if the trigger pulses are output from a Schmitt trigger, the pulses may be of variable duration and exceed the length of the monostable output pulse. The solution is to shorten up the trigger pulses to *spikes* with the assistance of the *simple differentiating circuit.* If we connect a capacitor in series with a resistor to form a high-pass filter (see Figure A7–13a), we can show that the output is approximately the derivative of the input voltage. In the illustration, the square wave first charges the capacitor in one sense, and the surge of current that flows produces a voltage spike across the output resistor. A time T later, the input voltage returns low and the capacitor discharges through R, producing a voltage spike in the opposite direction. Of course, the situation is simply an RC charging-discharging circuit. Exponential spikes of duration $\cong RC$ form across the resistance, and the sharpness is just determined by the RC time. Making either R or C small produces sharp pulses.

It may be that the negative spikes are not desirable or are even injurious to the electronic device that receives the spike pulses. They may be mostly clipped off (to about 0.6 V) by using diodes as shown in Figure A7–13b. Reverse the diode to clip in the other sense, as shown in Figure A7–13c. Actually, spikes going negative from $V+$ (or at least from some value above $1/3\ V+$) are required to trigger the 555 timer. The circuit of Figure A7–13c will do the job provided the cathode end of the diode/resistor combination is connected to $V+$. This raises the resting or quiescent level out of the differentiator to $V+$. Figure A7–14 shows such a circuit with values of R and C chosen so that the spike duration is less than 10 microseconds. This should accommodate nearly any situation since the manufacturer recommends that the minimum output pulse duration of the 555 timer be 10 microseconds. It is important to realize that the amplitude of the differentiator's output spikes will be the same as the amplitude of the pulses input to the

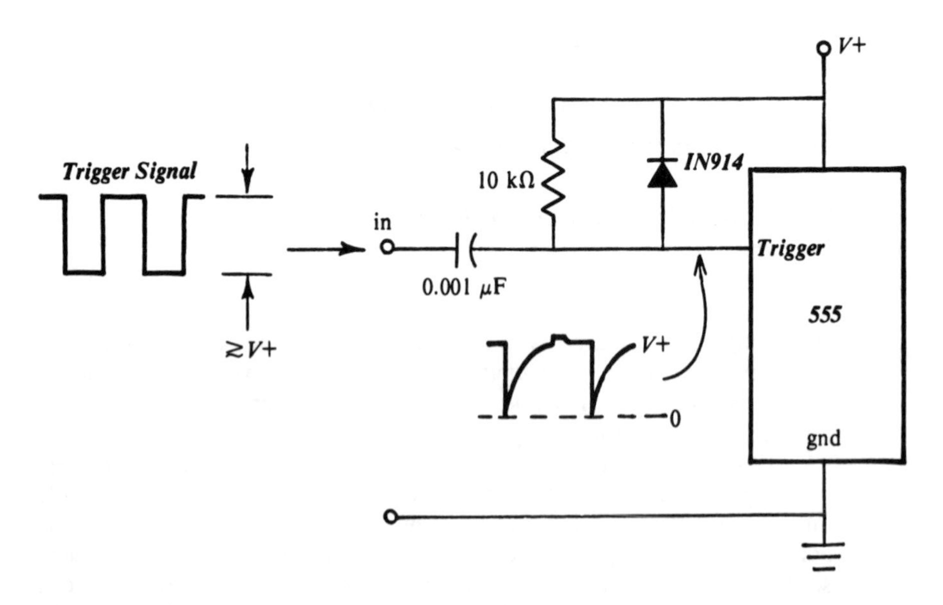

FIGURE A7–14. *Simple Differentiator Used to Produce Short Input Trigger Pulses to the 555 Timer*

differentiator. To achieve triggering, this means the pulse amplitude should equal $V+$, as shown in Figure A7–14.

The 555 *astable* multivibrator connection is shown in Figure A7–15. Here both the upper threshold and lower trigger comparator leads are connected to the "top" of capacitor C. The discharge switch connects to the junction between two resistors R_A and R_B, which are in series with capacitor C. Thus, if switch S is open, the capacitor charges through R_A and R_B toward $V+$. But if S is closed, the capacitor discharges through R_B toward ground.

Suppose S is closed (discharge shorted to ground). Then any charge on C will leak off through R_B to ground, with time constant R_BC. However, when the voltage across capacitor C gets down to $1/3\ V+$, the lower comparator sets the flip-flop; this causes the output to go high and also opens the discharge switch. Now capacitor C charges through R_A and R_B toward $V+$ with time constant $(R_A + R_B)C$. However, when the voltage on C reaches $2/3\ V+$, the upper comparator resets the flip-flop to low. Simultaneously the output goes low, the discharge switch closes so that C begins discharging again, and the cycle repeats. We have examined here the behavior of one cycle or period of the square wave.

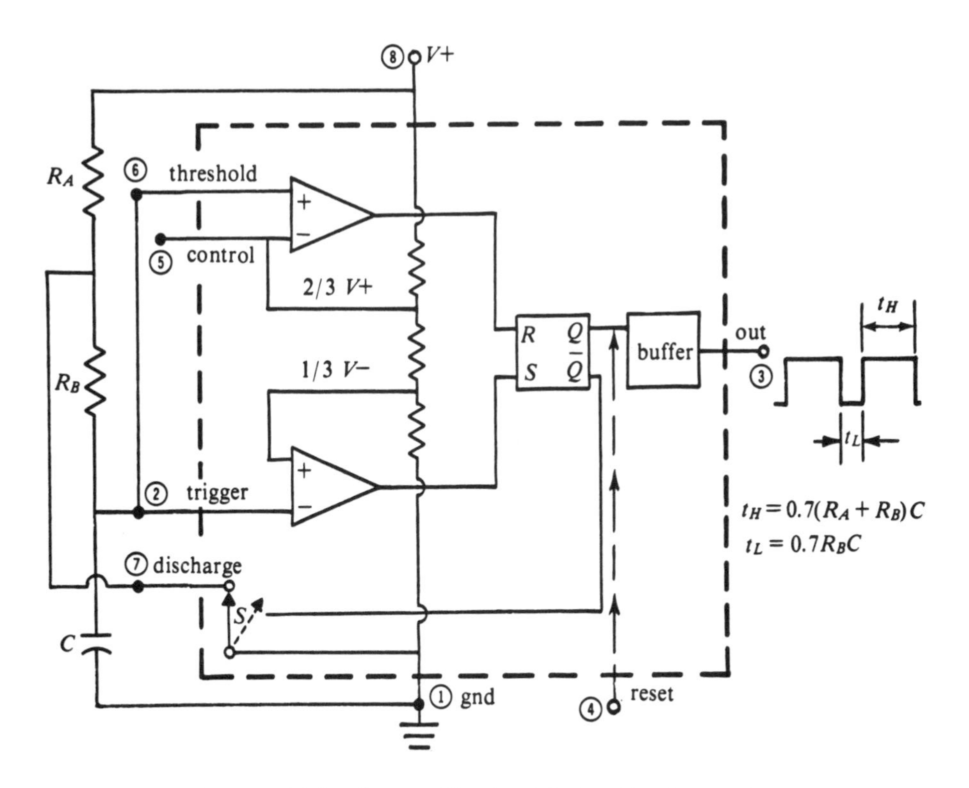

FIGURE A7–15. *555 Multivibrator Circuit with Astable Connection*

Note that since C charges through $(R_A + R_B)$ but discharges through R_B, the time constants for the output to be high and low are different. Thus, an asymmetric (in time) square wave is produced. The time for the output to be high t_H is given by

$$t_H = 0.7(R_A + R_B)C \quad \textbf{(A7–3)}$$

while the time to be low t_L is given by

$$t_L = 0.7R_BC \quad \textbf{(A7–4)}$$

The times t_H and t_L are illustrated at the output in Figure A7–15. The period of the square wave is given by

$$T = t_H + t_L = (1.4R_B + 0.7R_A)C = 1/f \quad \textbf{(A7–5)}$$

Manufacturers' specification sheets also show useful graphs of frequency versus resistance and capacitance that may be used to determine quickly the R and C required for a desired frequency.

It is possible to produce a *gated oscillator* or *tone-burst oscillator* from the 555 astable MV quite easily; that is, we can cause the oscillator to turn on for awhile and then off again. The 555 oscillator output will go low for as long as the reset (pin 4) is grounded—the oscillator itself is simply shut off. Thus, a positive pulse to the reset pin of the 555 astable MV will cause the output to turn on only for the duration of the pulse. Figure A7–16 illustrates the situation.

One application of a tone-burst oscillator is as a part of an audible monitoring system. For example, a heart-rate monitor can produce a pulse at each heartbeat; each pulse turns on or *gates* the 555 oscillator. If the 555 output is connected to a small loudspeaker, personnel in the area have an audible indication that the heart is functioning through the beep that is heard with each beat.

Changing the voltage of the control pin adjusts the upper threshold to values other than $2/3\ V+$; thus the time t_H of the square wave is changed or modulated if the 555 is operated in the astable connection. A *frequency-modulated square wave,* which is asymmetrical in time however, is generated in response to the voltage that is input to the control pin. For example, a siren sound can be generated. If the 555 is operated in monostable mode, the period of each pulse can be made changeable by changing a voltage input to the control pin; this is called *pulse width modulation.*

NOTE: If the control pin is not being used—as in the basic monostable and astable multivibrator usage—it is good practice to reduce noise problems by connecting a 0.01-microfarad capacitor between the control pin and ground.

The 555 timer can readily be used for a Schmitt trigger function. The threshold and the trigger pins may be connected together and used as the

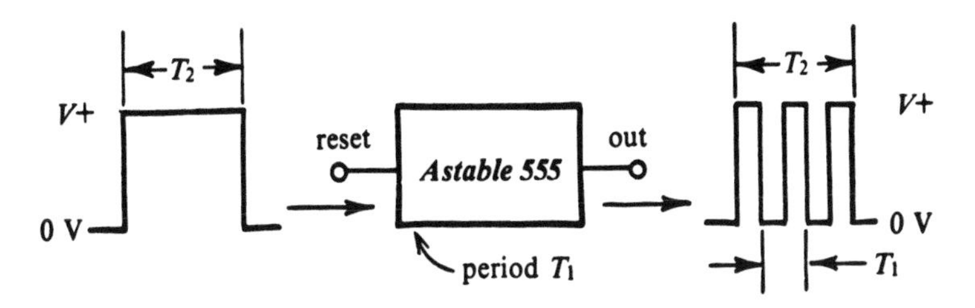

FIGURE A7–16. *555 Used as a Tone-Burst Oscillator*

device input; no additional resistors or capacitors need be connected, simply the power-supply voltage. Then the output of the timer moves high as soon as the input voltage exceeds 2/3 *V*+; the output moves low when the input falls below 1/3 *V*+. Notice that a Schmitt trigger with hysteresis width (for noise rejection) equal to 1/3 *V*+ is obtained. The two triggering levels can be adjusted—although not independently—by adjusting the voltage to the control input of the 555 timer.

For more information on these uses and many other 555 applications, the interested reader is referred to the books that are devoted to the 555; *IC Timer Cookbook* by W. Jung is a good representative.

Capabilities of the 555

The 555 can operate from supply voltage or *V*+ as low as 4.5 V and as high as 18 V. The output pulse height corresponds to the supply voltage used. The pulse width is quite stable against temperature variations: only 0.005% change per Celsius degree change. It is also stable against power-supply variations: only 0.1% change in pulse length per volt change in *V*+. The output is capable of supplying (as well as sinking) a substantial current: up to 200 milliamperes through an output impedance of about 20 ohms. This current is quite adequate to operate small light bulbs, LEDs, relays, and other small devices.

The shortest pulse width recommended for the 555 has already been mentioned: 10 microseconds. The longest pulse width is determined by the maximum desirable values for *R* and for *C*. Various small currents into the 555 limit *R* to about 10 megohms, and leakage current in capacitors of large size limits *C* to perhaps 10 microfarads, so that the maximum reliable pulse length is on the order of 100 seconds. If *R* and *C* are precisely known, inaccuracies in the 555 timer itself cause pulse duration precision of about 1 to 2%. Very long pulses are possible with various new types of IC timers. Thus, the LM322 has a wider range of pulse duration, while the Exar 2240 uses digital techniques to provide programmable pulse width of hours or days. The manufacturers' specification sheets should be consulted for limitations and recommendations for usage of these various devices.

Some Multivibrator Applications

Sequence Timing Control

Many situations in the laboratory as well as industry require that a sequence of operations or processes be controlled. Perhaps first a valve is to be opened for a certain interval to permit fluid to flow into a vessel. Then a mixing motor is to be operated for a certain time. After opening a drain valve for a time to permit emptying, the whole process is to be repeated some time later. The time-intervals for these processes should be individually adjustable, and a device to control the order and timing of the operations is called a *sequencer.*

Perhaps the simplest way to sequence processes in this manner is to use several monostable MVs in cascade. See Figure A7–17, where some "process A" and then "process B" are controlled. In this connection of monostable MVs, each stage quite literally becomes a timer, as its output is assumed to turn a process on during the time it is at the "high" voltage, and turn the process off while it is at the "low" voltage. As the output of a certain timer moves low to conclude a process, it triggers the next monostable timer into operation. This operation assumes negative pulse triggering as illustrated in Figure A7–7 (recall the 555 timer is triggered on a negative pulse). It is rather like a chain of dominoes with adjustable fall times. The entire sequence could be triggered by some initiating pulse (for example, from a push button) each time it is required. On the other hand it may automatically recycle, either by connecting the output from the last monostable MV to the input of the first, or by using the output from an astable MV to operate the first monostable. The latter technique is shown in Figure A7–17.

The various time intervals are set by adjustment of the *RC* time constant for corresponding monostable MVs, of course; this is illustrated by the adjustable resistors (potentiometers) in Figure A7–17a. The time interval of the overall recycling is controlled by the *RC* time constant(s) of the astable MV, and in fact operation similar to the 555 astable MV mode is assumed in the figure. Notice the *cause-effect* timing diagram of the three outputs in Figure A7–17b.

The output from a 555 monostable MV can control a process directly if it requires less than 200 milliamperes and less than 18 V_{dc}. However, in many cases the output will rather control an electromagnetic relay or a solid-state relay (see Chapters A3 and A8, respectively), which in turn switches the rather large ac current to a device such as an electric motor or a solenoid valve. Therefore, relays are parenthetically suggested as typical immediate destinations for the monostable MVs in Figure A7–17a. Note that if 555 timers are used for the monostable stages, a differentiating circuit such as shown in Figure A7–14 is required at each trigger input. Otherwise the low period from the previous monostable MV prevents the next monostable MV from completing its pulse.

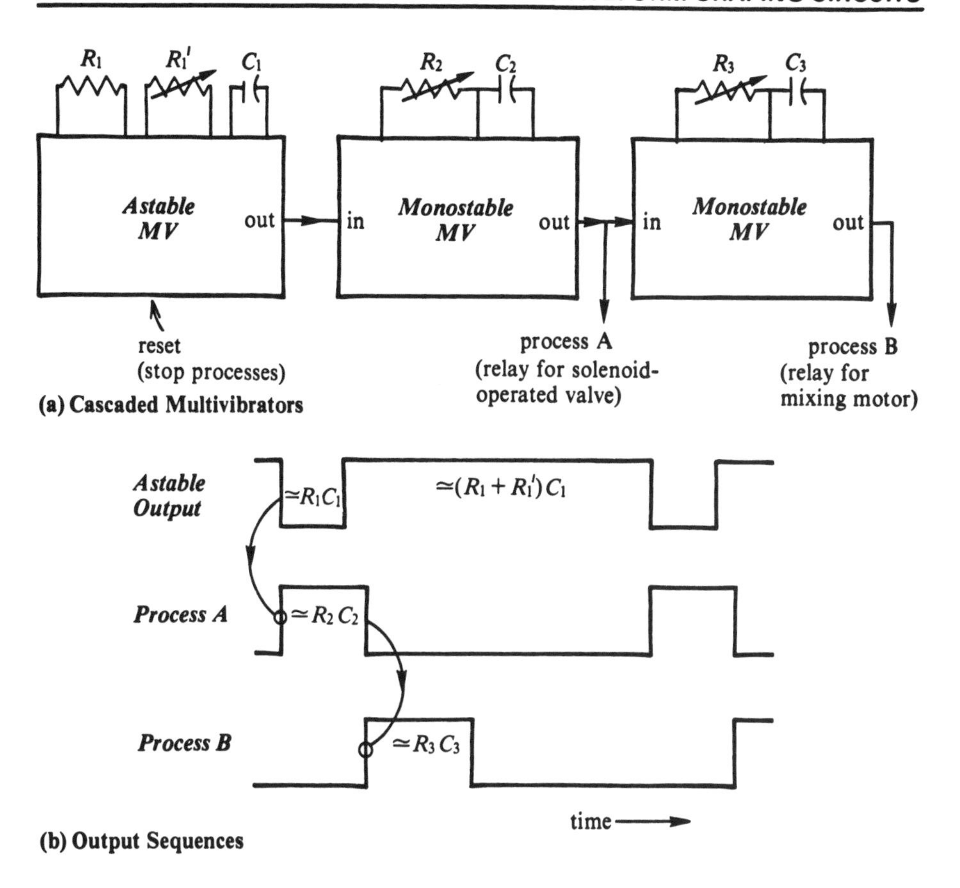

FIGURE A7–17. *Multivibrators Used to Obtain a Self-Cycling Sequencer*

Frequency-to-Voltage Converter or Ratemeter

There are a number of situations where a scientist or engineer wants to measure the frequency or rate at which something is occurring. We might want to monitor the heart rate of an animal, or the level of radioactivity, or the speed of a turning wheel. It is true that there are *digital* frequency meters that permit very precise measurement of the frequency (or the related period) of a repetitive electrical signal. However, there are circumstances where a voltage or *analog* indication of a rate is desired. We may want a quick visual indication of the rate through use of the needle of a meter, especially if it is somewhat rapidly changing; or we may want to monitor the rate over a period of time by recording it on a strip-chart recorder (these recorders typically accept a voltage as input). Also, this analog frequency meter that we are about to describe can be quite inexpensive and simple to construct and use.

The technique we describe here requires that each event in the repetitive process to be measured causes a voltage pulse of a definite length or duration and amplitude. Note that this suggests utilization of a monostable MV as a pulse-shaping device. As a specific example, assume a heartbeat ratemeter. Each beat could produce an electrical pulse of 0.25 seconds duration. If a pulse stream of this character is input to a *very* low-pass filter, the dc voltage out of the filter is proportional to the frequency of the pulses. This results from the fact that the voltage out of the low-pass filter is the "dc part" (or time average) of the input pulse signal. A similar behavior was presented for the *L*-section low-pass filter in Figure A4–17.

We will justify this output-voltage versus input-frequency relationship. If a voltage $v(t)$ is varying in an arbitrary way during a time T, the average (or dc part) of the voltage during that time is just the *area* under the corresponding $v(t)$ curve, divided by the time T (see Figure A7–18a). Mathematically,

$$v_{ave} = T^{-1}\int_0^T v(t)dt$$

To calculate the average voltage of our train of pulses, let each pulse be of amplitude V_p and time duration T_d (see Figure A7–18b). Let the frequency of the peaks be f; then there are f pulses per second. In a time T there will be fT, each with a v area of V_pT_d. The total area under the pulse curve in time T is the number of pulses multiplied by the area of each, or fTV_pT_d. To calculate the average voltage v_{ave} during time T, we finally divide by T and obtain $v_{ave} = fV_pT_d$.

For example, if the pulse amplitude is 5 V and pulse duration is 0.25 seconds, we have $v_{ave} = 1.25 f\,\text{Hz}^{-1}$ V. Then if the pulse frequency is 1 hertz, its average voltage is 1.25 V, while if the pulse frequency is 2 hertz, its average voltage is 2.5 V. This proportionality holds up to a maximum frequency of 4 Hz, for which the pulses abut each other, and the average voltage is clearly just the pulse amplitude voltage itself.

Figure A7–18c illustrates this example pulse-form input to a low-pass filter that is followed by a buffer amplifier and analog voltmeter. The time constant of the *RC* filter should be selected to be about 6 times longer than the *period* of the *lowest* frequency expected. Thus, the filter illustrated in Figure A7–18c will produce a reasonably well-filtered dc output that is proportional to frequency from about 0.3 hertz (or a 3-second period) to the upper limit of 4 hertz. The filter output is buffered with a voltage follower to isolate the high impedances of the filter from disturbance by the voltmeter load. Finally, a potentiometer is used as an amplitude adjustment. It permits adjustment of the meter to a convenient scale such as one volt per hertz. This adjustment is generally done during an initial calibration of the meter using an input pulse stream of known frequency from a function generator.

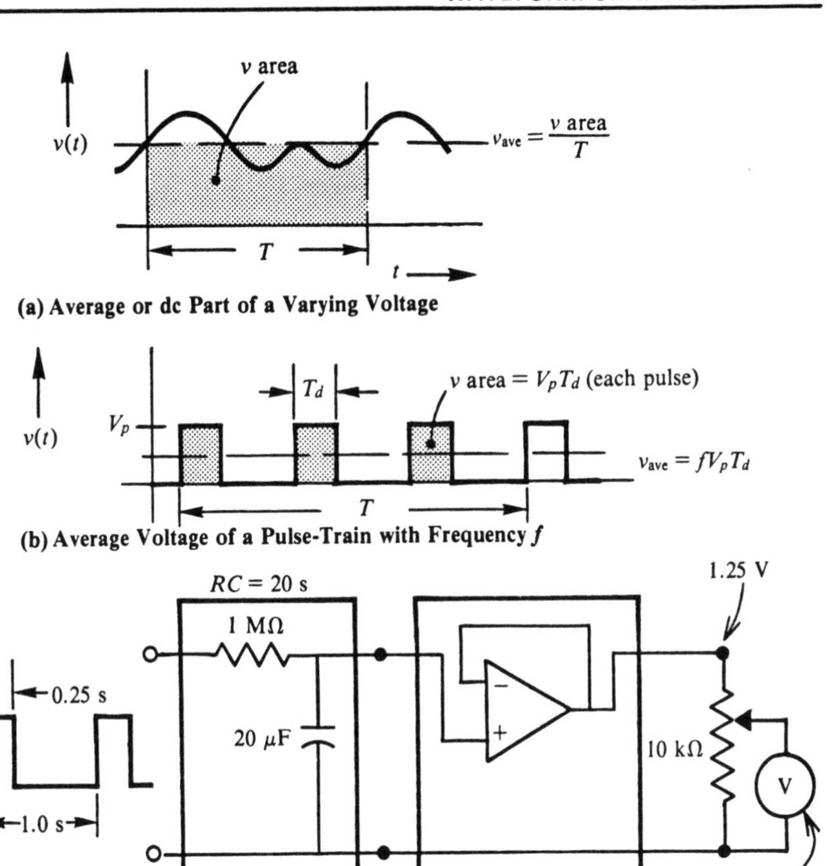

FIGURE A7–18. *Basic Functions in an Analog Frequency Meter or Ratemeter*

General Ratemeter Instrument

This example illustrates the application of many of the circuit types and principles, discussed in this and earlier chapters, to obtain a functional instrument. A block diagram of the example ratemeter instrument is presented in Figure A7–19, together with a representation of the signal as it passes through the instrument stages and is processed.

Block A of the instrument is the transducer, which senses the repetitive aspect of the experimental object. For example, it may be a photoresistor that senses darkening due to the spokes of a wheel as they pass or of the finger-

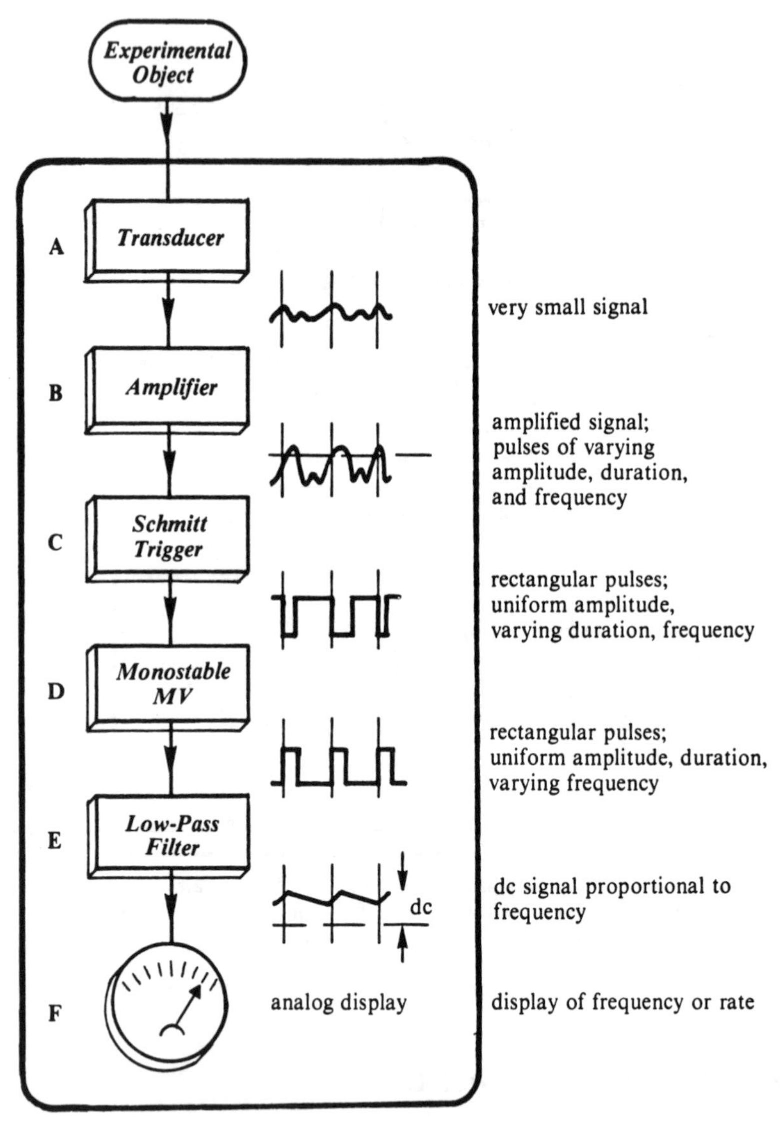

FIGURE A7–19. *A Ratemeter Instrument Constructed from Signal Processing Blocks*

tip each time blood moves into the capillaries. The photoresistor may be in a voltage divider circuit with capacitive coupling at the output to block the dc level but pass the voltage pulses. Note there is actually an ac signal from the repetitive process.

The ac signal from the transducer is generally quite small and must be processed next by an amplifier stage with voltage gain. The amplifier is shown as block B in the figure. However, the output from the amplifier will typically be pulses which vary somewhat in shape, amplitude, and duration. Also low-level noise pulses may be present in the signal. Therefore, the next signal processing step is block C—a Schmitt trigger that is set to discriminate against small spurious pulses and noise and produces rectangular output pulses of definite amplitude.

However, the Schmitt trigger output pulses may vary in duration, due to variations in the input pulse shape and its relationship to the threshold. Recall that the low-pass filter rate meter requires pulses of uniform duration as well as amplitude. Therefore, the Schmitt trigger output is presented to block D, a monostable multivibrator stage, which outputs a single uniform pulse for each input pulse.

Next the pulse train, which is now uniform in all respects but frequency, is presented to the low-pass filter in block E. The slowly varying dc voltage output from this stage is finally presented to an analog display such as a d'Arsonval meter. This meter may be marked with a convenient scale such as heartbeats per minute. Alternatively, a strip-chart recorder may be connected to obtain a convenient permanent record of the frequency versus time for many minutes, hours, or even days.

REVIEW EXERCISES

1. (a) A comparator box has 3 input/output connections. Tell what each is for.
   (b) Give another name for comparator.
2. A comparator constructed from an op amp uses which of the following: A. negative feedback, B. no feedback, C. positive feedback.
3. Suppose an op amp is used as a comparator and operates from power-supply voltages of ±9 V. The comparator output is taken directly from the op-amp output. Someone asserts that a DVM connected to the comparator output will read any of a range of voltages between −9 V and +9 V while the comparator is being used. Criticize the statement.
4. (a) Consider the comparator circuit shown in Diagram A7-1. The output voltage will be close to (choose one): A. −12 V, B. −6 V, C. 0 V, D. +6 V, E. +12 V.
   (b) Which diode is forward biased and conducting: A. D_1, B. D_2, C. both, D. neither.
   (c) Repeat question (a) if the 1-V battery is reversed.
   (d) Repeat question (b) if the 1-V battery is reversed.

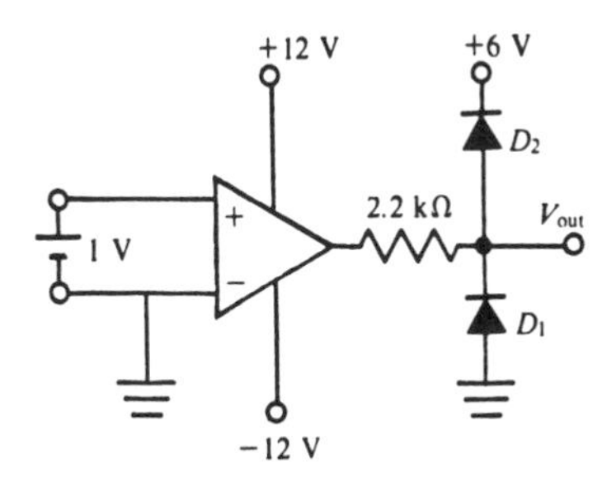

DIAGRAM A7–1.

5. What is the purpose of each diode in the above comparator circuit?
6. (a) What is the important difference between a comparator and a Schmitt trigger in terms of their behavior? What is the characteristic term used for the Schmitt-trigger behavior?
   (b) What feature of input signals often dictates use of a Schmitt trigger rather than a comparator?
7. A Schmitt trigger constructed from an op amp uses which of the following: A. negative feedback, B. no feedback, C. positive feedback.
8. The purpose of the voltage divider in the feedback circuit of the op-amp-based Schmitt trigger is to (choose one of the following): A. reduce the input signal before entering the op-amp input, B. clip the output signal at the ground, C. protect the op-amp output circuitry, D. produce two thresholds to the op-amp comparator.
9. A Schmitt trigger constructed from a single op amp has an inverting character. Explain this statement.
10. Consider the Schmitt-trigger circuit pictured in Diagram A7–2.
    (a) Which resistors are primarily responsible for the hysteresis of the trigger circuit?
    (b) Which resistors are primarily responsible for the average triggering level?

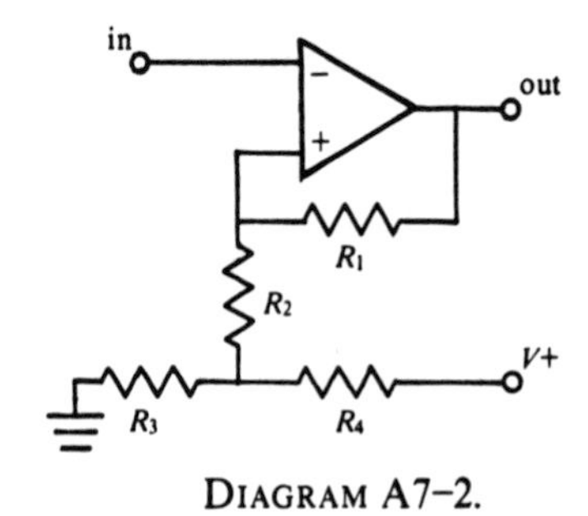

DIAGRAM A7–2.

11. Select a multivibrator type for each of the following tasks.
    (a) A pulse-output oscillator is needed.
    (b) A certain long-duration pulse out is required for each short-duration pulse in.
    (c) The frequency of a pulse train should be divided by 4 to obtain a lower-frequency pulse train.
12. Which of the following is *not* inside a 555 timer integrated circuit? A. comparator, B. resistor-capacitor combination, C. flip-flop, D. transistor switch.
13. What are the two important multivibrator types that may readily be obtained using a 555 timer IC?
14. A 555 timer is used with a 1-kΩ resistor and a 1-μF capacitor to make a one-shot. What is the expected duration of the output pulse?
15. A 555 timer may be used with a capacitor and two identical resistors to make which type of oscillator? A. sine wave, B. asymmetrical square wave in voltage and time, C. symmetrical square wave in voltage and time.
16. What are the two important threshold voltage levels inside the 555 timer IC (in terms of the supply voltage used)?
17. Why is a simple *RC* differentiating circuit sometimes required when using the 555 timer?

18. (a) Draw a simple *RC* differentiating circuit.
    (b) Describe in qualitative terms why two short pulses occur at the *RC*-circuit output in response to a single long pulse input.
    (c) What determines the duration of these short pulses?
19. (a) What is a sequencer?
    (b) Describe how a sequencer may readily be constructed from multivibrators.
20. Suppose the signal pictured in Diagram A7–3 is input to a good low-pass filter. What is the dc voltage expected from the filter?

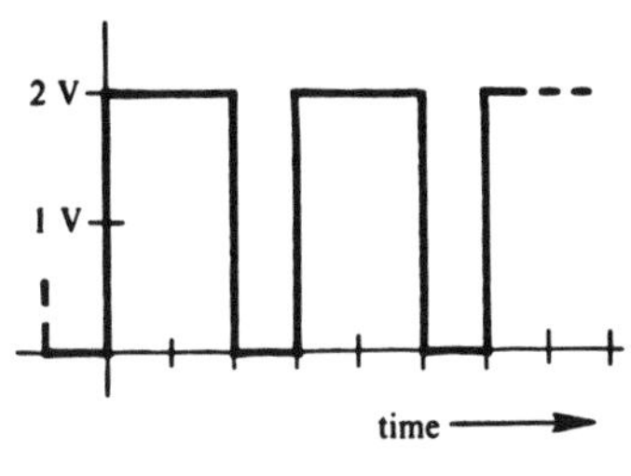

DIAGRAM A7–3.

21. (a) What is the difference between the output of the frequency information from a ratemeter as compared to a digital frequency counter?
    (b) Why is a Schmitt trigger often required in a ratemeter instrument?
    (c) Why is a monostable MV almost always required in a ratemeter instrument?

PROBLEMS

1. An op amp is operated from ±6-V supplies to form a comparator circuit as shown in Diagram A7–4. Give the output voltage for the following input voltage V_{in}:

   (a) −2 V (d) +1 V
   (b) −1 V (e) +2 V
   (c) 0 V

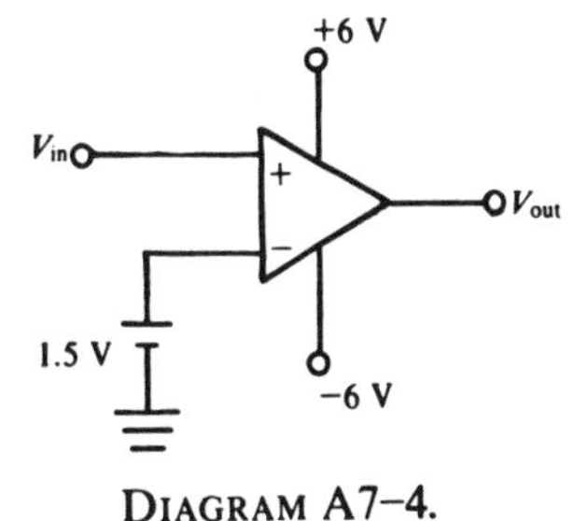

DIAGRAM A7–4.

2. The op amp in the circuit shown in Diagram A7–5 is to serve as a voltage level detector relative to a reference level of −9 V. Describe why the circuit will *not* work. Suggest a change so that it may function properly.

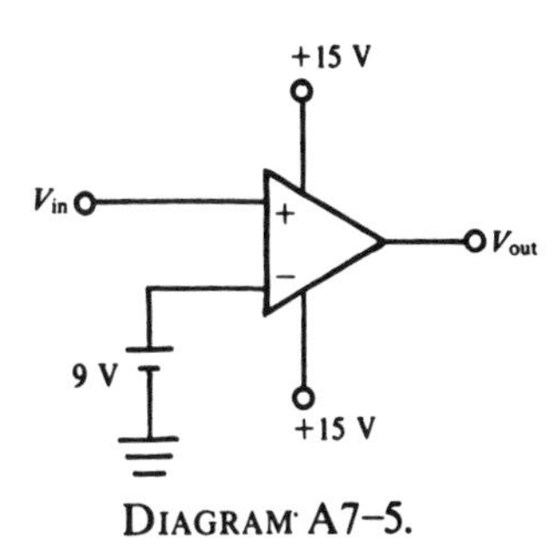

DIAGRAM A7–5.

3. Design a comparator that produces a high voltage out equal to +6 V when the input voltage is greater than +3 V and a low voltage out of −6 V when the input voltage is less than +3 V. The comparator is to be constructed from an op amp that is operated from ±12 V power-supply voltages. Two designs may be developed.
   (a) Do a rather simple design that uses only diodes and appropriate batteries along with the op amp and ±12-V power supply.
   NOTE: Assume self-current-limiting by the op amp.
   (b) Diagram a design that requires no batteries; diode and resistors are to be used (along with the op amp and ±12-V power supply). Arrange to have about 20-mA current drain from the ±12-V supplies and to have the op-amp output current limited to 2 mA.
4. Suppose we want a comparator that switches its output from 0 V to +5 V in 0.5 μsec. Discuss why the 741 op amp could *not* do the job. Suggest an op amp that might and justify your choice.
5. Consider the Schmitt trigger shown in Diagram A7–6. Assume the output switches between +10 and −10 V. The reference voltage is +3 V as shown. Find the two switching levels and also evaluate the hysteresis width for the unit. Show that it agrees with the formula in the text. Is the center of the two levels at 3 V?

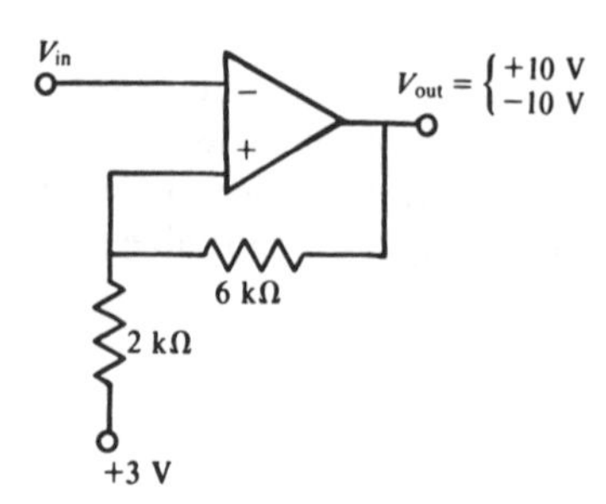

DIAGRAM A7–6.

6. Block-diagram a circuit that performs as a square-wave oscillator. It should have a precisely symmetrical square-wave output in terms of time, with some frequency *f*. Use two devices from the multivibrator family.
7. Assume that in response to a certain negative input pulse, a process *A* is to be turned on for one of three possible time periods: 1 second, 2 seconds, or 3 seconds. These time periods are to be switch selectable by an operator. At the conclusion of process *A*, there is to be a 5-second delay, after which process *B* is turned on for 10 seconds. Diagram a design that assumes a block model and timing relationships such as those shown in Figure A7–7 for the monostable MVs used.

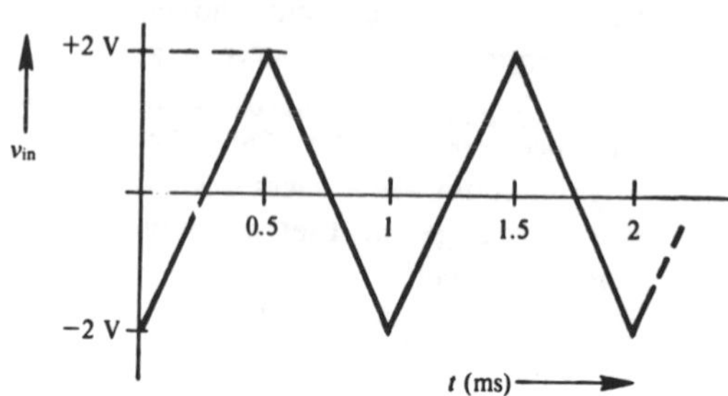

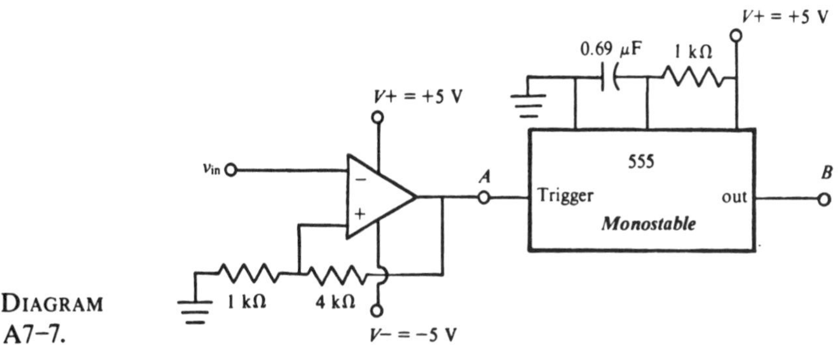

DIAGRAM A7–7.

8. (a) A symmetrical triangle-wave voltage that is 4 V_{p-p} with 1-kHz frequency is illustrated in Diagram A7–7. Suppose this voltage is input as v_{in} to the circuit shown. Draw with some accuracy the time graphs of the voltages v_A and v_B appearing at points A and B in the circuit.
   (b) What problem (perhaps unanticipated) occurs with the signal at B if the capacitor for the 555 timer is made less than about 0.25 μF in this application?
9. We have a design problem where pressing push button PB1 should cause a 6-V_{dc} motor to begin running. Subsequently, pressing push button PB2 causes the motor to stop. A design based on the 555 timer IC is shown in Diagram A7–8.

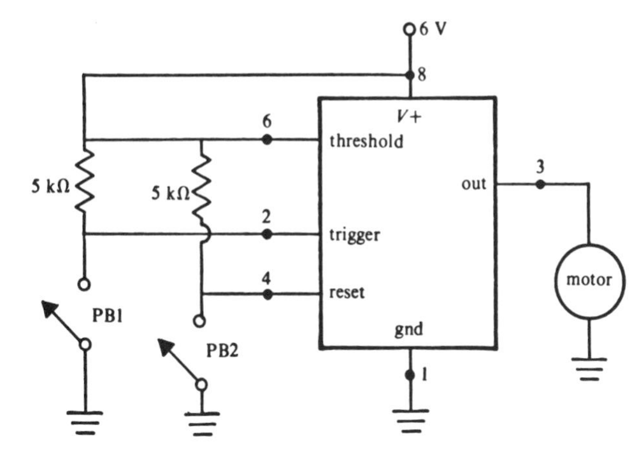

DIAGRAM A7–8.

   (a) Justify that a negative pulse to ground is input to the particular 555 input when a push button is pressed. (The trigger and reset inputs have high resistance.) Which push-button type should be used, NO or NC?
   (b) Draw the circuit, filling in the interior model for the 555 timer. Justify that the circuit will perform the desired function. What current limitation is placed on the motor?

(c) Suppose the threshold input is to be used rather than the reset (threshold and trigger have very high input impedance, similar to an op amp). Illustrate and justify briefly the push-button circuit then required.

10. We want a small lamp in a rat's cage to turn on in response to the rat pressing a sensitive snap-action switch. The light is to turn on for 2 seconds. Assume that we conveniently have at hand lamps that operate from 6 V at 100 mA. Develop a design using a 555 timer.

11. Assume that a certain timing sequence is required for three processes *A, B,* and *C.* When a push button is pressed, process *A* is turned on for 2 seconds; process *B* is also immediately turned on, but for 5 seconds. At the termination of process *B,* process *C* is turned on for 20 seconds. Use 555 timers, and a 1-MΩ timing resistor in each case. (Low leakage tantalum electrolytic capacitors are assumed.) It is not necessary to draw in the interior functions in the 555 timers, but connections to their input/output pins should be shown and the pins labeled. Note that one differentiating circuit is required. Explain why.

12. A 555 timer is used with a 1-μF capacitor and two 1-kΩ resistors to make an oscillator. The power supply voltage is 6 V.
    (a) What is the frequency of the output signal?
    (b) Make a moderately accurate graph of the output signal as a function of time.

13. Research demonstrates that rats can generate and sense ultrasonic sounds. We want a 30-kHz tone of 2-second duration to be produced near a rat's cage in response to the rat pressing a sensitive snap-action switch. Assume an ultrasonic sound transducer that operates from a signal level of 6 $V_{p\text{-}p}$ and has an input impedance of 200 Ω. Develop a design using 555 timers.

14. The *duty cycle* of an asymmetrical square wave (in time) is usually defined as the ratio of the time that the signal is high to the overall period of the wave.
    (a) Show that for the 555 timer, the duty cycle is given by $(R_A + R_B)/(R_A + 2R_B)$.
    (b) Argue that the duty cycle may be adjusted between 50% and 100%. Characterize the ratio R_A/R_B for these two extremes, respectively.

15. Consider the 555 timer operated in astable mode.
    (a) Justify the statement (actually somewhat careless) that reducing R_A to zero causes the output to be a symmetrical square wave in terms of time (50% duty cycle).
    (b) Demonstrate that the statement is careless because, if $R_A = 0$, the discharge switch (a transistor) within the 555 would be destroyed. What would destroy it?

    NOTE: The minimum value for R_A shown on the manufacturer's graph is 1-kΩ.

16. (a) The bias current into the threshold input of the 555 timer is specified as typically 0.1 μA (with a maximum of 0.25 μA). Assume that a typical 555 device is used in conjunction with a 6-V supply to make a long-pulse monostable. Show that if the timing resistor *R* exceeds 20 MΩ, the pulse cannot terminate.

NOTE: Thus the bias current limits the use of very large timing resistors. The LM322 monostable typically has a 0.3-nA bias current.

(b) One electrolytic capacitor family is specified to have a minimum leakage resistance given by $R = 50/C$ ΩF, where C is in farads. This resistance appears in parallel with the ideal capacitor. For example, if $C = 10$ μF then $R = 5$ MΩ. Note that large capacity implies smaller resistance and more leakage current. Assume that a 1-MΩ timing resistor is used with a 555 timer as a monostable. Show that if the timing capacitor C exceeds 25 μF for this type of electrolytic family, the pulse cannot terminate.

NOTE: Selected standard electrolytic capacitors can have a factor-of-10-lower leakage current and therefore higher leakage resistance than this example; selected tantalum electrolytics can be a factor of 100 better.

17. Rectangular pulses with 12-V amplitude and 2-msec duration are input to a low-pass filter that is part of a ratemeter.
    (a) What is the maximum operating frequency for this ratemeter, and what is the voltage out of the filter at this frequency?
    (b) If the frequency of the pulses is 50 Hz, what is the dc voltage out of the filter?

18. (a) What is the disadvantage of making the time constant of the low-pass filter in a ratemeter very long; for example, 10 minutes?
    (b) What problem exists if the time constant of the filter is made short; for example, equal to the period of the incoming pulse train?

CHAPTER A8

Discrete Electronic Devices

We now begin the study of *discrete active devices;* that is, basic single electronic elements such as a vacuum triode, bipolar transistor, or field-effect transistor that can be used in a circuit to increase the power level of a voltage signal. Note that the additional power does not come from the device itself, but rather from the power supply.

The low price of IC operational amplifiers has revolutionized much of analog electronics, almost to the point of wondering why bother with transistors or tubes (discrete devices) at all? We do in fact scale down their consideration greatly in this book; a brief overview is given, however, to better understand the limitations and requirements of integrated circuits, as well as to permit some understanding of older circuitry when it is encountered. Also, in some situations a transistor is quite simple to apply and is all that is required. For example, when simple current amplification is required, the bipolar transistor is a convenient device. If electronic switching of rather high current or high voltage is required, transistors—or in ac circuits, the SCR perhaps aided by a transistor—are very useful devices. This book develops a proficiency in transistor circuits to this level of application; it does *not* consider the broad field of transistor amplifier design.

General Amplifier Model

General Electronic Current-Controlling Element

The various electronic devices that are used in amplifier circuits to cause the circuits to do their "magic" of increasing power level all operate on one principle: They control the current passing through them in response to a low-power input voltage/current. A simple analogy is the water flowing through a garden hose (see Figure A8–1). By squeezing on the sides of the hose, we can control the stream of water through the hose. If we use pliers or some clamp with mechanical advantage, it is possible to control a powerful stream of water passing through the hose. The water pressure comes from an external pump, of course.

A vacuum tube or transistor behaves like such an element; it controls the current that a battery or power supply attempts to force through it. In fact, the British use the descriptive term *valve* rather than vacuum tube. Only a small voltage or current input (like the pliers force) is needed to effect the control.

Figure A8–2 illustrates a general active or current-controlling device with a control input lead to which voltage V_{in} is applied and two main current-carrying wires that carry the current to be controlled I_T. (T may represent transistor or tube here.) Note that three basic connections are involved. The most important parameter characterizing an active device is its *mutual transconductance* g_m. The mutual transconductance is simply the ratio of the *current change produced* to the *input voltage change* that caused it. Let ΔI_T be this change in device current, while ΔV_{in} is this change in control input voltage V_{in}. We may write

$$g_m = \frac{\Delta I_T}{\Delta V_{in}} \tag{A8–1}$$

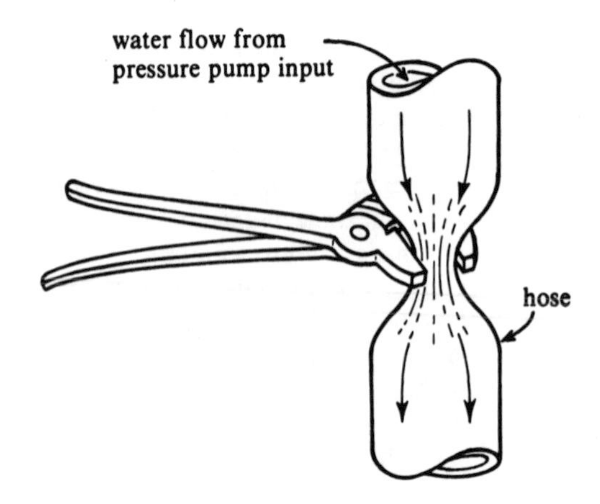

FIGURE A8–1. *Analogy of an Active Electronic Device*

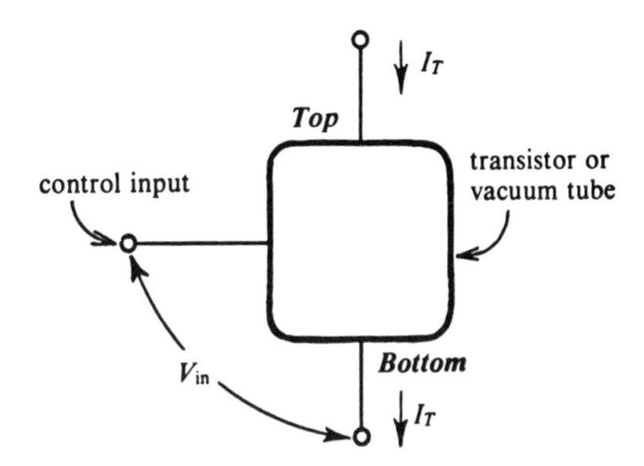

FIGURE A8–2. *General Electronic Active Device*

Notice that the dimensions of this quantity are amperes/volts = 1/ohms, or reciprocal ohms. The reciprocal of resistance is called *conductance,* and the unit for conductance has long been called the *mho.* Transconductance for transistors and tubes is typically given in millimhos or micromhos.

General Discrete Amplifiers

An amplifier is constructed by use of a resistor and power supply as shown in Figure A8–3. The dc voltage supply causes a current I_T to flow through R_L, which is in series with the transistor (or tube). The *bias voltage* V_b at (or current into) the control lead establishes the value of this current. A voltage then develops across R_L and also across the transistor. V_b is usually chosen so that about half the supply voltage appears across R_L and half across the transistor; this is the *steady or operating situation.*

When an ac voltage is applied to the input (shown as v_{in} in Figure A8–3), it passes through the *coupling capacitor* C_{in} to the control lead for the transistor; consequently, the transistor permits more or less current to flow through it. A variation in I_T—that is, ΔI_T—occurs that is proportional to v_{in}. From Equation A8–1 the variation is given by

$$\Delta I_T = g_m v_{in} \qquad \textbf{(A8–2)}$$

This varying current through R_L causes a variation in the voltage across R_L in the amount

$$\Delta V_{R_L} = \Delta I_T R_L = g_m v_{in} R_L \qquad \textbf{(A8–3)}$$

But the voltage at the top of R_L is fixed at the supply voltage, V_{supply} (relative to ground). Therefore, the voltage at point T (bottom end of R_L and top of transistor/tube) must *vary* by the amount given in Equation A8–3. We see shortly that this voltage becomes the ac voltage out of the amplifier.

A graph of V_T relative to ground (the voltage across the transistor/tube) and of v_{in} are shown in Figure A8–4. Notice that there is an ac voltage

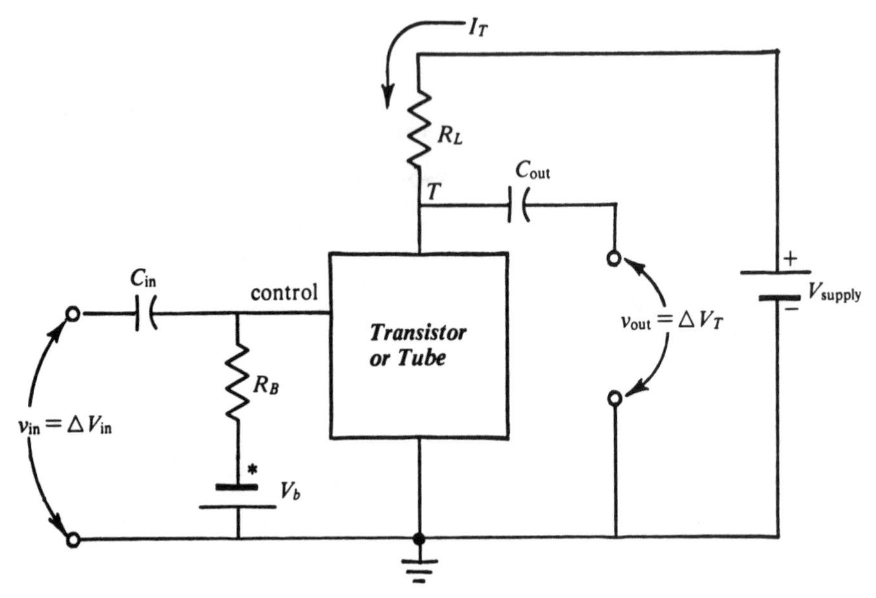

*(+) if bipolar transistor; (−) if junction FET or tube

FIGURE A8–3. *General Discrete Component Amplifier*

superposed or added to a large dc voltage. Hopefully, the ac voltage is a larger replica of the input voltage. A *blocking capacitor* C_{out} is connected between point T and the amplifier output as shown in Figure A8–3. This blocks the dc component of the voltage at T, but passes the ac to the output. The amount of the output voltage (peak-peak) is noted in Figure A8–4.

Note that V_T swings *down* while v_{in} swings *up*. This is because in the typical configuration of the tube or transistor, the current through the transistor/tube (as well as R_L) increases when $V_{control}$ increases. Then the increasing current through R_L causes an increasing voltage across the resistor R_L and a decreasing voltage across the transistor. Consequently, there is a *180° phase reversal* between the input and output signals.

We have the ac voltage out of the amplifier from Equation A8–3. It is

$$v_{out} = -(g_m R_L) v_{in}$$

In this relation for v_{out}, a minus sign has been inserted to take into account the 180° phase reversal or inversion of the output voltage relative to the input voltage. Then

$$A_v = \frac{v_{out}}{v_{in}} = -g_m R_L \qquad \textbf{(A8–4)}$$

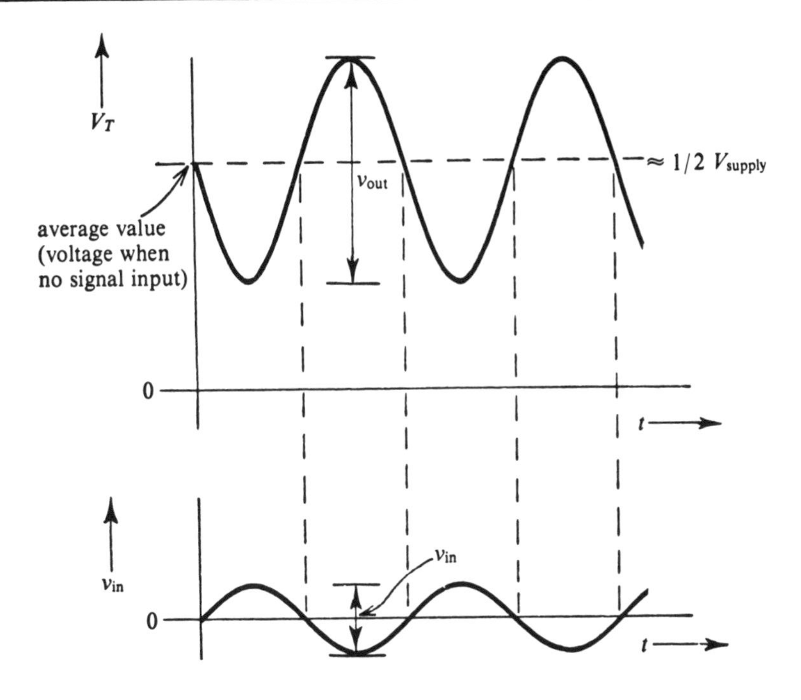

FIGURE A8–4. *Graph of Active Device Voltage and Input Voltage for a Typical Amplifier*

The ratio of the output voltage to the input voltage of an amplifier is called the *voltage amplification* or the *voltage gain* A_v of the amplifier, as discussed in general terms in Chapter A5. We therefore have the following general rule: *The amplification of a basic amplifier is simply the transconductance of the device times the load resistance used.*

Because of various small effects that are present in real discrete devices, this rule is only a good approximation to the voltage gain. Nevertheless, it is quite sufficient for many scientific purposes and certainly for the purposes of this book. The rule applies equally well to the operation of an important class of amplifier constructed from one of the various types of field-effect transistors, a bipolar transistor, or a pentode tube; the rule holds less accurately if a triode tube is used as the active device. The general mode of operation for these various active devices is presented in the following sections. Specific basic amplifier designs are also briefly illustrated and discussed.

We know from Chapter A5 that the input and output impedances of an amplifier are very important amplifier properties, and they will be discussed here. The input impedance of the basic amplifier shown in Figure A8–3 depends largely upon whether the active device is a field-effect transistor, bipolar transistor, or vacuum tube; therefore, this property must await specific device discussions. However, the output impedance of the basic amplifier pictured in Figure A8–3 is largely independent of the active device used, for devices other than the triode tube. We have the rule: *The output impedance of the basic amplifier type is essentially equal to the value of the resistor* R_L *used in the amplifier.* This rule shows that there is a conflict between the desires for high voltage gain and for low output impedance from an amplifier, because both of these amplifier properties are proportional to R_L.

The rule that the output impedance of the basic amplifier is simply R_L can be seen from the following argument. When a *true load* $R_{L'}$ is connected to the amplifier output, it effectively appears in parallel with the amplifier resistor R_L. The parallel character may be seen by imagining $R_{L'}$ to be connected to the output terminals in Figure A8–3 and using the fact that the coupling capacitor and the supply voltage source should have negligible impedance for the ac currents that are involved; that is, C_{out} and V_{supply} may be replaced by wires for the analysis of ac signals. However, then the effective parallel resistance $R_L \| R_{L'}$ must be used in place of the value R_L in the voltage-gain formula $A_v = g_m R_L$. Consequently, the voltage that appears across the load is proportional to $R_L \| R_{L'}$. If $R_L = R_{L'}$, the value of $R_L \| R_{L'}$ is reduced to just half of the value R_L, and the voltage across the load is half of the unloaded output value. Consequently, from our discussion of Thevenin's Theorem in Chapter A1, we see that the output impedance of the amplifier must be R_L.

NOTE: The term *basic* amplifier has been used for the circuit configuration shown in Figure A8–3. The important feature of this configuration is that the *bottom* lead of the active device pictured in Figure A8–2 is shown connected to *ground* in Figure A8–3. Notice that the bottom lead is the reference lead for the voltage V_{in} as it is presented to the control input. The formal term for this basic amplifier type is *grounded-X amplifier,* where *X* is the name for the bottom lead of the specific active device used. We shall see that *X* is *cathode* for vacuum tube, *source* for field-effect transistor, and *emitter* for bipolar transistor.

Vacuum Tubes

Vacuum Triode

The first electronic amplifying device, the *vacuum triode,* was invented by Lee De Forest in 1906. Although now used very seldom in new designs, it remains the easiest current-controlling element to understand. A wire spiral is introduced between the cathode and plate of the diode in Figure A4–3a to produce the *triode*. Figure A8–5a shows the triode structure, while Figure A8–5b gives the conventional schematic symbol.

Plenty of space is afforded by the wire of the grid for electrons to pass readily through it to the plate. Indeed, if the grid is connected directly to the cathode, the current through the device as a function of the *plate voltage* connected between the plate and the cathode shows just the same behavior as that for the vacuum diode. If the grid is made negative relative to the cathode, however, the current flowing through the tube can be greatly reduced because the electrons tend to be repelled back toward the cathode. Figure A8–5c shows the electron behavior in a graphic way. (Note that the filament connection is omitted from the drawing; this omission is conventional.) Thus, the grid becomes the current-controlling input so essential for an amplifying device. For further discussion of the triode, see the first supplement to this chapter.

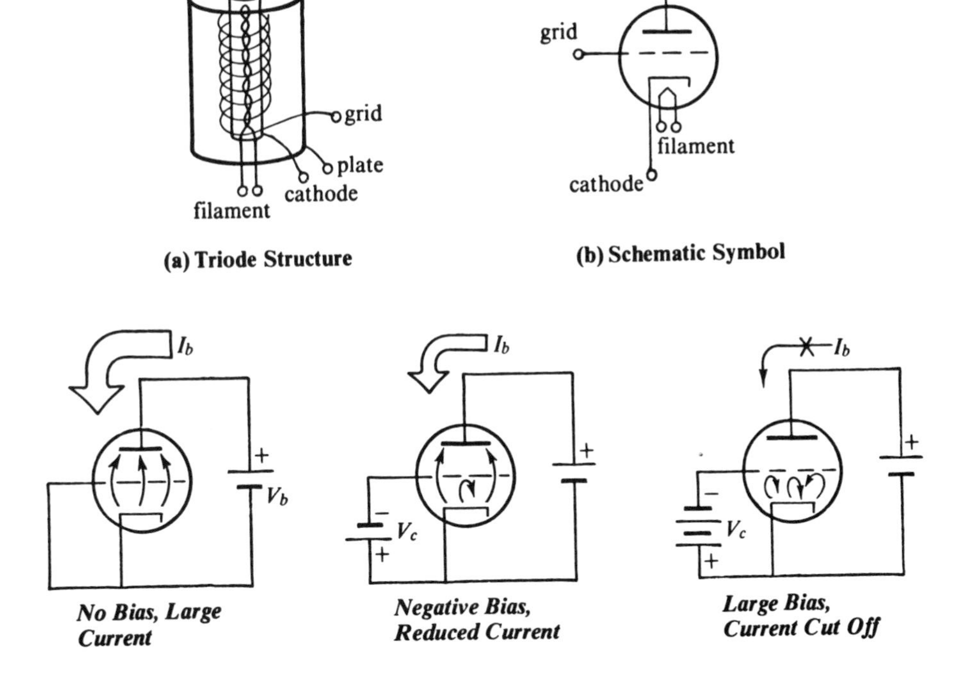

FIGURE A8–5. *Vacuum Triode*

Vacuum Tetrode and Pentode

Because of the capacitance between the control-grid and the plate structures, the triode is generally not well suited to radio-frequency amplifier applications; therefore, a *screen grid* was early introduced between the control grid and the plate. Then 4 important structures exist in this tube, and it is called a *tetrode.* However, the electrical characteristic for the tetrode has a problem that tends to produce unwanted oscillations. Therefore, still an additional grid structure was introduced between the screen grid and the plate; the resulting tube is called a *pentode.* Although vacuum tubes have been largely superseded by transistors, the pentode has continued to be used in high-frequency applications that involve large amounts of power. Only recently has there appeared a type of field-effect transistor that threatens to displace the pentode in this area. Indeed, the electrical characteristics and circuitry for the FET and the pentode are very similar, although only the FET is discussed in some detail in this book. Notice that the cathode-ray tube is, in fact, a special-purpose vacuum tube.

Field-Effect Transistors and Amplifiers

The simplest transistor type to understand is the *field-effect transistor* (FET). Although it became widely available in the mid-1960s, some years after the bipolar transistor, it had actually been proposed much earlier; however, semiconductor fabrication techniques had not been adequate at that time. The FET combines many of the best features of the vacuum tube and the transistor. It has a voltage-controlled input with a high effective resistance like the vacuum tube; on the other hand it is compact and lacks the vacuum tube's heat-producing and power-hungry filament (the bipolar transistor also has no filament).

N-Channel Junction FET

The schematic construction of the *N-channel junction field-effect transistor* is shown in Figure A8–6a. We can think of taking a bar of N-type semiconductor—which has a modest number of electrons as charge carriers—and constructing a region of P-type semiconductor on its side. Recall from Chapter A4 that a P-type semiconductor has holes that behave like plus charges as the majority charge carrier. The body of P-type material in contact with the body of N-type material constitutes a PN junction; hence the name junction FET, or JFET. Recall also from Chapter A4 that a junction of P- and N-type semiconductors such as we now have constitutes a diode

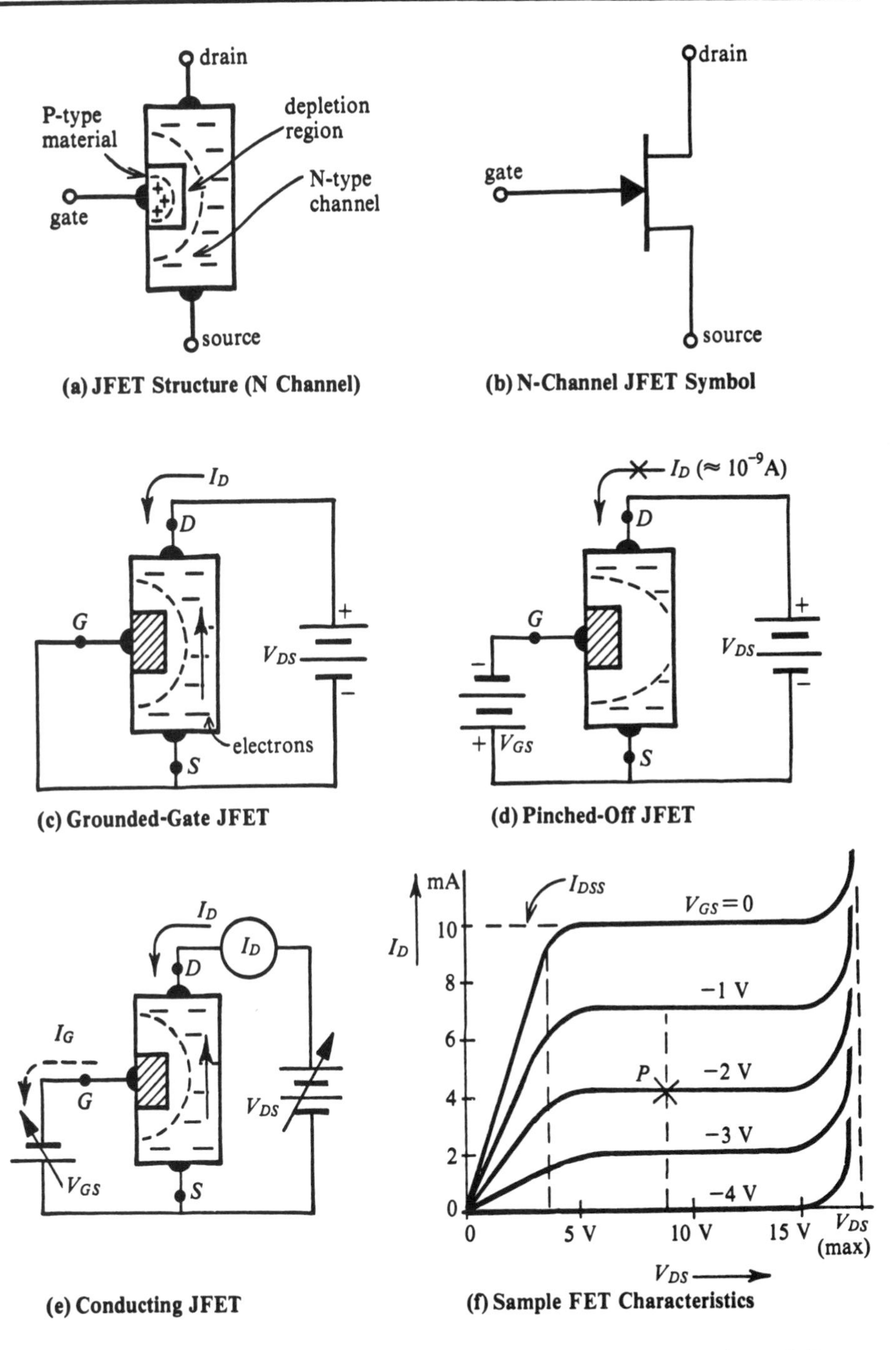

FIGURE A8–6 *Junction Field-Effect Transistor (JFET) for the N-Channel Case*

junction and that a depletion zone (substantially lacking any charge carriers) forms in the region of the junction. The region with free electron carriers in the material is indicated by the little minus signs in Figure A8–6a, while the depletion region between the dashed lines lacks the − carriers in the N bar and the + carriers in the P-gate material.

Wires or *leads* are attached to the top and bottom of the N-type bar and are called the *drain* (*D*) and the *source* (*S*), respectively. The third current-controlling *gate* (*G*) wire is connected to the P material on the side. The symbol for this 3-terminal device is shown in Figure A8–6b; note that the gate is signified with an arrow that represents the easy-current-flow direction of the diode that is formed. (Note also that the drain is not particularly distinguishable from the source, and indeed on some JFETs they can be used interchangeably.)

NOTE: Although the formal abbreviation for the junction field-effect transistor should perhaps be JFET, in common usage the J is often omitted. Accordingly, in the discussion that follows, both JFET and FET may be encountered. Nevertheless, there is a clear nomenclature distinction between FET and MOSFET; the MOSFET-type FET is discussed later in this chapter.

General Electrical Behavior of N-Channel JFET

If a voltage V_{DS} (the V_{DS} battery) is established between the drain and the source as shown in Figure A8–6c, a *drain current* I_D flows quite easily since there exists essentially a bar of conducting material—called the *channel*—between the drain and the source. Electrons move from the negative terminal of the V_{DS} battery to the source wire, where they become the "source" of electrons in the bar, and then move up to the drain, where they are literally drained from the bar as they move toward the positive terminal of the battery.

If a rather large voltage V_{GS} is established to make the gate *negative* with respect to the source as shown in Figure A8–6d, the gate-channel diode becomes very *reverse biased.* Then only a tiny leakage current flows from the gate (typically 10^{-9} A = 1 nA), but the depletion region is enlarged greatly and extends all the way across the bar. With no carriers existing in this region, the bar becomes an insulator. The conducting width or channel of the bar has effectively been pinched to zero, and essentially no drain current flows through the device; the FET is said to be *pinched off.* Notice the analogy to the pinched-off hose of Figure A8–1. The minimum gate-source voltage to achieve pinch-off is called the *pinch-off voltage* $V_{GS,\text{off}}$ for the device. This voltage varies from about −2 V to −15 V, depending upon the device construction, and is even quite variable (up to ±50%) for FETs with the same type number from a given manufacturer.

Electrical Characteristics of N-Channel JFET

To study the characteristics of an FET in conditions between the *conducting-hard* and the *cutoff* situations, a circuit to establish variable gate-source and drain-source voltages and to measure the resulting drain current is constructed as shown in Figure A8–6e. First suppose the gate-source voltage V_{GS} is set to zero (basically the gate is shorted to the source as in Figure A8–6c) and the drain-source voltage V_{DS} is slowly increased from zero. Observing the drain-current meter, we find that at first I_D increases quite linearly with increasing V_{DS} (just as current increases with the voltage across a resistor) until V_{DS} approaches the value of the pinch-off voltage for the unit. The I_D behavior is plotted in Figure A8–6f as the $V_{GS} = 0$ line that rises at an angle from the origin. Notice that the current I_D is plotted versus the drain-source voltage V_{DS} across the device.

When the drain-source voltage increases further, however, something rather surprising happens: The drain current levels off and remains virtually constant with increasing V_{DS}. The reason is a voltage drop (IR drop) along the resistive N-channel material. If the gate is at ground potential (relative to source) and the channel region near the top of the bar is at a voltage V_{DS}, the gate is negative relative to the channel by about this voltage. Thus, the gate-channel diode junction is *reverse biased* by the amount V_{DS}, and a large depletion zone is produced that tends to pinch off the channel. However, the current does not stop because the IR drop would then vanish and the channel would open again; instead, a narrow channel forms that becomes longer with increased V_{DS} so that the current flow remains practically constant.

Notice for the example FET of Figure A8–6f that for $V_{GS} = 0$ the drain current levels off at 10 milliamperes when the drain-source voltage is greater than about 4 V. This current value is called the *saturation current* I_{DSS} for the particular FET and is the maximum current that the device will pass. If the gate-source voltage V_{GS} is put equal to -1 V, the current increases more slowly from zero due to the smaller channel and levels off at a lower current of about 7 milliamperes (again when the drain-source voltage V_{DS} exceeds about 4 V). Similar behavior is seen for other V_{GS} voltages until finally, when $V_{GS} = -4$ V, the channel is pinched off and no current flows through the device. In normal amplifier operation the FET is operated in the region of the graph where the current curves are almost horizontal; that is, I_{DS} is independent of V_{DS}.

JFET Operating Limitations

An important practical limitation of any transistor is illustrated in the FET characteristic curves of Figure A8–6f. Notice that the current increases sharply as the drain-source voltage approaches $V_{DS,\max}$. This large current can quickly destroy the transistor, and the FET must be operated with V_{DS} less than this value. The particular FET type will therefore have a maximum safe operating V_{DS} specified by the manufacturer. In addition, the device should not dissipate power above a certain maximum, specified by the manufacturer as the *maximum device dissipation.* If this limit is ignored, the FET will quickly overheat and become ruined. The power dissipated in the FET is simply *the current through the FET times the voltage across it,* or $I_D V_{DS}$. Small FETs have typically been capable of only 1/2- to 1-watt dissipation and a maximum of 30 V to 50 V for the drain-source voltage.

However, recently high-power and high-voltage types have become available (for example, up to 600-V V_{DS} and 100-watts dissipation). These devices are actually a particular type of FET called the MOSFET, which is introduced later in this chapter. The high-power/high-voltage type of MOSFET is constructed with a *vertical* geometry over an area of semiconductor chip; hence, it is often called VMOS. Since the drain current is dispersed over greater area than in the *planer* technology of conventional transistors, greater current and power is possible.

N-Channel JFET in an Amplifier

Figure A8–7a shows a basic *grounded-source amplifier.* This circuit uses a separate *bias battery* V_{GG}, as well as a *supply battery* V_{DD}. The V_{GG} bias voltage is usually chosen to set the current flowing through the FET to a value such that the voltage across R_L ($V_{R_L} = I_D R_L$) is about half of the supply voltage V_{DD}. This bias rule is in accord with typical amplifier practice discussed in connection with Figure A5–6, and again with Figure A8–4. Then about half the supply voltage also appears across the FET as V_{DS}. Notice that the bias-battery voltage is arranged so that the gate of the N-channel FET is made *negative* relative to the source because we require the FET's gate-channel diode to be reverse biased. The operation of this amplifier now follows the general discussion given for Figure A8–3 and Figure A8–4.

The basic amplifier pictured in Figure A8–7a has some typical values written in for resistors, voltages, and capacitors. We discuss the amplifier by answering a number of questions using the illustrated circuit values and by assuming that the FET used has the characteristics shown in Figure A8–6f.

Example: Working from Figure A8–6f and Figure A8–7a:

1. What is the gate-source bias for the amplifier?
2. What is the drain current I_D, the voltage across R_L, and the V_{DS} voltage for the FET?
3. What power is dissipated in the FET?
4. Find the *operating point* for the FET on its characteristic graph.

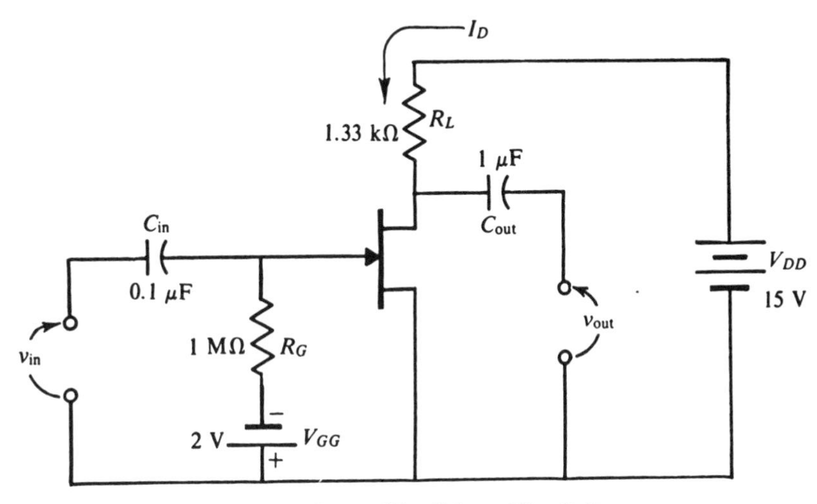

(a) Grounded-Source JFET Amplifier Using a Bias Battery As Well As Supply Battery

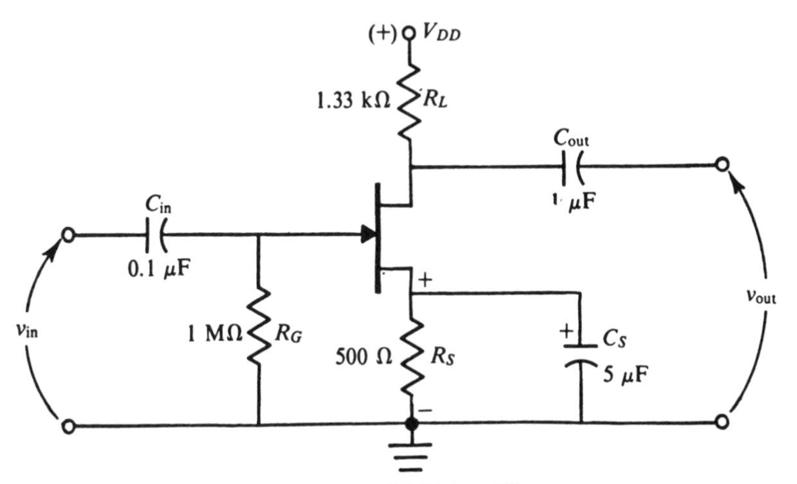

(b) Self-Biased, Grounded-Source JFET Amplifier

FIGURE A8–7. *JFET Grounded-Source Amplifier*

5. What is the mutual transconductance for the FET?
6. Calculate the expected voltage gain for this amplifier. If 0.05 V peak-peak is applied to the input, what voltage is expected at the output?

Solution:

1. The gate-source voltage is established by the bias battery V_{GG}. Be-

cause there is a gate current of only perhaps 10^{-9} A, the voltage across R_G is very small (10^{-9} A $\times$ 10^6 $\Omega = 10^{-3}$ V), and the gate-source voltage is virtually the same as the V_{GG} battery voltage, namely -2 V.

2. If $V_{GS} = -2$ V, from Figure A8–6f we expect the drain current to be 4 mA (assuming V_{DS} is more than 4 V). Then the voltage across the R_L resistor is $I_D R_L = 4$ mA $\times$ 1.33 kΩ. To find the drain-source voltage V_{DS}, we realize that the supply voltage equals the voltage across R_L plus the drain-source voltage across the FET: $V_{DD} = V_{R_L} + V_{DS}$ or 15 V = 5.3 V + V_{DS} so that $V_{DS} = 9.7$ V.
3. The power dissipated in the FET is the voltage across it times the current through it: $V_{DS} I_D = 9.7$ V $\times$ 4 mA = 39 mW. The heat-dissipation capabilities of the FET must be greater than this, but should be no problem here since even low-power units can manage 500 mW.
4. Since $V_{DS} = 9.3$ V and $V_{GS} = -2$ V (and $I_D = 4$ mA), the operating point can be plotted on the graph of Figure A8–6f as being at point P.
5. The mutual transconductance is calculated from Equation A8–1; for an FET $\Delta I_T = \Delta I_D$ and $\Delta V_{\text{in}} = \Delta V_{GS}$. We work from the characteristic graph. For calculating the voltage gain of the amplifier, g_m should be calculated about the operating point P. Looking at the FET characteristic in Figure A8–6f, we select V_{GS} curves above and below the operating point. As V_{GS} ranges from -3 V to -1 V, we see that I_D ranges from 2 mA to 6.7 mA (along a vertical line through point P). Then

$$g_m = \frac{\Delta I_D}{\Delta V_{GS}} = \frac{(6.7\text{ mA} - 2\text{ mA})}{(-1\text{ V}) - (-3\text{ V})}$$

$$= \frac{4.7\text{ mA}}{2\text{ V}} = 2.3 \times 10^{-3}\text{ mhos}$$

Basically, the mutual transconductance is a measure of the vertical spacing of the V_{GS} curves on the characteristic curves for the device.

6. The voltage gain for the amplifier is given by Equation A8–4.

$$A_v = -g_m R_L = -2.3 \times 10^{-3}\text{ mhos} \times 1.33\text{ k}\Omega = -3.0$$

If an input signal of 0.05 V peak-peak is applied to the input, we expect an output signal of

$$V_{\text{out}} = A_v v_{\text{in}} = 3 \times 0.05\text{ V} = 0.15\text{ V}_{p-p}$$

A few comments about the behavior of this amplifier are in order. Notice that the accuracy of the voltage-gain calculation relies on an accurate knowledge of g_m. Often, this is not accurately known for FETs "merely picked from the drawer"; this is a significant advantage of the op-amp design approach, where the gain is known as accurately as the circuit resistance values are known. However, the FET circuit can be expected to operate to high radio frequencies (perhaps 100 megahertz) that cannot generally be expected from op amps.

Notice, too, that the voltage gain is really rather low. This gain could be improved in either of two ways. First, an FET with a higher transconductance could be used; values to 1.5×10^{-2} mhos are quite common. Second, a larger R_L value could be used, but there is a penalty. As we saw in the general amplifier discussion earlier, the output impedance of the amplifier is essentially equal to R_L; it is usually desirable that this value not be too high.

Self-Biased JFET Amplifier

Next we examine the *self-biased amplifier* of Figure A8–7b. It is a nuisance to need two batteries or power supplies as shown in Figure A8–7a. The essential requirement in biasing the FET here is that the gate be somewhat negative relative to the source. This is usually accomplished by connecting the one end of resistance R_G to ground and introducing a resistance R_S between the source and ground as shown in Figure A8–7b. The current I_D flowing through R_S causes the source to be positive relative to ground as shown in the figure. However, the gate is nearly at ground (very little leakage current flows through the gate and through R_G to ground). Thus, the gate is negative relative to the source as required. The value of R_S is chosen by the fact that $V_{RS} = I_D R_S$ equals the gate-bias voltage V_{GS}.

The capacitor C_S smooths out the ac voltage that tends to develop across R_S when amplifying an ac voltage (we want a dc voltage across R_S). The resulting amplifier is called a *grounded-source amplifier,* even though the source is *not* at dc ground (as it is in Figure A8–7a). However, it *is* at ac ground if a large capacitor C_S is used because the small reactance of the capacitor effectively shorts ac currents to ground.

The input impedance of the FET amplifier is quite evident from Figure A8–7a. The resistance presented to the input voltage is just R_G (here 1 megohm) because the parallel resistance through the back-biased gate-source FET diode junction is very high. We might think that the input impedance could be made very high just by choosing a very large value for R_G or even leaving it out completely, but this is not true; a path must be provided for the leakage current through the gate-source diode. Typically, 10 megohms is the maximum recommended value.

P-Channel JFET

The FET type pictured in Figure A8–6 and discussed thus far has a channel for carrier flow constructed of N-type material. The complementary possibility is to use P-type material for the channel and a region of N-type material on the side as the gate, as shown in Figure A8–8a. The resulting device is called a *P-channel FET.* The easy-current-flow direction for the gate-channel diode junction is now *out* of the gate, and the arrow drawn on the gate of the N-channel symbol reflects this (see Figure A8–8b). When used, the gate-source diode must always be reverse biased, so the gate must be made positive relative to the source for the P-channel type; then the drain is made negative. The consequence is that the polarities of the supply voltage must be *reversed* for the P-channel JFET as compared to the N-channel JFET; otherwise, the circuit design considerations are identical, and the representative form for the grounded-source-type amplifier is shown in Figure A8–8c.

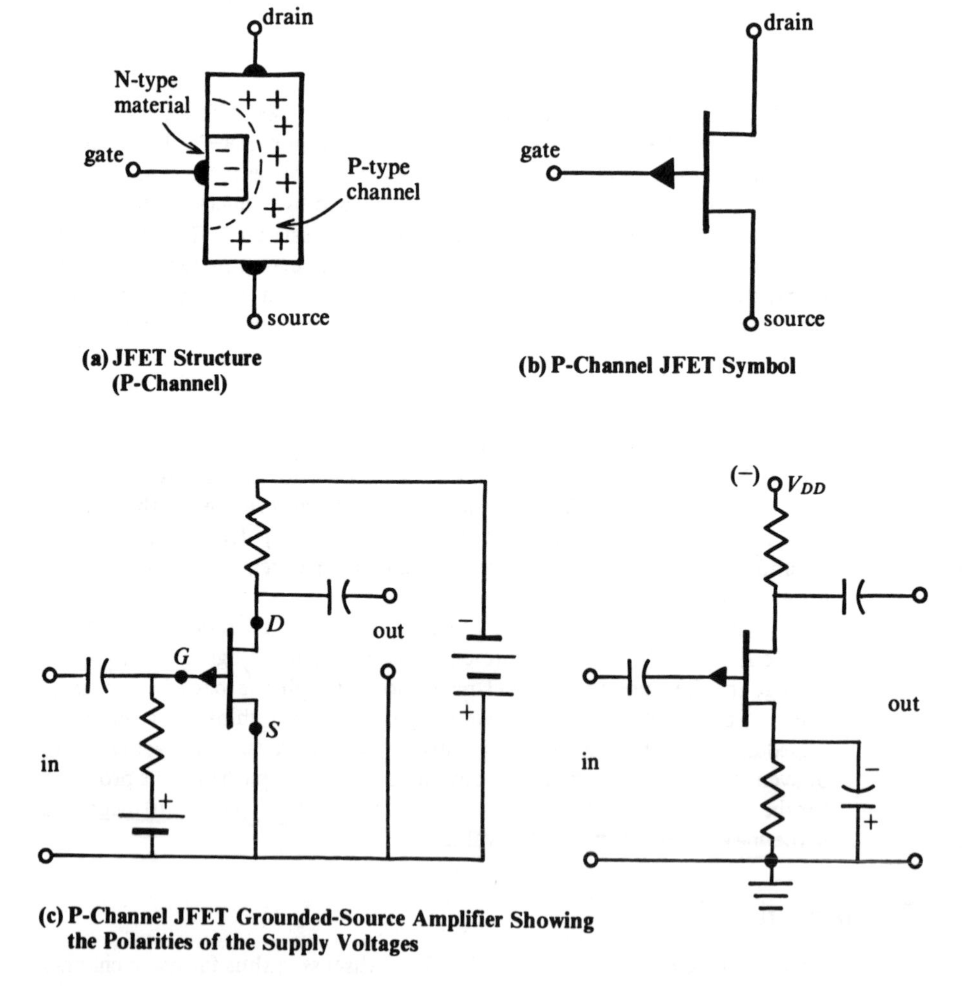

FIGURE A8–8. *P-Channel JFET and Circuit Polarities*

It should be emphasized that the gate is made quite positive relative to the source in order to pinch off the P-channel JFET, while we saw that the N-channel JFET requires a quite negative gate-to-source voltage. This complementary behavior can be useful in logic or switching circuits.

Metal-Oxide-Silicon FET (MOSFET)

While the input current is very small for the JFET amplifier (only the gate-channel-diode reverse-bias leakage), it is made astonishingly small for the *metal-oxide-silicon field-effect transistor* or MOSFET. The scheme is basically to put a layer of electrical insulation—typically silicon dioxide—between the metal gate plate and the silicon channel. The order of materials here is the reason for the name of this device. Of course the insulation restricts the current through the gate to very small values, typically 10^{-14} amperes.

Two different structures exist for MOSFETs, giving the depletion type and the enhancement type. The operation of the *depletion-type MOSFET* is very similar to the JFET: A negative gate-source voltage is required to pinch off the current through the N-channel version, while a positive V_{GS} is required for the complementary P-channel version. The symbols for these two versions are shown in Figure A8–9a. Notice that the gate is drawn as not electrically connected to the drain-to-source channel because it is insulated from it.

The *enhancement-type MOSFET* actually has a physical break in the channel so that no current flows through the device unless its gate is properly biased with respect to the source. To obtain device current, the gate must be positively biased relative to the source for the N-channel version, while the gate must be negatively biased relative to the source for the complementary P-channel version. The circuit symbols for the enhancement-type MOSFET are shown in Figure A8–9b; notice the breaks in each symbol's line joining the drain to the source that indicate the channel construction. For a more detailed discussion of the construction and mode of operation for MOSFETs, see the second supplement to this chapter.

Bipolar Transistors and Amplifiers

The invention of the transistor in 1948 at Bell Telephone Laboratories has revolutionized electronics. Its chief advantages relative to the vacuum tube are: (1) a much smaller size, (2) no heat-producing and power-consuming filament, and (3) a long, useful life. In a vacuum tube—similar to a light bulb—the filament eventually burns out; also the cathode loses its electron-emitting capability.

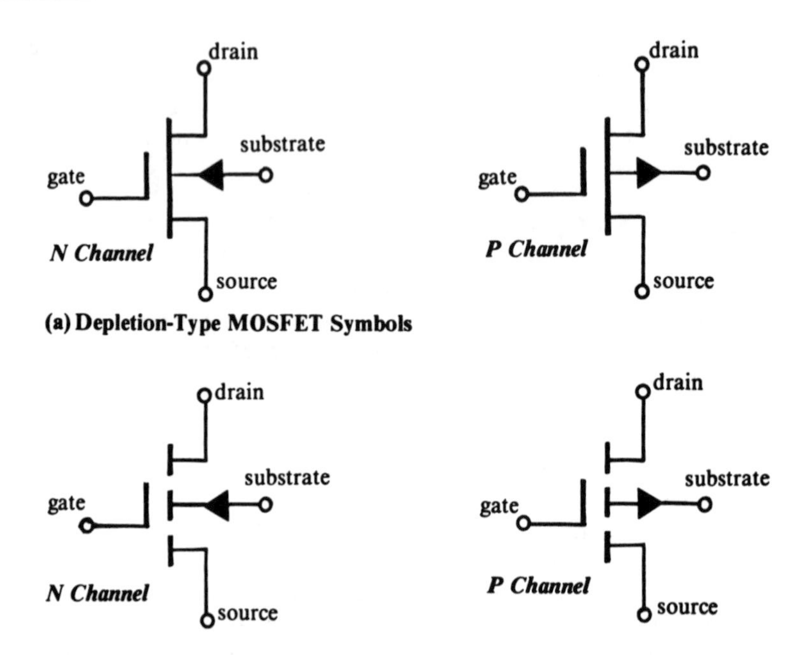

FIGURE A8–9. *MOSFET Symbols*

NPN Bipolar Transistor

The basic construction of an *NPN bipolar transistor* is quite simple: A *thin* layer of P-type material is sandwiched between two blocks of N-type material as shown in Figure A8–10a. Wires are connected to the three elements and are referred to as *collector* (*C*) (of electrons, like the plate), *base (B),* and *emitter (E)* (of electrons, like the cathode). The base is the input control element. However, in normal operation the base is somewhat positive relative to the emitter, whereas the grid was negative relative to the cathode and the gate was negative relative to the source in the JFET.

Because there are two PN junctions in the NPN transistor, the transistor consists of two diodes with their anodes inherently connected together, as modeled in Figure A8–10b. This diode structure is important to the operation of the bipolar transistor. Note, however, that merely connecting two diodes together will *not* make a transistor; as is shown in the next section, a *thin* base (the P material here) is needed. The circuit symbol for the NPN transistor is given in Figure A8–10c. The arrow is in the slant line corresponding to the emitter and points in the easy-current-flow direction of the base-emitter diode.

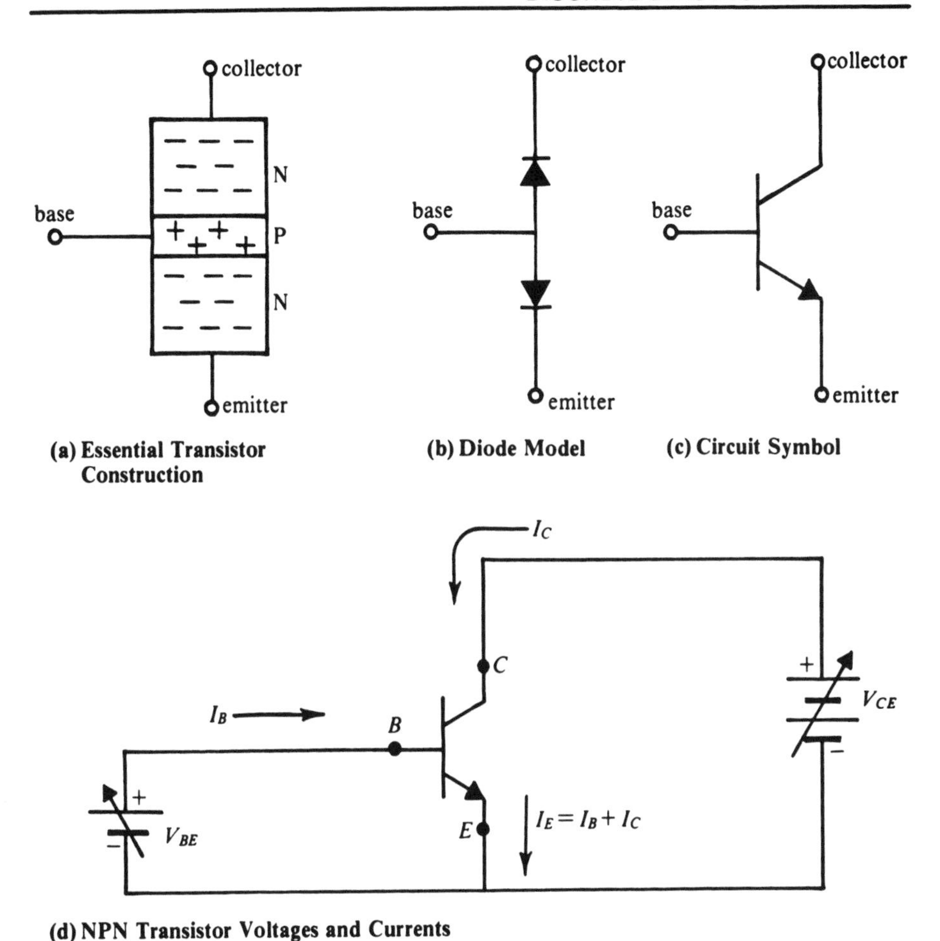

FIGURE A8-10. *NPN Bipolar Transistor*

NOTE: A very useful and simple test for a good or bad bipolar transistor is to make ohmmeter connections between the various pairs of transistor leads and observe the resistance. Thus, between the base and emitter, the ohmmeter should register a low resistance if the base is made positive relative to the emitter, but a very high resistance if the reverse connections are made. (For example, the Simpson VOM uses the red lead as positive for ohmmeter use; *not* all VOMs use this convention, however.) A bad transistor often exhibits a low resistance in both senses (a shorted junction) or a high resistance in both senses (an open junction). If either of these conditions is present, the transistor is certainly defective. A similar test can be made between the base and collector junctions. Between the collector and the emitter, the ohmmeter should register a high resistance for either polarity of connection.

General Electrical Behavior of NPN Transistor

Consider the operation of the transistor when connected to a pair of variable voltage supplies V_{BE} and V_{CE} as shown in Figure A8–10d. The directions of conventional current I_C flowing into the collector and the current I_B flowing into the base are shown. First notice that if the battery V_{BE} were *not* connected, essentially *no* current I_C would flow through the transistor. This situation is clear from the diode model in Figure A8–10b. The *C-B* diode is reverse biased and blocks the attempted current flow from V_{CE}.

The battery V_{BE} is connected in the forward-bias sense across the *B-E* diode. Then electrons begin to flow from the emitter (N block) into the base block as shown in Figure A8–11. The electrons can be thought of as being repelled from the negative terminal of the V_{BE} bias battery. However, only a small fraction of the electrons go out the base lead and constitute base current. The largest fraction of electrons zips across the *thin* base region and is quickly under the attraction of the large positive voltage from V_{CE}. The electrons are then collected by the collector and constitute the collector current. (The following paragraph gives a more rigorous description of the bipolar transistor mechanism for those readers who are interested.)

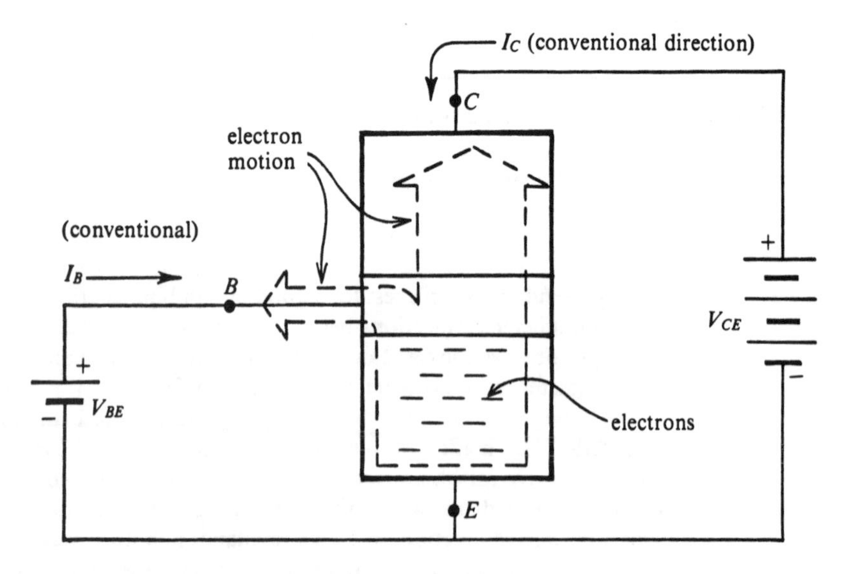

FIGURE A8–11. *Representation of Electron Currents in an NPN Transistor*

The forward bias imposed by the V_{BE} battery on the base-emitter diode junction reduces this particular junction barrier. Electrons then diffuse across the junction from the emitter block into the base block. However, the base block is made very thin and is also lightly doped so that there is little recombination of the electrons with the holes in the P-type base, which would draw carriers from the base lead. (Further, the light doping in the P-type base implies that few holes exist here to flow toward the B-E junction to give base current.) The electrons thus diffuse through the base block over to the base-collector diode junction. This junction is *reverse biased* by the rather large V_{CE} battery. But the large electric field existing in the depletion zone at this junction has the proper direction to push the electrons into the N-type collector block. Here they move to the positive collector terminal. Thus, the main current flow is between the collector and the emitter of the transistor and consists of electrons in the NPN transistor. Some holes are involved in the P region, however, so the device uses both types or poles of carriers; hence the term *bipolar* transistor.

Electrical Characteristics of NPN Transistor

The ratio of electrons moving through the transistor to those going out the base lead is quite constant since it depends primarily on geometrical construction. This ratio of I_C/I_B is called the *beta (β) or forward-current transfer ratio* h_{fe} of the transistor; values of 50 to 150 are quite common. That is,

$$\beta = h_{fe} = \frac{I_C}{I_B} \qquad \textbf{(A8–5)}$$

The electrical characteristics of the transistor can be represented on two graphs, considered the input characteristic and the output characteristic. The *input* or *base characteristics* are a plot of base current I_B versus the applied base-emitter voltage V_{BE}. It is simply a diode-like curve, but with a smaller current for a given V_{BE} than may be expected for a simple diode (because most of the current goes to the collector rather than to the base). See Figure A8–12a. Figure A8–12b is a plot of collector current versus applied collector-emitter voltage V_{CE}; this plot is called the *collector characteristics* or *output characteristics* of the transistor.

The collector characteristics show that the current flowing through the transistor is indeed determined almost entirely by the base current (which is initiated by V_{BE}). *Curves of constant base current* are drawn and for the transistor pictured $I_C \simeq 100\ I_B$ for all but quite small values of V_{CE}; that is, the β for this transistor is 100. Notice that the collector-emitter voltage has little effect on the collector current (except in the *saturation region;* if V_{CE} is very small, there is not much current flow).

The transistor is quite clearly a current amplifier in regard to the collector current compared to base (input) current. However, the base current

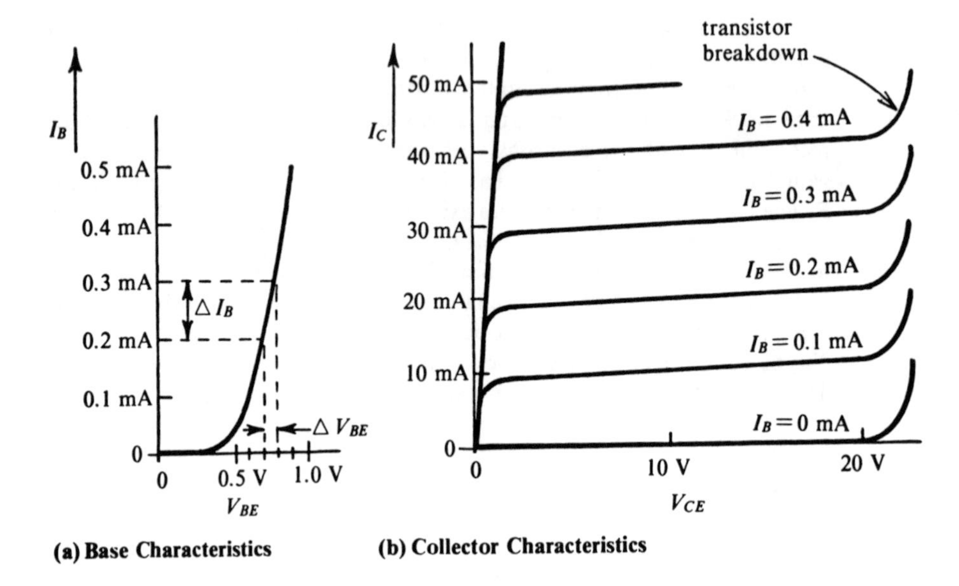

(a) Base Characteristics **(b) Collector Characteristics**

FIGURE A8–12. *Voltage-Current Behavior of the Bipolar Transistor*

is always caused by a base-emitter voltage. A small variation in V_{BE} causes a modest change in base current I_B, which in turn "causes" a large change (β times more) in collector current. Thus, the transistor follows the model of a voltage controlling a current that was shown in Figure A8–2.

To calculate its transconductance from Equation A8–1, we need

$$g_m = \frac{\Delta I_C}{\Delta V_{BE}} \tag{A8-6}$$

because $V_{\text{in}} = V_{BE}$ and $I_T = I_C$ in the case of the transistor. But ΔV_{BE} causes a certain change ΔI_B in the base current (see Figure A8–12a). We can write

$$\frac{\Delta V_{BE}}{\Delta I_B} \equiv r_{BE} \tag{A8-7}$$

where this ratio of voltage to current has dimensions of resistance and may be called the *ac input resistance* of the transistor. Further, the change caused in I_C due to a change in I_B is related through the β of the transistor (or h_{fe}). That is,

$$\Delta I_C = \beta \Delta I_B \tag{A8-8}$$

Then

$$\Delta I_C = \beta \Delta I_B = \frac{\beta \Delta V_{BE}}{r_{BE}} = \frac{\beta}{r_{BE}} \Delta V_{BE} \tag{A8-9}$$

Substituting this expression for ΔI_C into Equation A8–6 then gives an expression for transistor transconductance:

$$g_m = \frac{\beta}{r_{BE}} \qquad \textbf{(A8–10)}$$

The value for r_{BE} depends on the value of base current because it is the reciprocal of the slope of the input characteristic curve (which is exponential). However, as an *order-of-magnitude value,* we can assume that $r_{BE} \simeq 1000\ \Omega$ for low-power transistors. In fact, this is the situation diagrammed in Figure A8–12a, where the I_B current increases by 0.1 mA (from 0.2 mA to 0.3 mA) as the V_{BE} voltage increases by 0.1 V (from 0.7 V to 0.8 V) so that $r_{BE} = 0.1\ \text{mA}/0.1\ \text{V} = 1\ \text{k}\Omega$.

Example: What is the transconductance of the transistor with characteristics given in Figure A8–12 if operated with a collector current of 25 mA?

Solution: The base current for a collector current of 25 mA is 0.25 mA for this transistor. This is the point about which ΔV_{BE} and ΔI_{BE} equal 0.1 V and 0.1 mA, respectively. Then $r_{BE} = 1\ \text{k}\Omega$ as argued above. Next the transconductance is calculated from

$$g_m = \frac{\beta}{r_{BE}} = \frac{100}{1000\Omega} = 0.1\ \text{mhos} = 100{,}000\ \mu\text{mhos}$$

since we saw earlier that $\beta = I_C/I_B = 100$ for this transistor.

NOTE: The interested reader may refer to Problem 26 at the end of this chapter for a useful theoretical relation for g_m and r_{BE} for a transistor.

Basic Grounded-Emitter Amplifier

A *grounded-emitter amplifier* that uses a simple base-bias scheme is shown in Figure A8–13. As mentioned in the general amplifier case earlier, the transistor should be biased so that a continuous idle current I_C flows through the transistor even if no input signal is present. The R_B resistor supplies a base current I_B to the transistor input so that a proportional collector current I_C is flowing. In selecting the value of R_B, the idea is to adjust the idle-current value of I_C through the transistor so that $V_{R_L} \simeq 1/2\,V_{CC}$, where V_{R_L} is the voltage across resistor R_L. But $V_{R_L} = I_C R_L$, and if V_{CC} and R_L are given (or assumed), the value of I_C is determined as $I_C = V_{CC}/2R_L$. Next, the base current I_B must be less than this value by the factor β of the transistor. However, I_B is established by choosing the proper R_B. In Figure A8–14 we see that

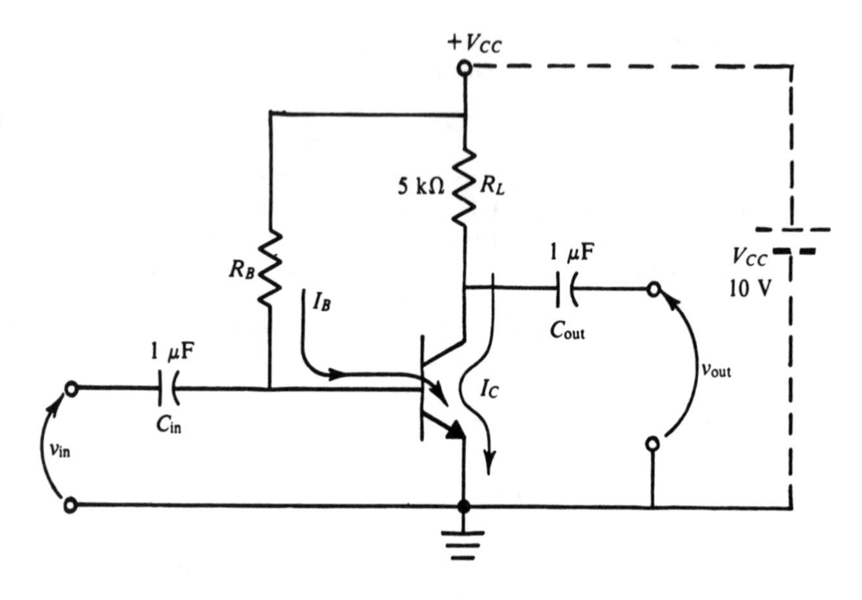

FIGURE A8–13. *Grounded-Emitter Amplifier Using Simple Base-Bias Current Scheme*

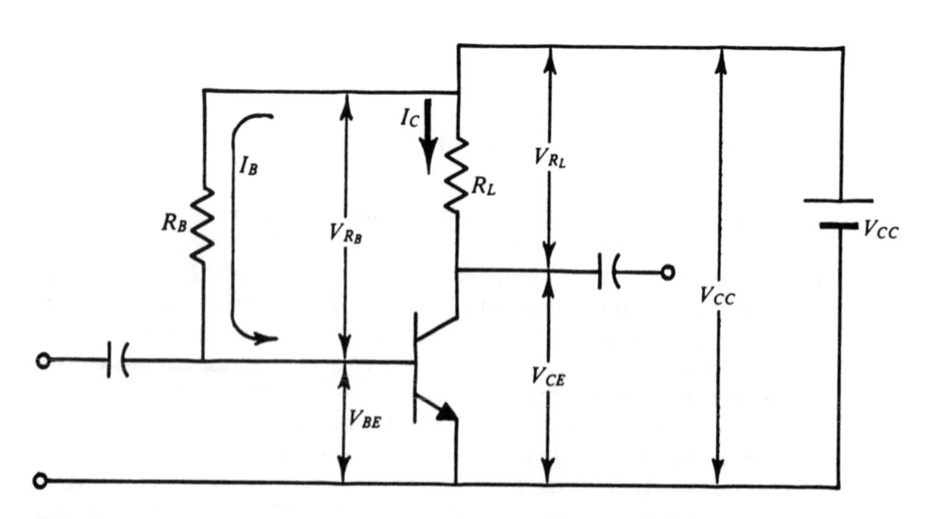

FIGURE A8–14. *Voltages Involved in Establishing the Base-Bias Resistor R_B*

$V_{CC} = V_{R_B} + V_{BE} = I_B R_B + V_{BE} \simeq I_B R_B + 0.6$ V. Usually V_{CC} is much larger than 0.6 V so that we have $V_{CC} \simeq I_B R_B$. Then

$$R_B = \frac{V_{CC}}{I_B} = \frac{V_{CC}}{I_C/\beta}$$

determines the bias resistor R_B.

Example: Suppose $V_{CC} = 10$ V and $R_L = 5$ kΩ for a transistor amplifier as shown in Figure A8–13. Assume the β of the transistor is 100. Find R_B. Also estimate the voltage amplification.

Solution: For normal operation we assume

$$V_{R_L} = \frac{V_{CC}}{2} = \frac{10 \text{ V}}{2} = 5 \text{ V}$$

Then

$$I_C = \frac{V_{R_L}}{R_L} = \frac{5 \text{ V}}{5 \text{ k}\Omega} = 1 \text{ mA}$$

Next

$$I_B = \frac{I_C}{\beta} = \frac{1 \text{ mA}}{100} = 0.01 \text{ mA}$$

Finally

$$R_B = \frac{10 \text{ V}}{0.01 \text{ mA}} = 1{,}000 \text{ k}\Omega = 1 \text{ M}\Omega$$

The voltage amplification is given by

$$A_v = g_m R_L$$

Here $R_L = 5$ kΩ. Further

$$g_m = \frac{\beta}{r_{BE}} \simeq \frac{100}{1{,}000 \ \Omega} = 0.1 \text{ mhos}$$

Then

$$A_v = 0.1 \text{ mhos} \times 5{,}000 \ \Omega = 500$$

(Here the rule-of-thumb estimate was used for r_{BE}.)

Input and Output Impedance of Grounded-Emitter Amplifier

The input impedance of the grounded-emitter amplifier of Figure A8–13 is generally determined by the ac resistance appearing between the base and emitter of the transistor. This is simply the r_{BE} resistance that has already been discussed and is typically on the order of 1 kilohm. This value is indeed low compared to what is often desired and comes about because the input is across a *forward-biased* diode junction. The JFET presents a much higher input impedance ($\sim 10^7$ ohms) because its controlling input is across a reverse-biased diode, while the MOSFET presents an extremely high input impedance ($\sim 10^{14}$ ohms) because its input is to an insulating layer.

As we mentioned earlier in the discussion of the general basic amplifier, the output impedance of the amplifier is generally close to the value of the R_L resistor used in the design. Although this resistor may be made small to obtain a low output impedance, notice that the voltage amplification is proportional to R_L so that a price of lower amplification is paid.

Transistor as a Switch

In digital circuits we are interested in yes/no, on/off, or open/closed switch situations. Also we are sometimes in need of an electronic switch. A transistor can be used as a voltage-controlled switch. Consider the circuit shown in Figure A8–15a. If the in lead is connected to ground as in Figure A8–15b, no base bias is present and no base current I_B flows. Then I_C is zero as well and the transistor behaves like an open switch. In spite of V_{CC}, no current flows through the load R_L. The out voltage must be at V_{CC} because no voltage drop exists across R_L either. Although the input voltage is zero (low), the output voltage is high (at V_{CC}). The input and output are seen to be opposites.

On the other hand, suppose the in lead is connected to V_{CC} as shown in Figure A8–15c. Also, assume $R_B = R_L$. Then a base current I_B flows that is given by

$$I_B = \frac{V_{R_B}}{R_B} \simeq \frac{V_{CC}}{R_L} \qquad \textbf{(A8–11)}$$

Next a collector current *tries* to flow that is β times as much as this base current. Actually, the maximum possible collector current is also the value of Equation A8–11. It is limited by the fact that the supply voltage must equal the voltage across the load resistor plus the voltage across the transistor (see Figure A8–14); that is,

$$V_{CC} = V_{R_L} + V_{CE} = I_C R_L + V_{CE} \qquad \textbf{(A8–12)}$$

V_{CC} is fixed and the minimum that V_{CE} can be is zero; then maximum collector current possible is $I_C = V_{CC}/R_L$. The result is that the voltage appearing across the transistor goes virtually to zero (as does the out voltage). The current flowing through R_L is not limited by the transistor at all; it behaves like a closed switch. Also notice that the output voltage is again the opposite of the input voltage: Although the input is connected to a high voltage (V_{CC}), the output voltage is low (nearly zero).

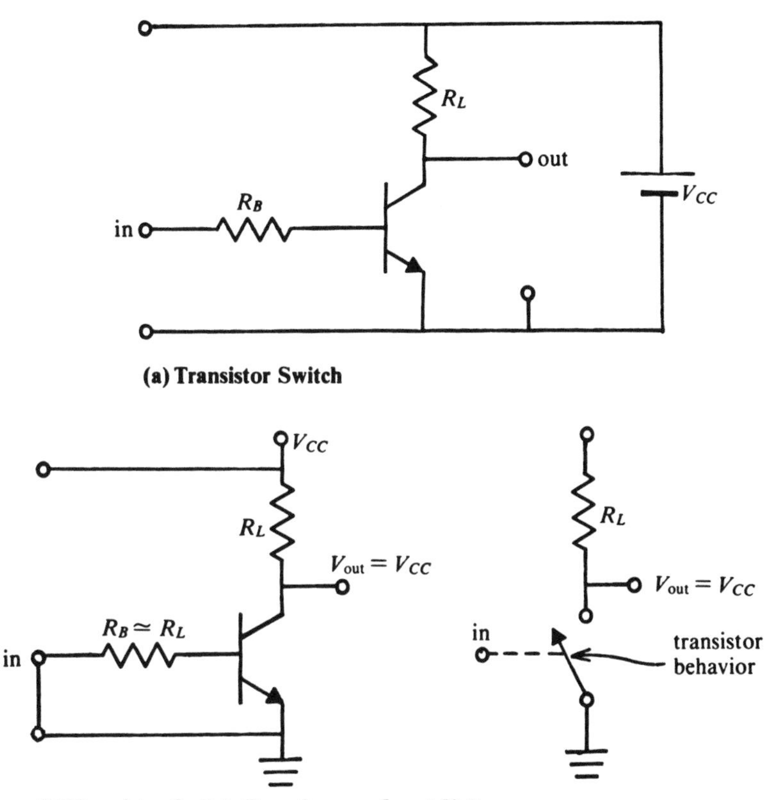

(a) Transistor Switch

(b) Transistor Switch Open because Input Voltage Is Zero (or Unconnected)

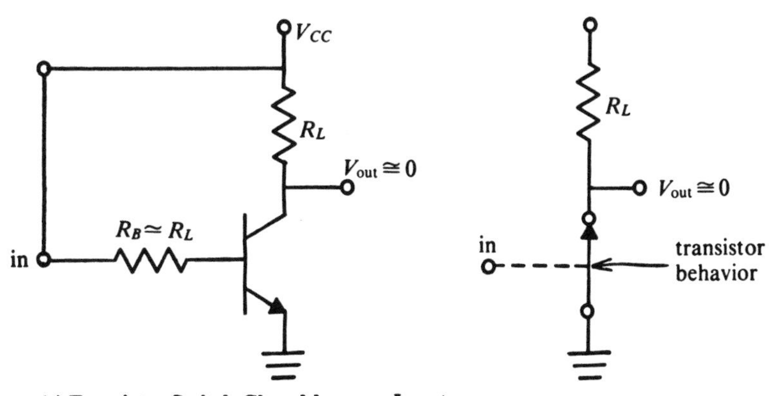

(c) Transistor Switch Closed because Input Voltage Is High at V_{CC}

FIGURE A8–15. *Transistor as a Switch*

The transistor switch is controlled by applying a suitable voltage to the input. Zero volts should be applied to open the switch, and a high voltage should be applied to close the switch. The load resistance R_L can be any resistive element; for example, a light bulb, relay coil, dc motor, and the like.

Graphical View of Transistor Amplifier and Switch

It can help to clarify transistor circuit operation to use a graphical representation on the collector characteristics. Equation A8–12 contains two variables: I_C and V_{CE}. Suppose we solve for I_C in terms of V_{CE} and the other constant quantities V_{CC} and R_L. Then

$$I_C = \left(\frac{-1}{R_L}\right) V_{CE} + \frac{V_{CC}}{R_L} \qquad \textbf{(A8–13)}$$

This is an equation of form $y = mx + b$ and therefore clearly represents a straight-line form if I_C is plotted versus V_{CE}. The slope of the line is $(-1/R_L)$ and the I_C intercept is V_{CC}/R_L. The variables are just the same as those for the collector characteristic, and we can plot the line on such a graph for the specific transistor used. Figure A8–16 illustrates the situation for a numerical example. The line plotted, as well as the collector characteristics, is for the circuit with transistor as diagrammed at the left in the figure. It should be noted that $I_C = 0$ in Equation A8–12 or Equation A8–13 implies $V_{CC} = V_{CE}$; this determines the V_{CE} intercept as V_{CC}. The line is called the *load line*. It really represents the requirement imposed by Kirchhoff's Voltage Law on the resistor R_L in series with the transistor; that is, by Equation A8–12. The state of transistor current and voltage *must* fall somewhere along this line. If the base current I_B is now given, a unique state is determined by the intersection of the given required base-current curve with the load line. For example, in Figure A8–16 if the base current is known to be 0.2 milliamperes, the operating point must be at the little circle pictured. Then I_C is about 10 milliamperes and V_{CE} is about 5 V. Furthermore, the voltage V_{R_L} across the R_L must be 10 V since $V_{CE} + V_{R_L} = 15$ V. This V_{CE} and V_{R_L} relationship to V_{CC} is diagrammed below the V_{CE} axis in Figure A8–16.

To use this transistor as a switch, the base current should either be zero, or more than 0.3 milliamperes. If $I_B = 0$, V_{CE} and also the output voltage is 15 V. If I_B is 0.3 milliamperes *or more,* all these I_B curves come down nearly vertically at about 0.5 V for V_{CE}. Then $V_{CE} = V_{out} \simeq 0.5$ V and the transistor is conducting hard at about 14 milliamperes. The transistor switch is closed and the output voltage is the *transistor saturation voltage* of about 0.5 V.

Finally, the operation of the transistor as an amplifier can also be appreciated from this graph. Suppose an input voltage causes I_B to vary between 0.1 milliamperes and 0.2 milliamperes. Then $V_{CE} = V_{out}$ can be seen

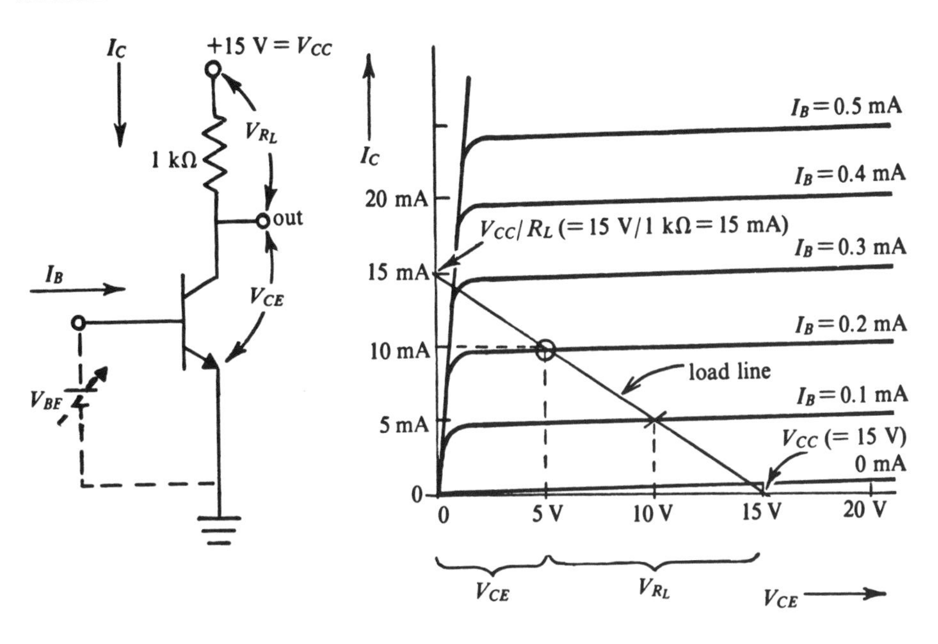

FIGURE A8–16. *Load Line on Collector Characteristics for a Transistor*

to vary between 10 V and 5 V, giving a peak-peak voltage of 5 V. But the input voltage that causes the 0.1-milliampere base-current variation is certainly considerably less than 1 V since the diode curve rises so steeply. It is of the order $V_{in} = \Delta V_{BE} = 0.1 \text{ mA} \times 1{,}000\ \Omega = 0.1$ V. Then the voltage amplification is about 5 V/0.1 V = 50.

Some Practical Transistor Considerations

Transistors are specified by a certain type number usually prefixed by 2N; for example, 2N708 (a switching transistor), 2N3055 (a power transistor), and 2N2222 (small general-purpose transistor). They are rated according to maximum current I_C, maximum collector-emitter voltage V_{CE}, and range of β or h_{fe}. The maximum-allowed V_{CE} is determined by the collector-emitter voltage for which the current begins to increase excessively; this *transistor breakdown* situation is evident in Figure A8–12b for V_{CE} greater than about 20 V. Unfortunately, the β value can vary by 50% or more between transistors of even the same type from the same manufacturer. (Vacuum tubes are much more uniform.) Another important rating is the *maximum device dissipation.* The heating in a transistor is determined by the product of voltage and current: $P = V_{CE}I_C$. If a transistor overheats, it can be ruined quickly. (Tubes are more forgiving.)

Low-power transistors typically can dissipate 1 or 2 watts, and a common case configuration is the TO–5 case shown in Figure A8–17a. In this case the collector, base, and emitter leads usually form a triangle with *C, B,* and *E* designated as shown. A small tab is near the emitter. A handy memory aid is that when a TO–5-type transistor is viewed from the *bottom,* the wires lead out in the same direction as the leads drawn in the transistor symbol. See the bottom view and transistor symbol illustrated in Figure A8–17a.

The larger TO–3 case is often used for transistors capable of 10- to 150-watts power dissipation. Then the transistor is securely attached to aluminum heat fins (a heat sink) to dissipate the heat. One lead (often the collector) is connected to the case itself and may need to be insulated from the heat sink by a mica spacer.

PNP Bipolar Transistor

Although only the NPN transistor has been discussed thus far in this chapter, using semiconductor blocks of P, N, and P (rather than N, P, and N) gives the *PNP bipolar transistor.* Then the diode senses are reversed and the battery polarities must be reversed; that is, the collector as well as the base must be made negative relative to the emitter. Otherwise, the operation of the transistor is similar to the NPN transistor (see Figure A8–18).

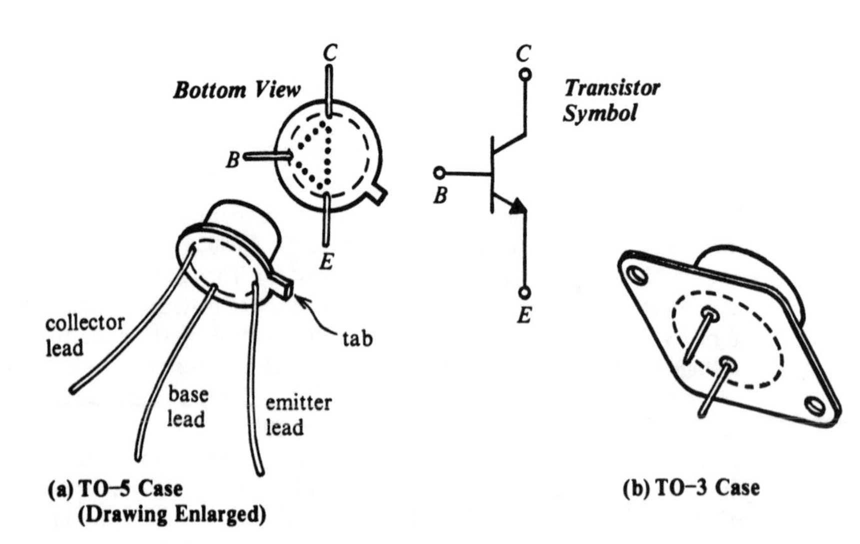

FIGURE A8–17. *Two Common Transistor Cases*

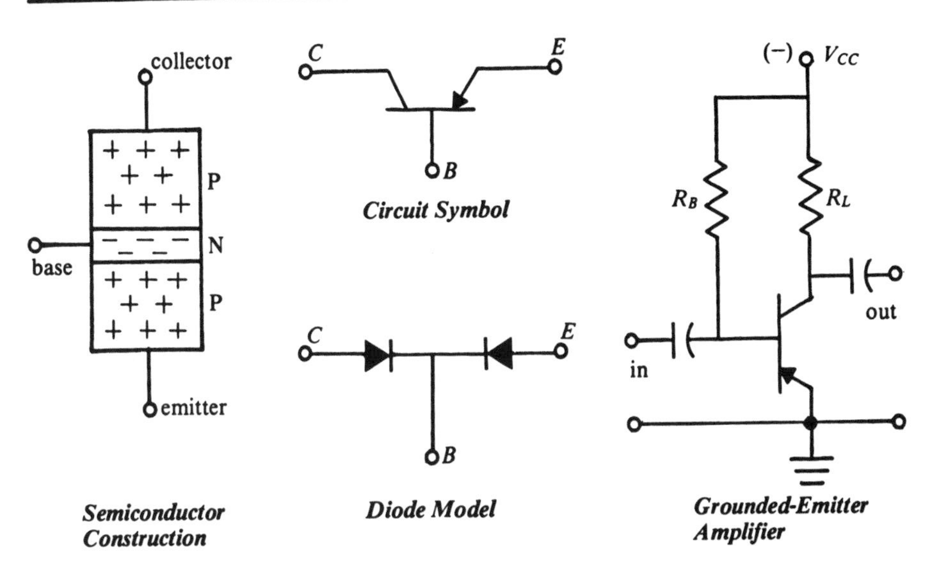

FIGURE A8–18. *PNP Bipolar Transistor*

Transistor Biasing

For pedagogical reasons, a simple base-bias scheme was used in discussing the bipolar-transistor grounded-emitter amplifier (for example, the circuit of Figure A8–13). In fact, we often encounter a more elaborate circuit because the transistor's β or h_{fe} typically varies greatly. As has been mentioned, β varies between devices with the same type number. Further, the β for a transistor increases significantly with temperature; it increases about 1% for each degree Celsius for silicon transistors and much more for germanium types. This is the reason germanium is now seldom used.

To largely overcome the problem of β variation, the better and more elaborate transistor bias circuit uses a voltage divider to establish the base current, along with a negative-feedback resistor between the emitter and ground. Problem 13 of the Problems for this chapter illustrates the form of the typical better-bias circuit; the interested reader may investigate this circuit further via that problem. Also, Figure A8–35b in the second supplement to this chapter shows a MOSFET in a better-bias circuit. The MOSFET circuit illustrated in the latter figure has the same structure and is used because the transconductance (or drain characteristic) for FETs generally varies greatly between unselected devices as well.

Transistor Current Amplifier/Switch Application

Suppose we want to construct a water-level sensing device that sounds a 120-V_{ac} buzzer (or activates some 120-V_{ac} device) when the level reaches a certain point. We propose to use a pair of wire probes and the fact that when water surrounds the probes, the resistance between them is 100 kilohms or less due to the conductivity of the water.

Assume we have available an electromagnetic relay with a coil resistance of 2 kilohms and that the coil requires 10 milliamperes to operate the relay. (Recall from Chapter A3 that the coil is a small electromagnet that attracts a small iron armature and thereby closes a pair of switch contacts if the coil current exceeds the pull-in current for the device.) The voltage that must appear across the coil to operate it is then 20 V minimum. We assume a 25-V_{dc} supply is available; this voltage provides a 25% operating margin for the coil.

A circuit to accomplish the task is shown in Figure A8–19. Notice first that current amplification is necessary. When 25 V is placed across the probes and the probes are in water, a current of only 25 V / 100 kΩ = 0.25 mA flows, which is much less than the 10 milliamperes required to operate the relay (and, of course, very much less than that required to operate a large lamp, motor, or other power-consuming device). Therefore the water probe is connected in series with the base-emitter junction of a transistor so that the water resistance effectively becomes R_B of Figure A8–15a. The transistor becomes a current-controlled switch placed in series with the coil in the 25-V circuit. When water is present around the probes, we desire the transistor switch to close and a coil current of at least 10 milliamperes to flow through the coil and transistor (as transistor collector current I_C).

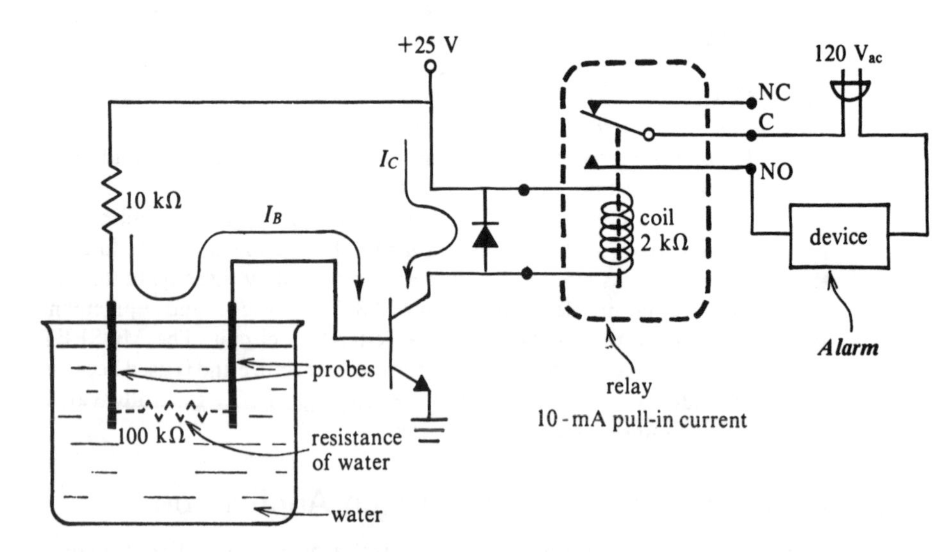

FIGURE A8–19. *Transistor Used to Drive a Relay-Controlled 120-V_{ac} Device*

A 10-kilohm protective resistor has been placed in series with the water probe in the base circuit to protect the transistor from direct connection of the base to 25 V in the event that the probes become accidently shorted. If water is present we calculate the base current from $V_{CC} = I_B R_B + V_{BE}$ or 25 V $= I_B$ (10 kΩ + 100 kΩ) + 0.6 V. Then $I_B \simeq 0.23$ mA. The collector current and therefore coil current is proportional to this by $I_C = \beta I_B$. We desire $I_C > 10$ mA; then $\beta I_B > 10$ mA, or $\beta > 10$ mA/0.23 mA = 43. The transistor chosen should have a β (forward-current transfer ratio h_{fe}) of at least 43 to multiply the base current up to the 10 milliamperes required to operate the relay. (A larger β does no harm since the transistor is simply driven further into saturation.) Furthermore, the transistor must have a maximum collector-emitter voltage rating $V_{CE,\max}$ of more than 25 V because when the transistor is not conducting, it acts as an open switch. The 25-V supply voltage will then appear across the transistor from collector-to-emitter. A transistor with β or $h_{fe} = 75$ or more and $V_{CE,\max}$ of 35 V or more would be a safe selection. The $I_{C,\max}$ specification for the transistor should be no problem since almost any contemporary transistor is capable of 100 milliamperes or more. Power is no problem either since V_{CE} is low when I_C is high, and vice versa.

Notice that the 120-V_{ac} circuit makes no electrical connection to the probe circuit, as the insulation between the armature and coil should be quite good. This electrical isolation of 120-V circuits from controlling electronics is often desirable. A final practical element is placed in this circuit. Notice the diode, placed with anode to collector and cathode to power supply. This diode arrangement protects the transistor from the large voltage that may be induced in the coil of the relay when the current through it is stopped suddenly by the opening of the transistor switch (recall inductive emf and Equation A1–41). A voltage transient can be produced that swings well above the supply voltage of 25 V and consequently exceeds the $V_{CE,\max}$ rating of the transistor. The diode becomes forward biased during such an excursion and clips off the voltage spike in a manner similar to that discussed in Figure A7–13 for the differentiating circuit.

Darlington Connection

The *Darlington connection,* which sounds like the name of a movie, is actually a rather simple and useful connection of two transistors. The transistors are connected so that the emitter current of the first transistor passes into the base of the second (see Figure A8–20). Then the base current I_{B1} into transistor 1 is amplified by the factor β of that transistor before it enters the base of transistor 2 as its base current I_{B2}. The collector current for transistor 2 is in turn greater than its base current by the factor of its β_2, so that the resulting device current is a factor $\beta_1 \times \beta_2$ greater than its input current. We can regard the transistor pair as a *super beta* transistor, and indeed such

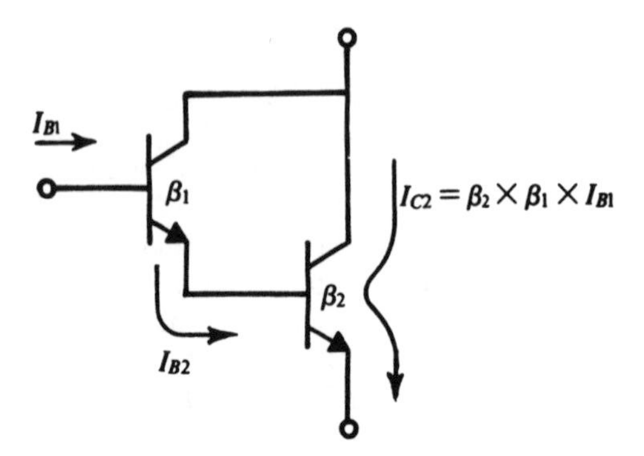

FIGURE A8–20. *Darlington Connection of Two Transistors*

devices are available in a single 3-terminal transistor package called Darlington transistors with a β on the order of $10^2 \times 10^2 = 10^4$. Notice that a Darlington transistor pair used in the water-sensing circuit of Figure A8–19 would increase the sensitivity greatly; the water resistance could be as high as 10 megohms and still give satisfactory operation.

Emitter Follower or Grounded-Collector Amplifier

The problem of low input impedance and generally high output impedance for the grounded-emitter amplifier was discussed earlier. The *grounded-collector amplifier* greatly improves on both of these ills, although at the price of providing only unity gain (actually slightly less than unity). The load resistor R_L is connected between the emitter and ground as shown in Figure A8–21 and the collector is connected to the positive power supply (for an NPN transistor). Although the collector is not at dc ground, it is in fact at ac ground if the power supply is of good quality and has a low impedance to ground. The name of the amplifier follows from this fact.

It is not difficult to see that if the input voltage goes up by a certain amount, the emitter voltage will increase by nearly as much. This is because the base-emitter voltage is basically that for a forward-biased diode, and the base-emitter voltage changes very little even though the current passing through the transistor changes a great deal. Therefore, the emitter voltage will quite faithfully follow the input voltage up and down, and the amplifier output will follow suit because it is connected (through a dc blocking capacitor) to the emitter; hence the grounded-collector amplifier's other name, *emitter follower*.

Because the emitter follows the base quite closely, there is a greatly reduced tendency for current to flow into the r_{BE} resistance that effectively appears between the base and emitter of the transistor. It is possible to show that the input impedance R_{in} of this amplifier is greater than merely r_{BE}; it is usually given to a good approximation as $R_{in} = r_{BE} + \beta R_L$. In addition,

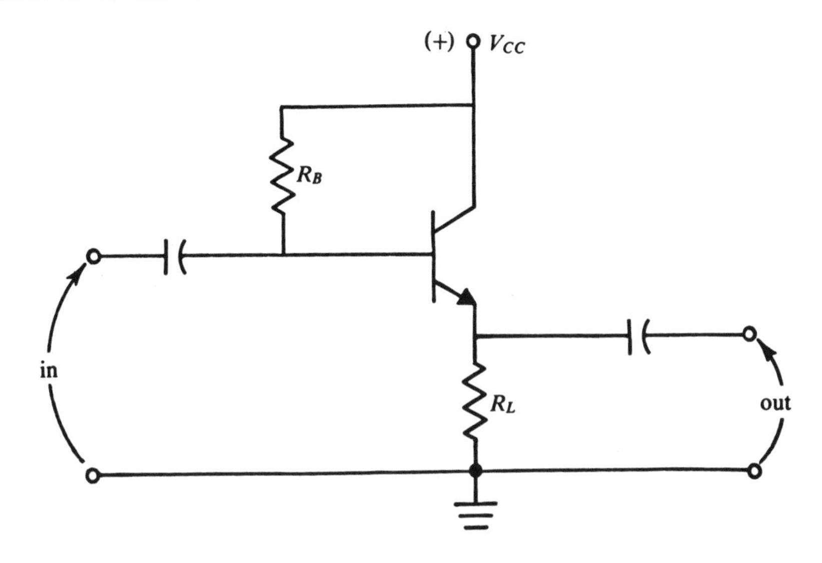

FIGURE A8–21. *Basic Emitter Follower or Grounded-Collector Amplifier*

the output impedance R_{out} is generally much less than R_L; it can be shown to be approximately given by $R_{out} \simeq [R_L \| (1/g_m)]$. Here g_m is the transconductance of the transistor, and its reciprocal has dimensions of ohms. For example, if $g_m = 0.1$ mho, $(1/g_m) = 10\ \Omega$. This value is generally much less than R_L so that the parallel resistance of R_L with $(1/g_m)$ is nearly just that of $(1/g_m)$, or 10 ohms. This is the typical output impedance of a low-power bipolar transistor emitter follower.

We can construct a follower amplifier with a triode, pentode, JFET, or MOSFET as well. They are termed *cathode follower* or *source follower* as the case may be. In all cases, the output impedance can be expected to be given by $R_{out} \simeq (1/g_m)$.

To obtain an amplifier with high input impedance, high voltage gain, and low output impedance, we can construct a multistage or cascaded amplifier with emitter followers as input and output stages in order to obtain desirable input/output impedances. In between may be placed one or more grounded-emitter stages to obtain the necessary gain.

Basic Differential Amplifier

Figure A8–22 illustrates a *basic differential amplifier* constructed from bipolar transistors. The amplifier essentially consists of two grounded-emitter–type amplifiers with the output taken between the two collectors. An input passes into the base lead of each transistor. Suppose each input changes

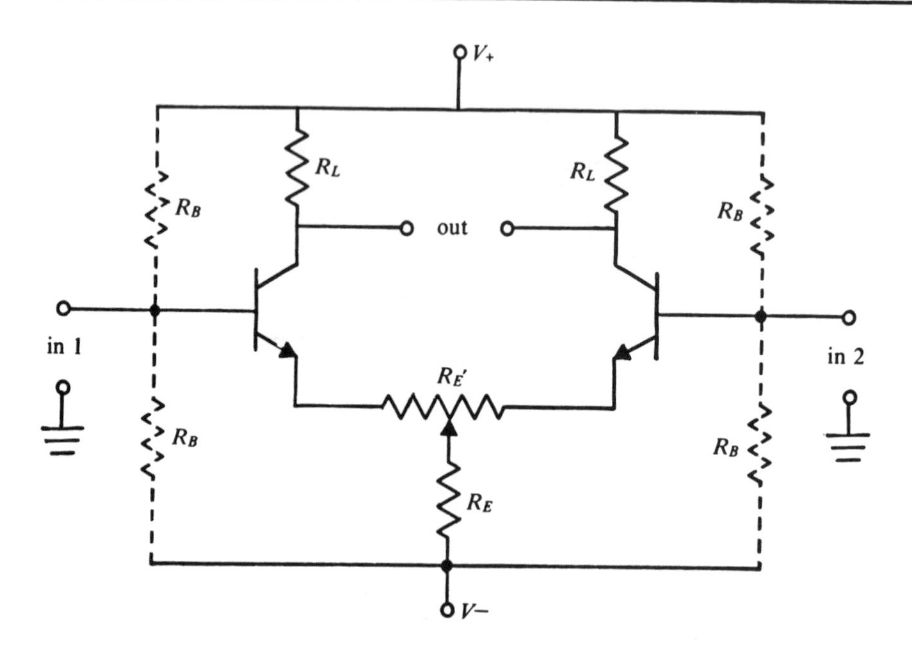

FIGURE A8–22. *Basic Differential Amplifier Constructed from Bipolar Transistors*

positive by the same amount. Then we expect the collector of each transistor to move negative by the same amount (though the amount is different from the input voltage) because of the symmetry of the circuit. The output voltage is really the *difference* between the two collector voltages, and it is clear that for perfectly matched resistors and transistors, the difference will be zero for the input case stated. Actually, this input case represents a common-mode input, and since the output is zero, we clearly have the common-mode rejection desired of a difference amplifier. The R_E' potentiometer serves to help balance the circuit against small mismatches in resistors or transistors.

If one input changes positive while the other changes negative, however, the collectors of the two transistors change in opposite senses by some amplified amount, and the output signal is then very substantial. Indeed, it is possible to show that the difference-voltage gain is approximately $g_m R_L$, where g_m is the transconductance of either transistor and R_L is the load resistance. To obtain further voltage gain, several difference amplifiers may be cascaded together. This is generally the method of choice to obtain high-gain dc amplifiers, and the examination of a dc oscilloscope or op-amp schematic invariably reveals several cascaded differential amplifiers. For example, to obtain very high input impedance in the RCA 3140 op amp, the first stage consists of a pair of MOSFETs.

When designing a differential amplifier, the base-bias current would generally be supplied by resistors such as those labeled R_B in Figure A8–22.

These resistors then appear in parallel with the effective input impedance of the transistor amplifier. In order to permit the largest possible input impedance, these resistors are *omitted* in virtually all operational amplifiers; therefore they are shown as dashed components in Figure A8–22. Consequently, the bias current required by the transistors must be supplied by the signal source connected to the op amp. This is precisely the warning made in Chapter A6 with regard to input bias current for op amps.

Other Semiconductor Devices

Phototransistor

The photodiode was discussed in Chapter A3. An interesting variant on this theme is the *phototransistor.* Such transistors are fabricated in a transparent case so that light may fall on the reverse-biased collector-base junction. Transistor action then essentially multiplies the photocurrent in the base region (which is generated by the light), and a device more sensitive than a simple photodiode results. The base lead may not be available from some phototransistors, so that they look rather like diodes. If the base lead is available, it is *not* used for the phototransistor application.

Figure A8–23 illustrates the symbol for a phototransistor and also the use of a phototransistor to operate a sensitive relay directly. Thus, if the incident light intensity is above 5 milliwatts/centimeter^2, the TIL81 photo-

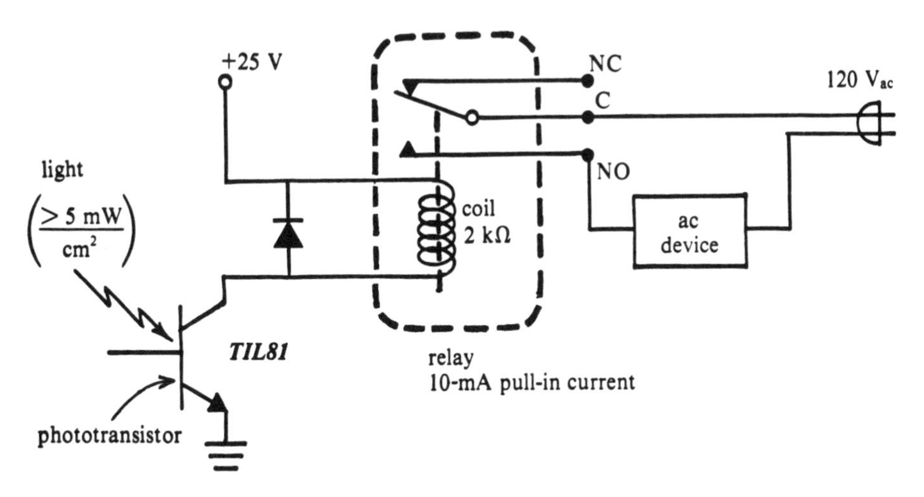

FIGURE A8–23. *Phototransistor Used in a Simple Relay Control Circuit*

transistor passes some 20 milliamperes for a V_{CE} greater than 3 V. The relay assumed for Figure A8–19 could then be operated by this phototransistor and is employed again in Figure A8–23.

Optocoupler

The optically coupled isolator or *optocoupler* is a useful, rather recent device for providing electrical isolation between systems operating with different voltages, different grounds, and so forth. It typically consists of an infrared *light-emitting diode* (LED) placed in a light-tight, electrically insulating package near a phototransistor. If a current is passed through the LED, light from the diode falls on the phototransistor, and current will pass through the transistor if a voltage exists across the transistor. The output current I_C through the transistor is approximately directly proportional to the input current I_f through the LED. If no current passes through the input LED, the output transistor behaves like an open switch. Figure A8–24 illustrates the essential construction and electrical characteristics for the 4N26 optocoupler. Optocoupler types also exist that have a Darlington transistor output stage; this output increases the dc current transfer ratio, as well as the maximum output capability. Thus, the 4N33 Darlington-transistor–type optocoupler has a current transfer ratio of 500 and a maximum current capability of 60 milliamperes. This may be contrasted with the corresponding respective values of 20 and 3.5 milliamperes that are listed for the 4N26 type in Figure A8–24.

Silicon-Controlled Rectifier (SCR)

The *silicon-controlled rectifier* or SCR is a semiconductor device that is very useful as an electronic switch for controlling rather large ac currents (on the

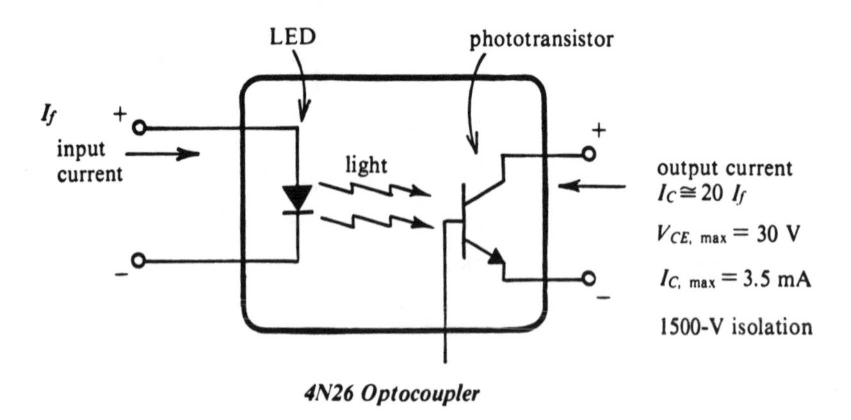

FIGURE A8–24. *An Optically Coupled Isolator*

order of amperes) to motors, light bulbs, electric heaters, and so on. It is a 4-layer PNPN device, with 3 leads attached called anode, gate, and cathode as shown in Figure A8–25a. The schematic symbol for an SCR is shown in Figure A8–25b. We see that there is a gate terminal added to the cathode of the familiar diode symbol. The diode arrow points in the possible easy-flow direction for current through the SCR.

The characteristic for the SCR is presented in Figure A8–26. For the purpose of this book, it is not necessary to justify this characteristic in detail in terms of the 4-layer semiconductor structure of the SCR, although some features can be understood in terms of elementary diode behavior. Rather, we consider the action of the SCR as presented by the experimental characteristic and discuss in general terms how it is commonly exploited.

Of course, the term anode implies that the terminal so named is normally connected to a voltage source in a manner to make it positive relative to the cathode. However, as we examine the possible current path from anode to cathode, we note a PN, an NP, and a PN junction; these are connected in series. The NP junction is *reverse biased* when the anode is connected positive.

Assume first that the gate is left unconnected or is negative relative to the cathode. Then, when a positive voltage is applied from anode to cathode, only a small *forward-leakage current* I_{FX} flows through the SCR until the anode is made very positive relative to the cathode. Examining the characteristic in Figure A8–26, we see that the SCR current finally begins to flow substantially when the voltage across the SCR exceeds the *breakover voltage*. This situation is represented by the SCR characteristic curve labeled $I_G = 0$ since no gate current I_G is flowing into the gate in this case. Of course, we expect that at some voltage the reverse-breakdown capabilities of the

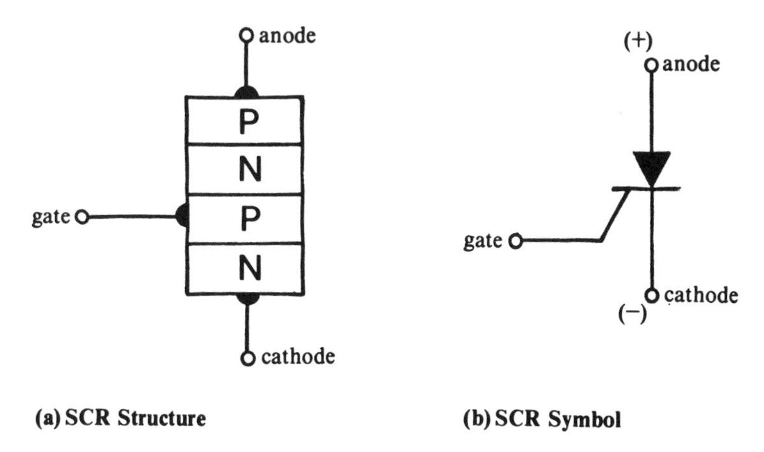

FIGURE A8–25. *Structure and Symbol for the Silicon-Controlled Rectifier*

blocking diode will be exceeded and current will flow. It is interesting that "once the dam is broken," so to speak, current flows easily through the device, with only a small voltage drop of about 1 V across the SCR (from anode to cathode). The SCR has changed from an open-switch mode to a mode much like a forward-biased diode or a closed switch.

Also notice in Figure A8–26 that very little *reverse-leakage current* I_{RX} flows when the anode is made negative relative to the cathode. This can be expected from the fact that the PN junctions in the SCR are reverse biased in this case (the other two junctions are forward biased). However, if the reverse voltage is made sufficiently large, reverse diode breakdown will occurs for the blocking junctions. This is just the same as the typical reverse breakdown shown for diodes in Figure A4–7. The voltage for reverse breakdown is about the same as the SCR's forward-breakover voltage (for $I_G = 0$). As was true for normal diodes discussed in Chapter A4, the reverse-breakdown situation can ruin the diode through VI—voltage × current—heating if the current is allowed to become large. The SCR is *not* similarly ruined in the forward-direction breakover case since $V_{\text{anode-cathode}}$ quickly drops to about 1 V (the VI product is then small). Typical SCRs can pass a current of 1 to 15 amperes in the forward direction without harm, depending on the SCR rating and whether proper heat sinking has been employed.

Generally an SCR is operated with the anode-cathode voltage always *less* than some *maximum forward-blocking voltage* V_{FM} and also less than some *maximum reverse-blocking voltage* V_{RM} in the reverse sense. For example, the V_{RM}, V_{FM} values for the 2N2322, 2N2323, 2N2324, 2N2325, and 2N2326 SCR devices are, respectively, 25 V, 50 V, 100 V, 150 V, and 200 V.

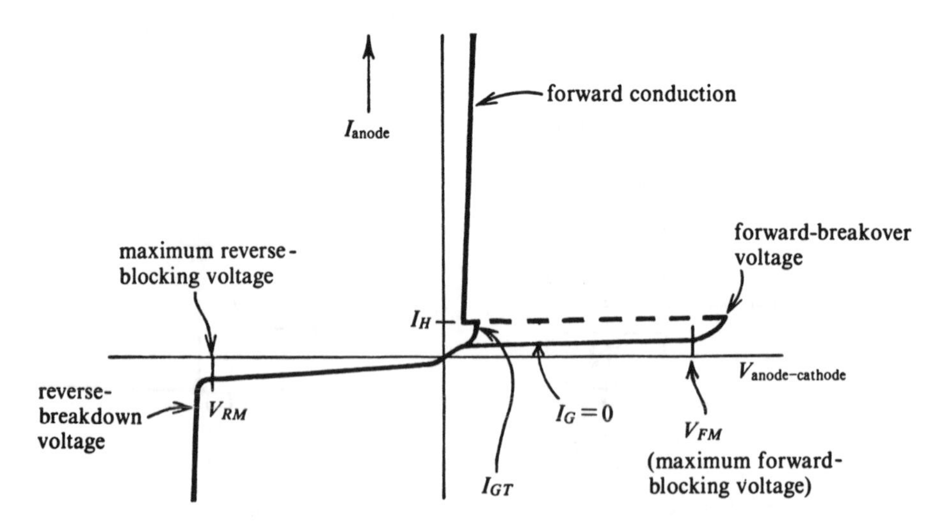

FIGURE A8–26. *Silicon-Controlled Rectifier Characteristic*

These SCRs can have an average on-state current of 1.0 amperes if their TO–5-style case is maintained at less than 85°C. The 2N1842/3/4/5/6/7/8/9/2N1850 SCRs have a blocking voltage ranging from 25 V for the 2N1842 to 500 V for the 2N1850. This family can handle the rather substantial current of 16 amperes. A 2N184X SCR must be bolted to a heat sink to achieve this current, and the threaded stud is in fact the anode.

However, when a gate current is supplied to the SCR, the forward-breakover voltage (but not the reverse-breakdown voltage) is reduced in proportion to the gate current. If a gate current greater than some *gate-trigger current* I_{GT} is supplied to the gate, the device goes into the breakover or conduction state for a *small* positive anode-cathode voltage on the order of 1 to 2 V. The gate must be brought positive by only about 1 V relative to the cathode to cause the gate-trigger current I_{GT} to flow. Notice that this voltage forward biases the bottom PN junction and, therefore, gate current flows into the SCR. For the low-power 2N232X family of SCRs, I_{GT} is less than 0.1 milliamperes. For the rather high-power 2N184X family, I_{GT} is typically less than 15 milliamperes, although 80 milliamperes is the guaranteed maximum required.

Once in the conducting state, the SCR remains so even after the gate current/voltage is removed. The SCR remains in the "on" state until the anode voltage is reduced so that the anode current falls below a small *holding current* I_H (typically a few milliamperes). The SCR will certainly be turned off if the anode is actually made negative relative to the cathode. However, a negative voltage to the SCR gate will *not* turn off the SCR. The gate-trigger current I_{GT} to turn on an SCR must be supplied for some minimum time, ranging from about 1 microsecond for 1-ampere SCRs, to about 20 microseconds for 15- to 50-ampere SCRs. The turn-on time for an SCR is on the order of 1 microsecond, while the turn-off time after reverse bias is about 20 microseconds.

In order to experiment with an SCR, a simple circuit can be constructed with an SCR, a light bulb, and a switch. The circuit is illustrated in Figure A8–27a and shows how an SCR can electronically control the current through the bulb. Assume first that 120 V_{dc} is applied to the series circuit. We require the blocking voltage capability V_{FM}, V_{RM} for the SCR to be greater than 120 V for these experiments. Then the light bulb remains dark as long as the gate to the SCR is connected to its cathode or is left open-circuited. The SCR behaves like an open switch. However, if the gate is brought positive by about 1 V relative to the cathode, gate current will flow that exceeds I_{GT} (provided the voltage source to the gate is capable of this small current of a few milliamperes). The SCR is triggered into the "on" condition, and the light bulb lights as current flows. The current continues to flow even if the gate voltage (and hence gate current) is removed. Thus, only a brief gate-current pulse is required to turn on the light permanently. This fact is indicated by the pulse form to the gate in Figure A8–27a. Of course, if the switch S is opened, the current to the light is interrupted and the light

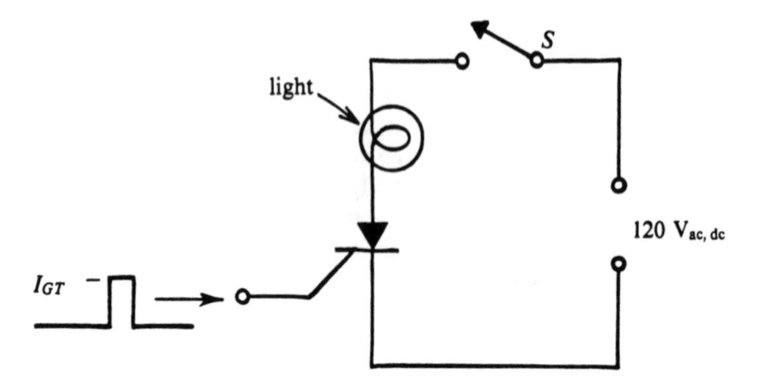

(a) Simple Control of a Light Bulb

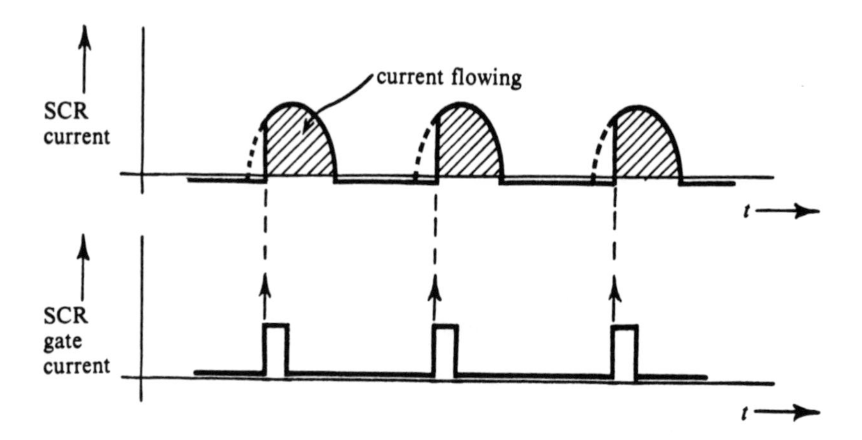

(b) Current Flowing in Nearly Full Half-Cycles

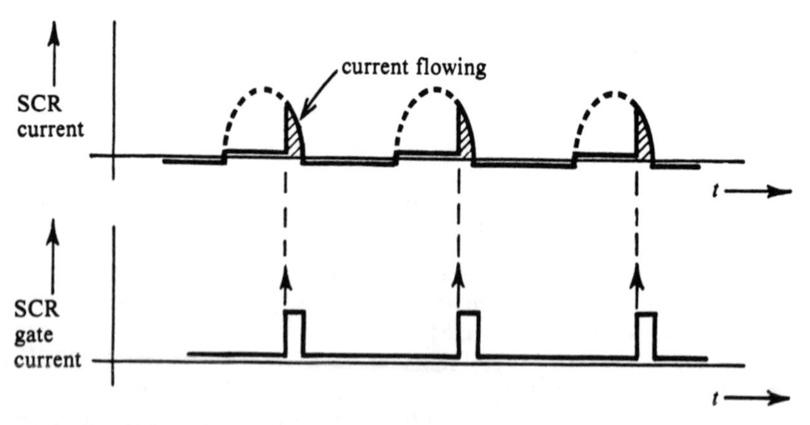

(c) Current Flowing for Small Portions of Each Cycle

FIGURE A8–27. *Control of Current (Especially ac Current) with an SCR*

bulb goes dark. It then remains dark after the switch is closed until another current pulse is input to the SCR gate. Another way to extinguish the bulb is to reduce the voltage to the SCR/bulb to zero or even negative. Then the bulb remains dark after the voltage is increased to 120 V_{dc} again.

Suppose on the other hand that 120 V_{ac} is connected to the light bulb and SCR series circuit in Figure A8–27a. Even if switch S is closed, the light remains dark until a gate-trigger current I_{GT} is supplied to the SCR. If this gate current is supplied *continuously* to the SCR, the light bulb glows since current flows each half-cycle that the SCR is positively biased. Half-rectified current (such as shown in Figure A4–9c) flows through the bulb, as the SCR behaves very much like a simple semiconductor diode. However, if the gate current is removed from the SCR, the light bulb extinguishes as soon as the supply voltage swings negative. The SCR current stops flowing and the SCR reverts to the "off" state. Therefore, the SCR behaves rather like an electrically controlled switch or relay in this application. The SCR can be a convenient solid-state switch for ac currents.

The SCR can be used to control the light bulb brightness in the ac case of Figure A8–27a by repetitively sending the current pulse I_{GT} to the SCR gate. One pulse is sent to the gate during each cycle of the 120 V_{ac}, as illustrated in Figure A8–27b and Figure A8–27c. By adjusting the timing of the gate pulses so that they occur early in the cycle (see Figure A8–27b), the light bulb shines quite brightly (at only half of its normal power since only half-wave current is flowing). However, if the gate pulse is applied late in each half-cycle (see Figure A8–27c), the current flowing through the bulb is reduced to only a fraction and the bulb shines dimly. Typically, this timing is adjusted through the use of an RC timing circuit with an adjustable R. (Such an RC circuit is illustrated later in conjunction with a similar triac circuit in Figure A8–29.)

Triac and Diac

The *triac* (or TRIAC) semiconductor device performs like two SCRs in parallel and oriented in opposite directions. The symbol for the triac is shown in Figure A8–28a. When sufficient current I_{GT} is supplied to the gate of a triac, it is triggered into conduction and conducts current in *either* direction through the device. This bidirectional current capability can be seen from its characteristic shown in Figure A8–28b. Therefore, it is well suited to use as an electrical switch in ac circuits. By proper timing of the gate-current pulses, the SCR current can be adjusted from full-wave ac (in contrast to the half-wave rectified case of Figure A8–27b) to small current or no current.

As example applications, triacs are commonly used in light-dimmer circuits and as motor-speed controls for universal ac/dc electric motors. The timing of the trigger current is made adjustable during each half-cycle through use of an RC circuit with adjustable R. If the triac is turned on later in the

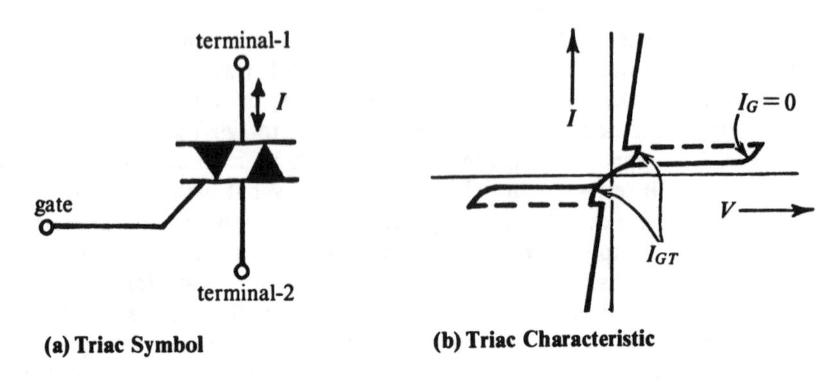

FIGURE A8–28. *Triac Semiconductor Switch*

half-cycle, the light becomes dimmer due to the reduced *duty cycle.* In the case of the motor, it runs more slowly with a reduced duty cycle. The schematic of a simple circuit to achieve a light dimmer with a triac is shown in Figure A8–29.

The circuit of Figure A8–29 uses still another device with a breakover characteristic called a *diac* or bilateral trigger. This device breaks over in *either* direction at a voltage of about 20 V to 35 V, depending on the type. It is like a triac without a gate, and its symbol in Figure A8–29 corresponds to this fact. A diac is generally placed in a triac circuit such as this to buffer the *RC* phase-shift (or timing) network from the SCR gate until the firing point

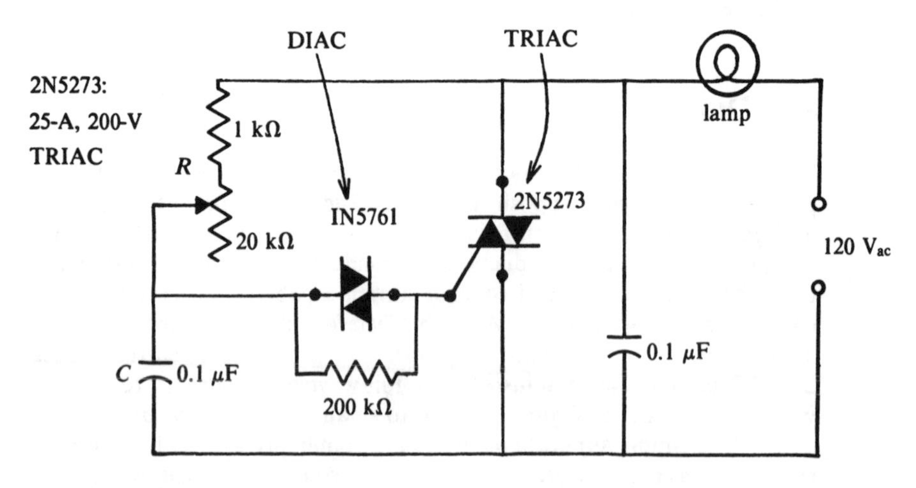

FIGURE A8–29. *Triac in a Light-Dimmer Control Circuit (Courtesy of Texas Instruments)*

is reached. Alternatively, it *is* possible to use a high-voltage transistor or FET follower between the *RC* circuit and the triac gate to provide the isolation desired.

In the vacuum tube era, a triode with a mercury vapor within the envelope served a switching function very similar to that of the modern-day SCR. This tube is called a *thyratron*. In a similar manner, an SCR or a triac solid-state device is referred to as a *thyristor*.

Solid-State Relay

In recent years, *solid-state relays* have been gaining wide usage in place of electromechanical relays. They have no moving parts, nor do they have the contact points that become pitted and inoperative after perhaps 10^5 or 10^6 operations in conventional relays. The typical solid-state relay represents a wedding between the optocoupler and a triac; Figure A8–30 illustrates its essential structure. If a certain minimum current is applied to the control input, the LED causes a photodetector in the device to trigger the triac, and an ac-capable switch effectively closes between the load output lines. As shown in Figure A8–30, the output lines should be connected in series with the ac device to be controlled, and the device will then be turned on for as long as current is supplied to the control input. The device indeed functions in nearly the same way as a relay; however, it will *not* switch a dc load, while a conventional relay can, of course. The triac is replaced with a power transistor in devices intended for dc loads.

A typical solid-state relay might operate with 15 milliamperes at 2 V to its input and control up to 10 amperes at 140 V_{ac}. Thus, a 0.03-watt input can control 1400 watts, with a life expectancy of 10^8 operations. These devices are very useful in electronic control of high-power devices; for example, when a microcomputer must control an electric motor.

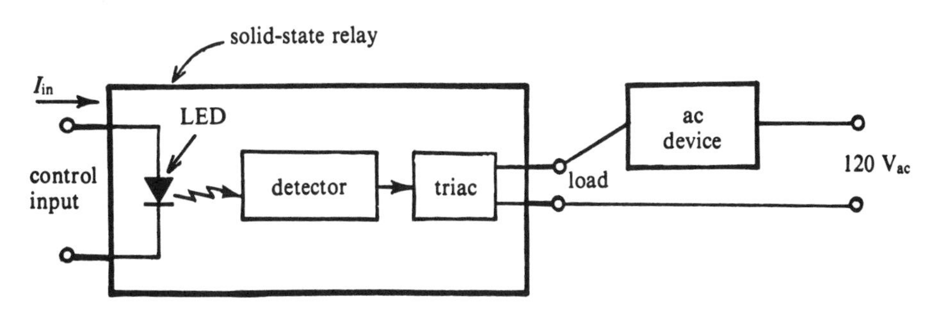

FIGURE A8–30. *Solid-State Relay*

Supplement to Chapter: Triode Characteristics and Basic Circuit

The essential construction, circuit symbol, and current-controlling action of the triode were given in the chapter.

There are three variables involved in the electrical characterization of the triode: the *plate current* I_b flowing from plate to cathode (conventional-current sense); the *plate-to-cathode voltage* V_b; and the controlling *grid-to-cathode voltage* V_c. A circuit to obtain the relationship between these variables is shown in Figure A8–31a, and a typical *plate-characteristics* graph is shown in Figure A8–31b. Notice that curves of constant grid voltage V_c are drawn. The curve for $V_c = 0$ is indeed like that for a diode, while making the grid more negative at a given plate voltage V_b does decrease the plate current I_b.

Figure A8–32 shows a basic *triode grounded-cathode amplifier* that uses a *bias battery* V_{CC} as well as *supply battery* V_{BB}. The V_{CC} bias is usually chosen such that the voltage across R_L ($V_{R_L} = I_b R_L$) is about half the supply voltage V_{BB}. Then about half of V_{BB} also appears across the tube.

It is inconvenient to use two batteries or supply voltages. The essential requirement in biasing the tube is that the grid be somewhat negative relative to the cathode; this is usually accomplished by connecting the resistance R_g to ground and introducing a resistance R_K between the cathode and ground as shown in Figure A8–33. Now the current I_b flowing through R_K causes the cathode to be positive relative to ground, as shown in the figure. However,

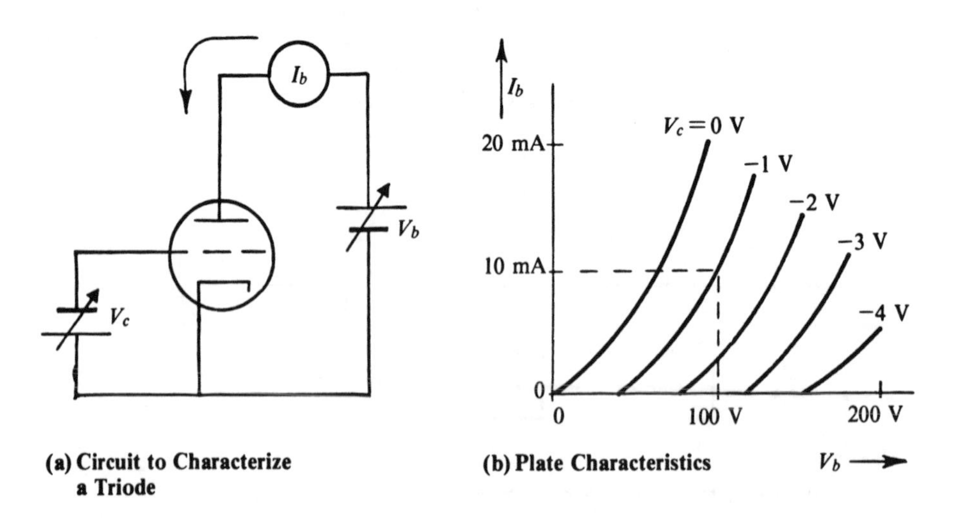

(a) Circuit to Characterize a Triode

(b) Plate Characteristics

FIGURE A8–31. *Characteristics of a Typical Triode*

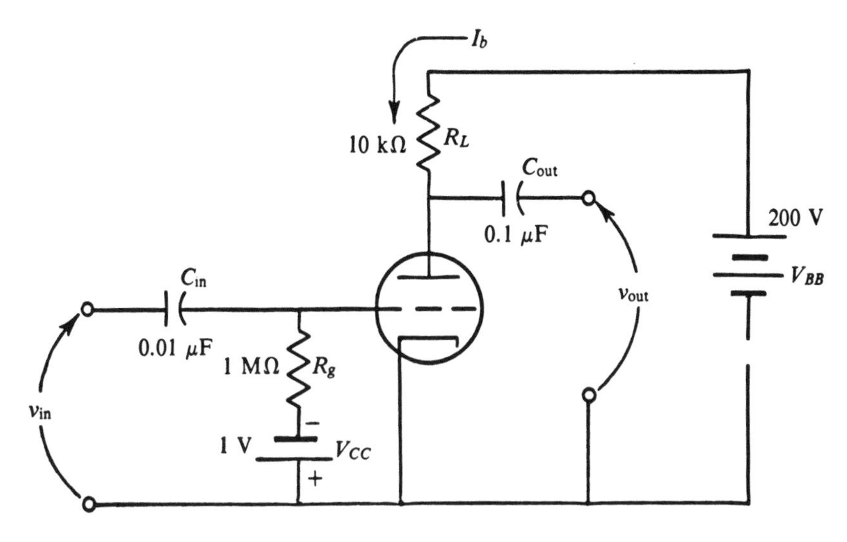

FIGURE A8–32. *Triode Grounded-Cathode Amplifier Using Two Batteries or Supply Voltages*

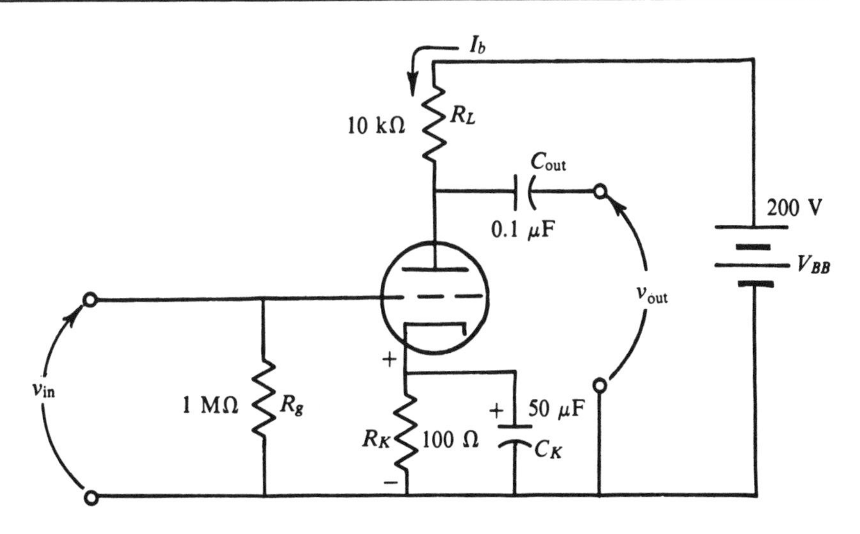

FIGURE A8–33. *Self-Biased, Grounded-Cathode Triode Amplifier*

the grid is nearly at ground because very little current flows to the grid and through R_g to ground. Thus the grid is negative relative to the cathode as required.

The capacitor C_K smooths out the ac that tends to develop across R_K when amplifying an ac voltage; we want a dc voltage across R_K, or else negative feedback will decrease the amplification obtained. The resulting amplifier is called a *self-biased, grounded-cathode amplifier.* The cathode is not at ac ground (as it is in Figure A8–32), but it *is* at dc ground if the large capacitor C_K effectively shorts ac currents to ground.

Supplement to Chapter: MOSFET Construction and Operation

Some of the general properties of the metal-oxide-silicon field-effect transistor (MOSFET), as well as circuit symbols, were given in the chapter. The central feature of this device is the fact that the gate structure is insulated from the channel; hence, the name *insulated-gate FET* or IGFET is also applied to these devices.

N-Channel Depletion-Type MOSFET

Two different schemes exist for constructing MOSFETs that result in the enhancement type and the depletion type. The *depletion type* operates rather similarly to the JFET and is discussed first. Figure A8–34a shows the essential features of the construction. For the N-channel version, drain, channel, and source regions are all constructed of N-type material, produced on a base or *substrate* of P-type material. Metal contacts are made to these three regions (as well as to the substrate), but a layer of insulation is established between the gate metal plate and the N-type semiconductor channel. Now suppose the gate is made negative relative to the source as shown. The negative gate plate attracts holes into the channel region near the plate. This configuration partially *depletes* the channel of some of the N-type carriers that conduct electricity between the drain and the source; hence the name *depletion* for this MOSFET type. The size of the channel is effectively reduced, and the device operation is very similar to that for the N-channel JFET. Its characteristic is therefore also similar to that drawn in Figure A8–6f.

However, the gate can also be made positive relative to the source without harm to the MOSFET (this *cannot* be done for the N-channel JFET since the gate diode becomes forward biased). The channel now is actually enlarged or *enhanced* into the P-material substrate, and more current will flow through the FET. Thus I_D versus V_{DS} curves exist also for positive V_{GS} values, and these simply occur in regular fashion *above* the negative V_{GS} curves. Figure A8–34b shows the characteristic curves for the N-channel depletion-type MOSFET.

The substrate P-type material must be *reverse biased* relative to the N-type channel so that it behaves like an inert base material. A connection

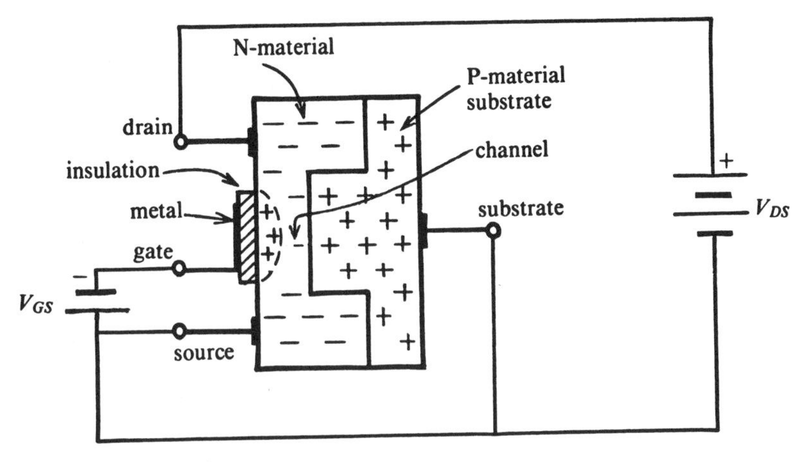

(a) Construction Scheme

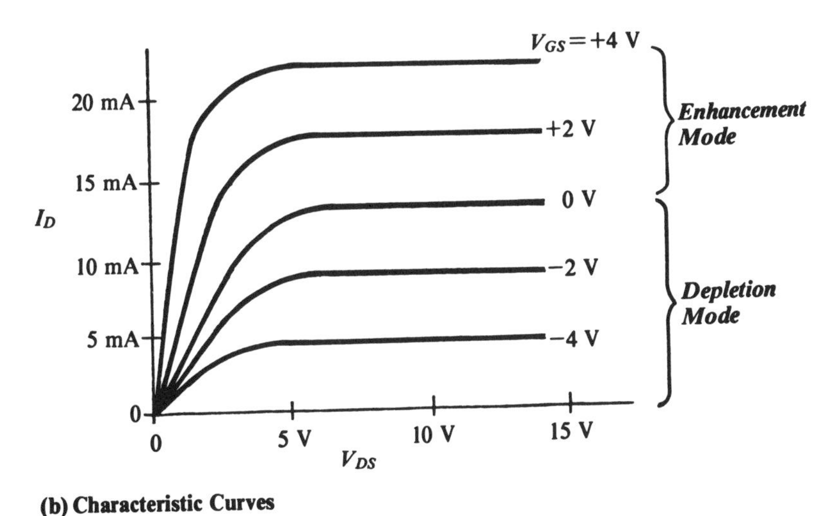

(b) Characteristic Curves

FIGURE A8–34. *N-Channel Depletion-Type MOSFET Construction and Characteristics*

exists to the substrate material that is often internally connected to the source lead inside the transistor package; if it does extend separately out of the package, the substrate lead is simply connected to the source lead or perhaps to ground.

The circuit symbol for the N-channel depletion-type MOSFET was shown in Figure A8–9a. Notice that the gate is drawn as quite separate from the channel to emphasize its insulated aspect. Note also that the N-channel version is designated by the arrow on the substrate lead, which shows the easy-current-flow direction through the substrate-to-channel diode junction.

Figure A8–35a shows a typical self-biased amplifier configuration for the N-channel depletion-type MOSFET. The arrangement of components and supply voltage is the same as for the N-channel JFET discussed earlier in Chapter A8. However, the gate need not be biased negatively relative to the source, but can even be biased positively. Thus R_S and C_S can be eliminated from the circuit of Figure A8–35a to give an amplifier operating with zero V_{GS} bias. To bias the gate positive relative to the source, a voltage divider operating from the $+V_{DD}$ power supply is generally used, as is illustrated by R_1 and R_2 in Figure A8–35b. If the R_S resistor is retained, this circuit has the advantage of accommodating FETs with considerable variation in characteristics; that is, variations in saturation current and pinch-off voltage.

N-Channel Enhancement-Type MOSFET

The essential construction features of the N-channel *enhancement-type* MOSFET are shown in Figure A8–36a. Notice that two separate N-channel regions are produced in the volume of P-type substrate, with *no* N-type material connecting the two; rather there is P-type substrate material be-

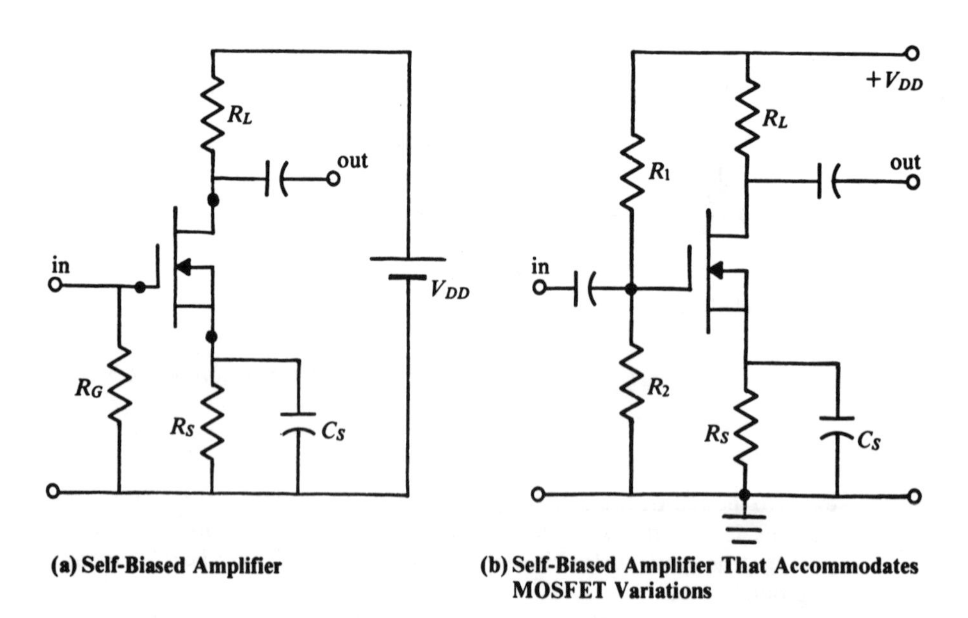

(a) Self-Biased Amplifier

(b) Self-Biased Amplifier That Accommodates MOSFET Variations

FIGURE A8–35. *N-Channel Depletion-Type MOSFET Grounded-Source Amplifier Circuits*

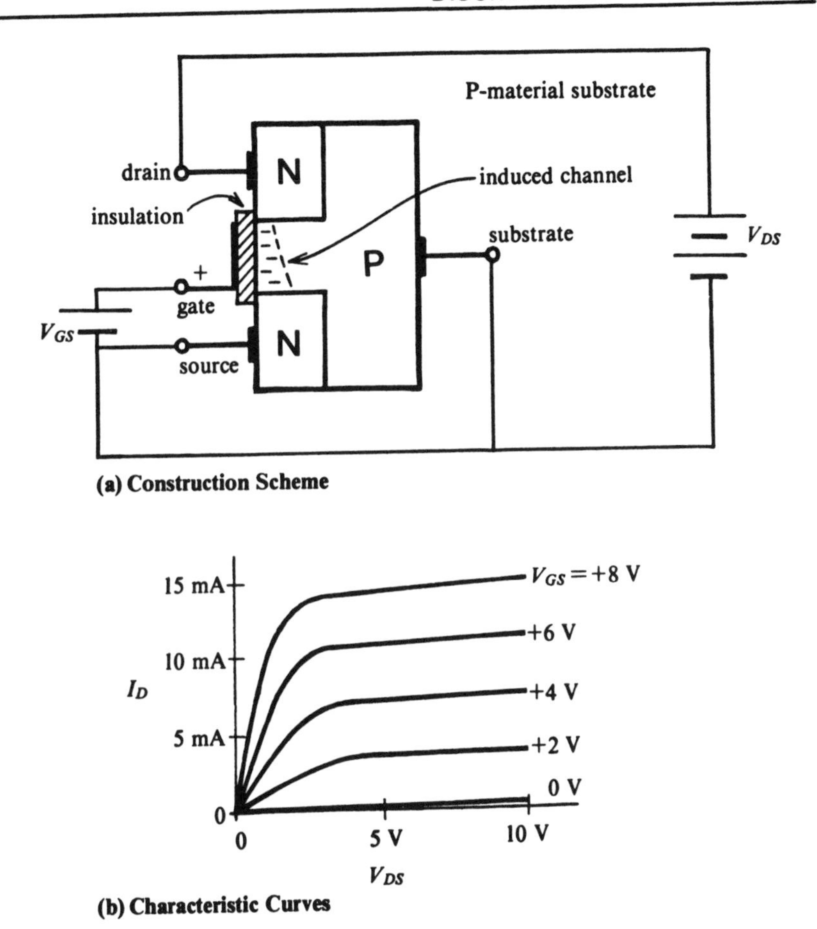

FIGURE A8–36. *N-Channel Enhancement-Type MOSFET Construction and Characteristics*

tween the two regions. These regions become the drain and source when metallic connections are attached. Insulating material is placed over the area between the drain and source regions, and a metal plate for the gate is placed on top of this, as shown in Figure A8–36a.

The NPN regions give two diode junctions rather like the bipolar transistor. Then if only a voltage V_{DS} is connected between the drain and the source, no current will flow because of the back-biased drain-substrate diode. However, if the gate is made *positive* relative to the source by some V_{GS}, (−) carriers (which are minority carriers in the P substrate) are attracted out of the substrate bulk to the region under the gate plate. Then an N-channel bridge is induced between the drain and source so that essentially a piece of

N-type material now exists between the drain and source leads and current will flow under the push of V_{DS}.

A large positive V_{DS} will cause a large channel and a large drain current will flow. The resulting characteristics are shown in Figure A8–36b. Notice that when $V_{GS} = 0$ there is only a tiny leakage current through the device; the MOSFET is turned off very much like the bipolar transistor is under similar circumstances. This behavior makes the enhancement-type MOSFET convenient for use as a switch in on/off or digital circuits. Indeed, it is very commonly used in microprocessors and other large-scale digital integrated circuits. The term *enhancement* comes from the fact that a channel is enhanced into existence in this type of MOSFET. The circuit symbol for this device graphically shows the break between the drain and the source (see Figure A8–9b).

An amplifier for the N-channel enhancement-type MOSFET must have the gate biased positively relative to the source, and a circuit such as that in Figure A8–35b is typically used.

P-Channel MOSFETs

We have examined the N-channel version of both the depletion and enhancement types of MOSFETs. Clearly the substrate could be made of P-type material rather than N type, and the channel should then be P type. This construction change produces a *complementary* or P-channel form of each MOSFET type, and the circuit symbols are shown on the right side of Figure A8–9.

The story is now very much like that for the complementary forms of the JFET and the bipolar transistors: All applied voltages should have reversed polarity. Thus if a P-channel MOSFET is used, the drain supply voltage V_{DD} in Figure A8–35 must supply negative voltage to the drain; turn the battery V_{DD} around in the figure. Also, the P-channel enhancement-type MOSFET requires the gate to be negative relative to the source in order to turn it on; that is, give drain-current flow.

REVIEW EXERCISES

1. (a) What is the difference between a vacuum diode and a vacuum triode?
   (b) Briefly, how does a vacuum triode control current?
2. (a) What is generally controlled by a discrete electronic device?
   (b) Give the definition of transconductance, both in words and as a mathematical relation.
   (c) What are the dimensions of transconductance?
3. (a) Diagram the (pedagogical) structure of an N-channel JFET. Label the three leads attached.

(b) Draw the corresponding JFET symbol. What aspect of the symbol indicates the N-channel character?
(c) What is the channel in the device?
(d) Why does a large negative gate-source voltage pinch off the current through an N-channel JFET?
(e) What is meant by the saturation current for a JFET? Sketch the typical FET characteristic; label the axes and the curves and point out the saturation current on it.

4. Name two safe operating limits specified by manufacturers for JFETs.
5. What is the reason for the interest in the new VMOS devices?
6. (a) What is meant by the bias or biasing in an amplifier?
   (b) How is a JFET amplifier typically self-biased?
7. (a) What determines the output impedance of a grounded-source amplifier?
   (b) What determines the input impedance of a self-biased grounded-source amplifier?
   (c) What determines the voltage gain of a grounded-source amplifier?
8. (a) Diagram the construction and the symbol for the P-channel JFET.
   (b) What determines the arrow direction for the gate in the P-channel JFET?
   (c) Briefly characterize the difference in the electrical usage of the P-channel JFET from the N-channel device.
9. (a) What do the letters in MOSFET stand for?
   (b) Which MOSFET type has an electrical behavior or characteristic very similar to the corresponding (N-channel or P-channel) JFET?
   (c) What distinguishes the other MOSFET type physically and electrically?
10. (a) Select the NPN transistor from the symbols shown in Diagram A8–1.
    (b) Select the N-channel FET from the symbols shown in Diagram A8–1.

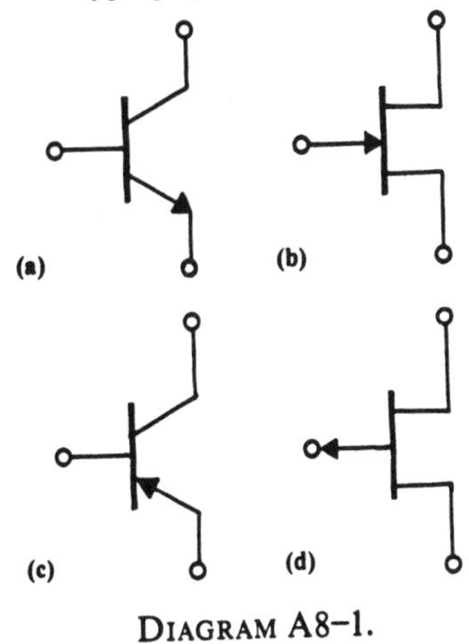

DIAGRAM A8–1.

11. (a) Diagram the (pedagogical) construction of an NPN transistor along with the schematic symbol; label the three connections for both.
    (b) Draw a diode equivalent for the NPN transistor.
    (c) Why is the collector current much larger than the base current in a normally operating transistor?

(d) Why can an NPN transistor not be constructed by soldering the anode leads of two diodes together?

12. What is the forward-current transfer ratio of a bipolar transistor? Give the definition both in words and as a formula.

13. (a) Why are two graphs required to show the electrical characteristics of a bipolar transistor?
NOTE: Only one was required for the FET or the vacuum tube.
(b) What determines the ac input resistance r_{BE} for a transistor? Answer with reference to an appropriate graph.
(c) Draw a collector characteristic for some typical transistor, and show how the β for the transistor may be determined from it.
(d) Relate the transconductance of a transistor to the above quantities.

14. (a) What determines the input impedance of the grounded-emitter amplifier?
(b) What determines the output impedance of the grounded-emitter amplifier?
(c) What determines the voltage gain of the grounded-emitter amplifier?
(d) What penalty is paid for reducing the output impedance of a grounded-emitter amplifier?

15. The transistor with SPDT switch S controls the light bulb in the circuit shown in Diagram A8–2.
(a) What setting of the switch turns off the light? Explain why briefly.
(b) What setting of the switch turns on the light? Briefly explain why.
(c) Why would the transistor probably be ruined if resistor R_b is omitted?
NOTE: Consider the diode model of the transistor.

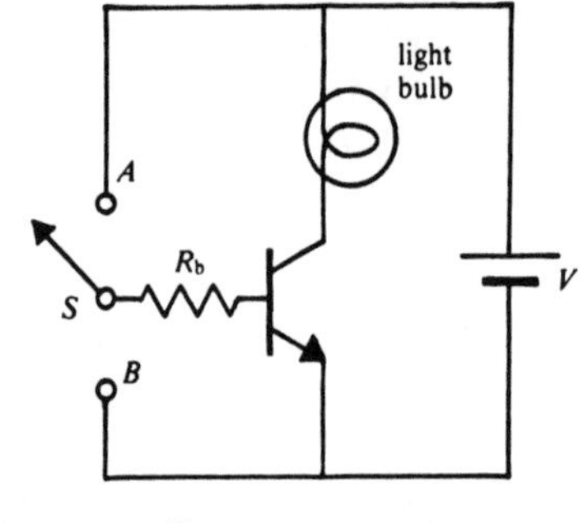

DIAGRAM A8–2.

16. What is meant by the saturation region for a bipolar transistor?

17. (a) Illustrate transistor breakdown behavior in a sketch of an appropriate transistor characteristic.
(b) What manufacturer transistor rating is related to this behavior?

18. Consider either a grounded-source or grounded-emitter amplifier. When the load resistor R_L is increased (choose one): A. The output impedance and the voltage gain both increase. B. The output impedance increases, while the voltage gain decreases. C. The output impedance decreases, while the voltage gain increases. D. The output impedance and voltage gain both decrease.

19. In which of the following is the input element—grid, gate, or base—biased positive with respect to the grounded element—cathode, source, or collector? A. triode, B. N-channel FET, C. NPN transistor.

20. Discuss why the input impedance of an FET amplifier is much greater than that

of a bipolar transistor amplifier. Invoke the construction of each device in your discussion.

21. (a) What is the universal voltage-gain formula for a grounded-source/emitter/cathode amplifier constructed from an FET/transistor/pentode?
    (b) Explain why it is difficult to construct an amplifier from a single transistor or tube with an accurately predictable voltage gain.
22. It is useful to have a constant-current source to obtain the conventional characteristic for (choose one): A. FET, B. triode, C. (bipolar) transistor, D. MOSFET.
23. When the temperature of a bipolar transistor increases, its h_{fe} (choose one): A. decreases, B. remains almost constant, C. increases.
24. (a) Diagram the (pedagogical) construction of a PNP transistor. Draw the corresponding PNP transistor symbol and diode model.
    (b) What should be the polarity of the base relative to the emitter to forward bias the PNP transistor?
25. What basic property of the bipolar transistor may be exploited when the current in a situation is not sufficient to operate a relay?
26. Suppose very large current amplification is required. Diagram a simple transistor circuit that provides it and give its name.
27. (a) Diagram a transistor amplifier that affords low output impedance and relatively high input impedance.
    (b) Give two names for this amplifier and justify each briefly.
28. (a) Diagram the basic amplifier type that is used for the input of an op amp (and in dc amplifiers generally).
    (b) What is missing from this amplifier when used for an op-amp input? What is consequently required of the circuitry connected to the op-amp inputs?
    (c) Argue why this amplifier type has high common-mode rejection.
29. What is the advantage of a phototransistor over a photodiode?
30. (a) Diagram an optocoupler. What is its purpose?
    (b) Discuss qualitatively why the output current might be expected to be proportional to the input current.
31. (a) What does SCR stand for?
    (b) Diagram the symbol for an SCR and label its leads.
    (c) What is the utility of the SCR?
32. (a) Diagram the symbol for a triac.
    (b) What is the advantage of a triac over an SCR?
33. (a) Diagram the components of a solid-state relay that utilizes a triac.
    (b) Why can't this relay type be used with dc loads?
    (c) Name two advantages of a solid-state relay over a conventional relay.
    (d) A conventional relay provides good electrical isolation between its input and output. How does a solid-state relay accomplish a similar desirable characteristic?

PROBLEMS

1. (a) Suppose that an ohmmeter is alternately connected in the two connections A and B as shown in Diagram A8–3 to a "good" N-channel FET. The + terminal of each ohmmeter connection is shown. Tell whether the ohmmeter reads a small resistance or a large resistance in each case.

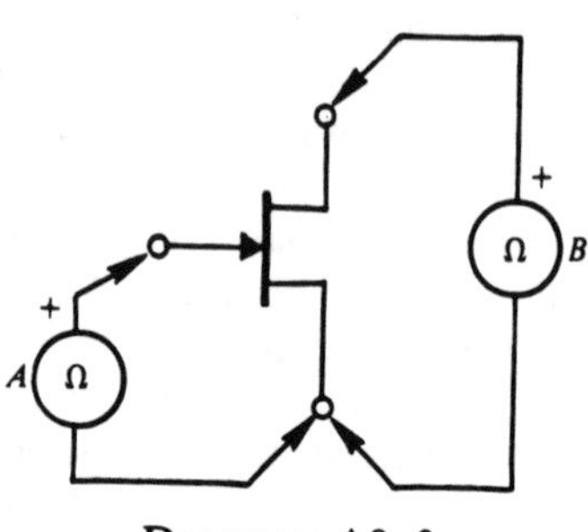

DIAGRAM A8–3.

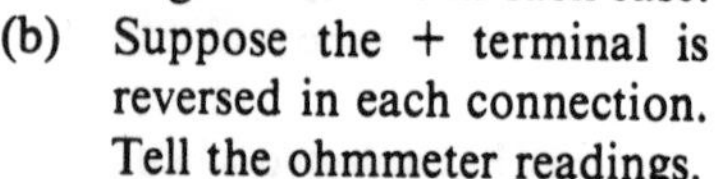

(b) Suppose the + terminal is reversed in each connection. Tell the ohmmeter readings.

2. (a) The saturation current for a certain FET is 20 mA, and its pinch-off voltage is −6 V. $V_{DS,\max}$ for the device is 30 V. Sketch the characteristic for this FET, making the *approximation* that the curves of constant gate voltage are equally spaced.
   (b) Calculate the transconductance of the FET in this approximation.
   (c) If the FET is used in a grounded-source amplifier with a load resistance of 10 kΩ, what voltage gain is expected?
   (d) If the FET is used in a source-follower circuit, what output impedance is expected?
3. Assume I_{DSS} for the JFET in the circuit pictured in Diagram A8–4 is 6 mA, and the pinch-off voltage is very small. What is the voltage V_{out} (relative to ground)?

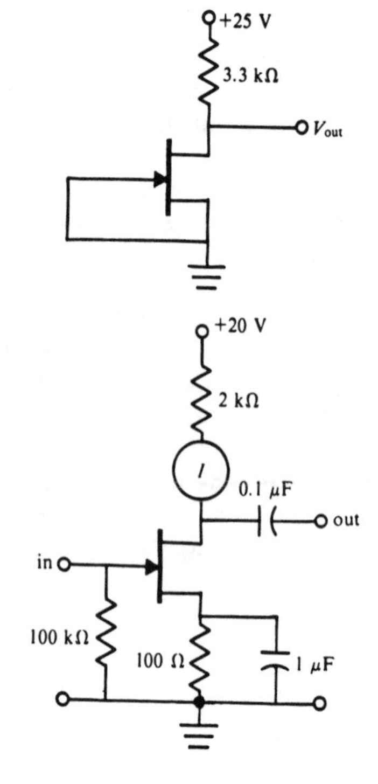

DIAGRAM A8–4.

4. (a) Consider the circuit shown in Diagram A8–5. Someone claims the ammeter I reads 20 mA. Explain why this is impossible if all other features are correct as drawn.
   (b) What input impedance is expected for this amplifier?

DIAGRAM A8–5.

(c) What output impedance is expected for this amplifier?
(d) If the transconductance of the FET is 10 mmhos (10^{-2} mhos), what voltage gain is expected for the amplifier?

5. (a) Suppose a JFET is operated with its drain-source voltage at less than 50% of its pinch-off voltage. Argue from its characteristic that between the drain and source, it functions like an Ohm's Law resistor whose resistance can be changed by changing the gate-source voltage V_{GS}.
(b) Argue that the circuit pictured in Diagram A8–6 will function as an adjustable gain amplifier where the control input changes the gain in response to a negative voltage. Suppose the control input is grounded. Is the gain a minimum or a maximum? Why?

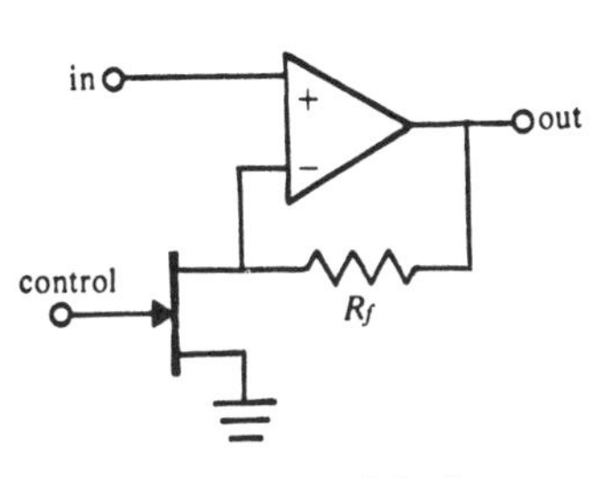

DIAGRAM A8–6.

6. Suppose an ohmmeter is connected respectively in the three connections shown in Diagram A8–7 to a "good" transistor. Tell whether the ohmmeter reads a small resistance or a very large resistance in each case. Note that the + terminal of the ohmmeter is indicated for each connection.

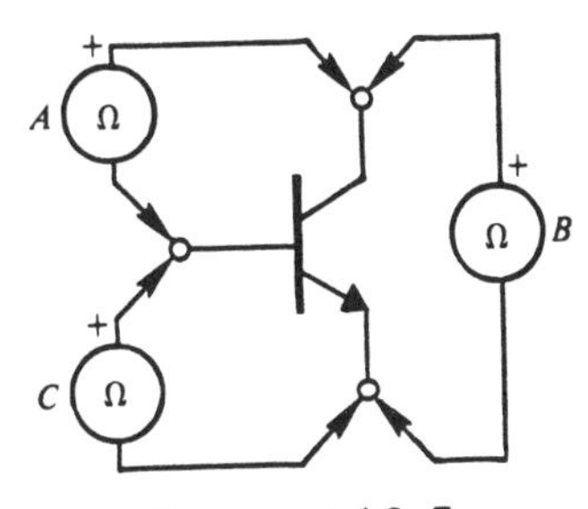

DIAGRAM A8–7.

7. (a) Sketch the appearance of a typical base characteristic for a bipolar silicon transistor. Include reasonable numbers of the horizontal and vertical axes.
(b) Illustrate how the resistance r_{BE} is determined from this characteristic.
(c) Sketch a collector characteristic for a transistor with a beta of 40. Use five I_B values, in increments of 50 μamps. Assume that the maximum V_{CE} allowed for the device is 40 V.

8. (a) Consider the transistor circuit illustrated in Diagram A8–8. If $V_{out} = 5$ V, what is the approximate h_{fe} of the transistor?
(b) What power is dissipated in the transistor?

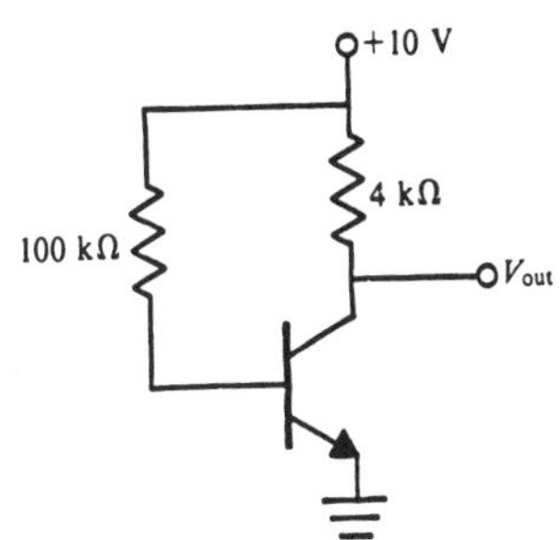

DIAGRAM A8–8.

9. The silicon transistor in the circuit shown in Diagram A8–9 has a β of 40. What are the readings expected for the voltmeters V_1 and V_2 and milliammeter I?

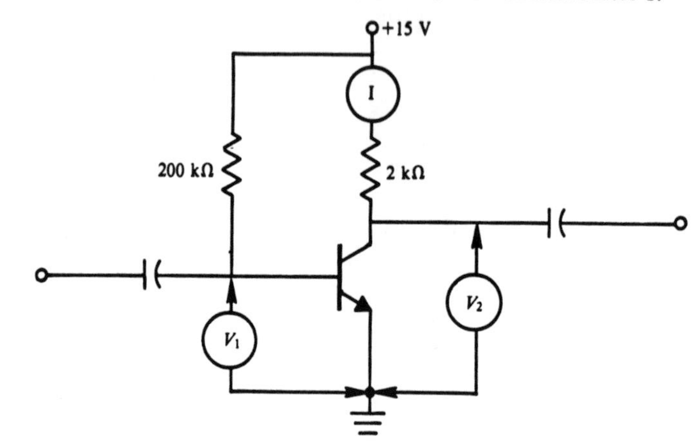

DIAGRAM A8–9.

10. Suppose a silicon transistor with an h_{fe} of 50 is used in each of the circuits (a) and (b) shown in Diagrams A8–10a and b. Find the voltage V_{out} relative to ground in each case.

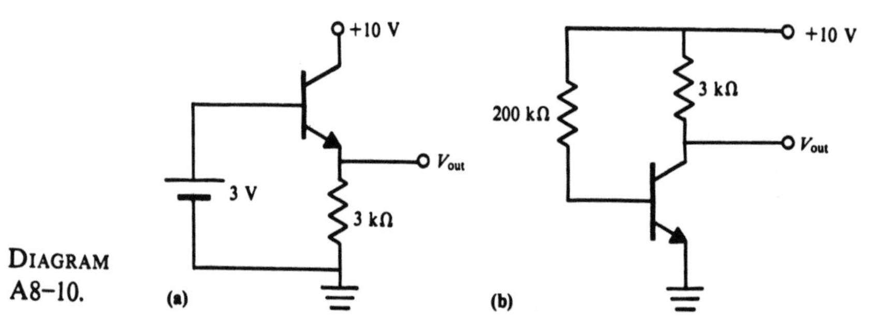

DIAGRAM A8–10.

11. A silicon transistor with a beta of 50 is used in the circuit shown in Diagram A8–11.
    (a) If $R_b = 5$ kΩ, what is the voltage V_{out} when the switch is at positions A and B, respectively?
    (b) What is the approximate maximum value of R_b for which the V_{out} voltages re-

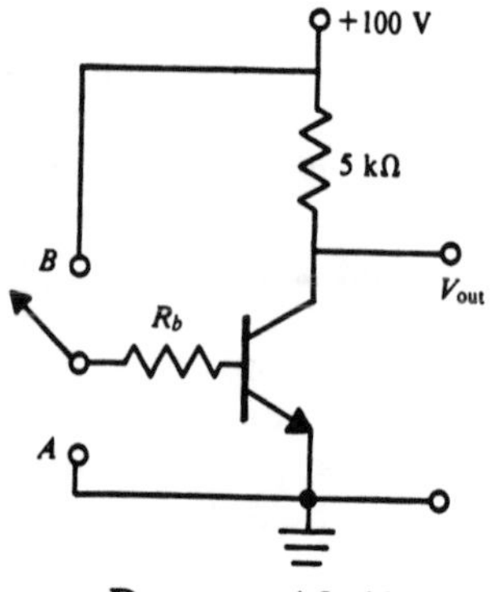

DIAGRAM A8–11.

main the same as in part (a) when the switch is operated?

(c) What minimum ratings for collector current and collector-emitter voltage capability are required for the transistor used?

12. The transistor in the amplifier pictured in Diagram A8–12 has $r_{BE} = 1200\ \Omega$ and $h_{fe} = 80$.
   (a) Find the collector current and collector-emitter voltage.
   (b) What is the approximate largest peak-peak output voltage without distortion from this amplifier?
   (c) What is the voltage gain for this amplifier?
   (d) Select C_1 and C_2 to have reactive impedances of 10% of the input impedance and output impedance, respectively, of the amplifier at a frequency of 500 Hz.
   (e) Discuss the reasons for the rule used in part (d); for example, why would we *not* want the reactance to be 10 times the input or output impedance?

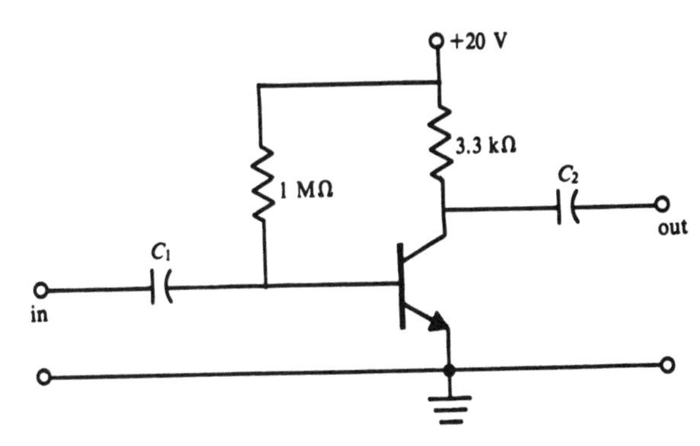

DIAGRAM A8–12.

13. Consider a silicon transistor in the better-bias circuit shown in Diagram A8–13.
   (a) Assume the base current to be very small (due to the large β of the transistor). Deduce the voltage relative to ground at points *A*, *B*, and *C*, respectively.

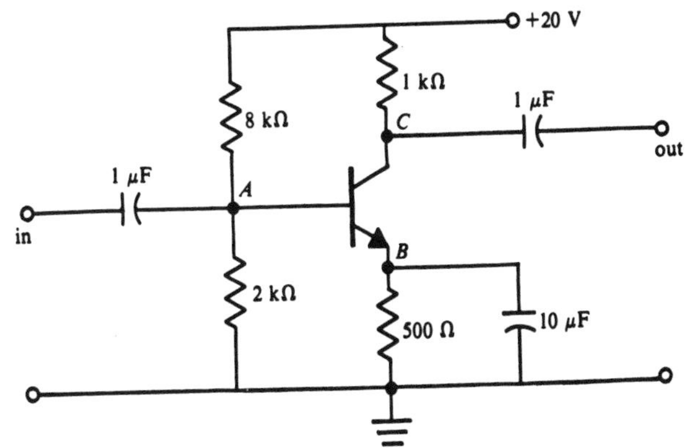

DIAGRAM A8–13.

NOTE: At A consider a voltage divider; at B consider the base-emitter characteristic for a silicon transistor; at C use Ohm's Law and the currents flowing.

(b) Suppose β increases by 25%. Discuss why none of the voltages in part (a) change. Thus, the circuit is desirable if the circuit experiences temperature variations. Explain this statement.

(c) Why is it evident that the amplifier pictured must be an ac amplifier?

14. (a) Justify the following statement about transistors in terms of their characteristics: The drain current for a given JFET cannot exceed a certain value, but the collector current for a bipolar transistor can be increased to any value (in principle).

(b) In practice, can the current through a bipolar transistor be made arbitrarily large? Why?

15. Consider the FET circuit shown in Diagram A8–14. Justify that this circuit is approximately a constant-current source to R_L, with current I_{DSS}. However, R_L must be less than V/I_{DSS}; explain why.

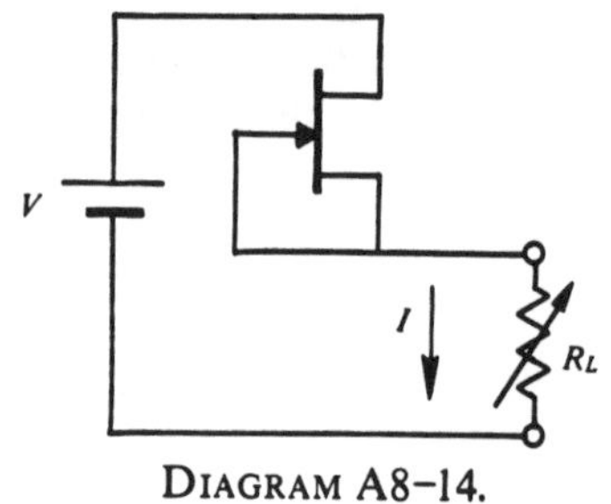

DIAGRAM A8–14.

16. Consider the bipolar transistor circuit shown in Diagram A8–15. Justify that this is approximately a constant-current source to R_L, with current $I = \beta V/R_b$ (if $V >> 0.6$ V). However R_L must be less than R_b/β; explain why.

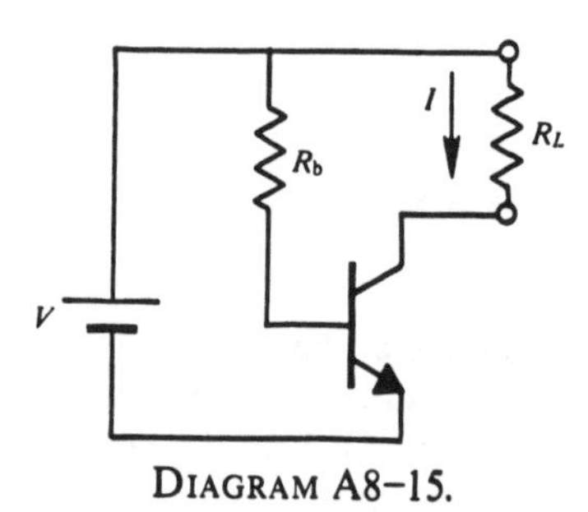

DIAGRAM A8–15.

17. Consider Diagram A8–16. The switching versus input character of JFET and bipolar transistors is contrasted.

(a) Suppose both inputs are grounded. What can be said about the two output voltages?

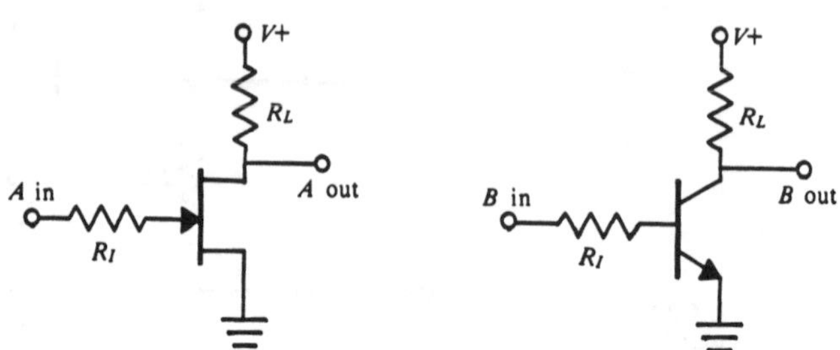

DIAGRAM A8–16.

(b) Does the answer to part (a) depend on the value of R_I in either case?
(c) Suppose *B* in is connected to *V*+. What can be said about *B* out? Does it depend upon the value of R_I?
(d) *A* in should *not* be connected to *V*+. Why?
(e) What voltage should be connected to *A* in so that the *A*-out voltage equals *V*+?

18. Consider the emitter-follower amplifier shown in Diagram A8–17. The transistor has an h_{fe} of 120 and r_{BE} of 800 Ω.
   (a) What should be the value of R_B so that the emitter voltage is 5 V relative to ground?
   (b) What then is the maximum undistorted peak-peak voltage that may be output (or input) for the amplifier?
   (c) What is the expected input impedance and output impedance of the amplifier?
   (d) Select C_1 and C_2 to have 10% of these respective impedances at 1 kHz.

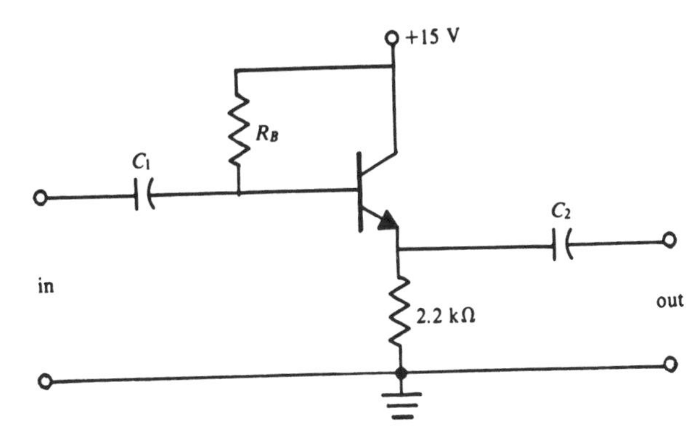

DIAGRAM A8–17.

19. (a) Suppose the voltage supply used with the basic differential amplifier of Figure A8–22 is not stable, but fluctuates. Argue that this should have no effect on the output from a good differential amplifier.
   NOTE: This favorable property is called good *supply voltage rejection.*
   (b) Suppose the temperature changes so that the beta of the two transistors change in the differential amplifier. Although this changes the current through each transistor, argue that no effect is expected at the output from a good differential amplifier.
   NOTE: This is a favorable temperature drift characteristic.

20. (a) It is quite clear that the grounded-emitter amplifier pictured in Figure A8–13 is not a dc amplifier. Why?
   (b) Discuss how the differential amplifier pictured in Figure A8–22 avoids the need for each of the two offending elements in part (a).

21. (a) Consider the ac device operated in the circuit with a phototransistor in Figure A8–23. Is the ac device turned on when it is dark or when it is bright at the TIL81 transistor? Explain why.

(b) Describe a simple change in the circuit to obtain the opposite behavior from (a).

22. Consider Diagram A8–18. We wish to count objects as they pass a certain point and have at hand an electromechanical counter that increments once for each 240-mA pulse of current through its 100-Ω coil. Thus, a 24-V pulse to the coil is required for each count. Assume that the objects interrupt a light beam and that the resistance R_p of a photoresistor increases from 1 kΩ to 100 kΩ when the light beam is blocked. A circuit to operate the counter is illustrated.
    (a) Assume the transistor beta is very large. Discuss how the circuit achieves the counting.
    NOTE: Use an emitter follower argument.
    (b) Assume that the transistor β is 100. Prove that the maximum current to the counter is only about 150 mA. Thus, the circuit will not operate.
    (c) What transistor connection could be used to obtain a super-beta transistor for application here?

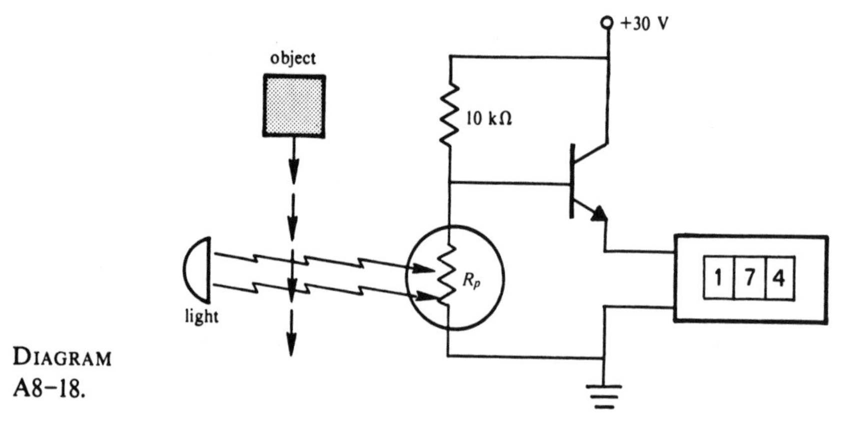

DIAGRAM A8–18.

23. By opening or closing the proper switches in the circuit shown in Diagram A8–19, we may obtain 500 W, 250 W, or 0 W from the electric resistance heater. What switch settings are required for each of the three heating conditions? Explain how the circuit operates in terms of the SCR and triac behavior. Assume I_{GT} = 10 mA for both the SCR and the triac.

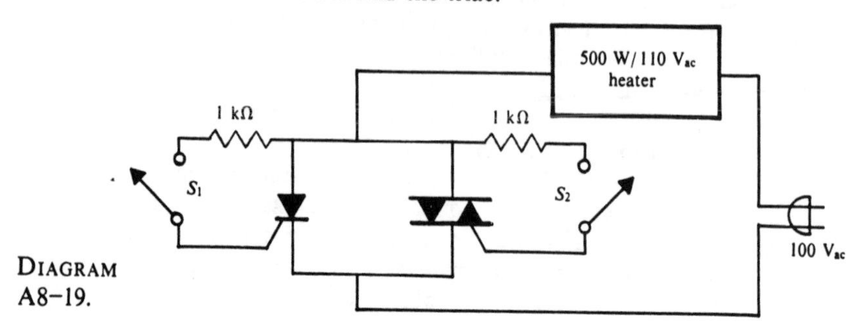

DIAGRAM A8–19.

24. The intrusion alarm shown in Diagram A8–20 is constructed from an SCR and an automobile horn that operates from 12 V at 1.5 A. A snap-action switch S_1 is positioned against a door so that the switch is depressed until the door opens.
    (a) Justify that the horn sounds when the door opens for a properly selected SCR. Will it stop sounding if the door is quickly closed again? What is the utility of switch S_2?
    (b) What forward-blocking-voltage, on-state-current, and trigger-current characteristics are required for the SCR in this circuit?

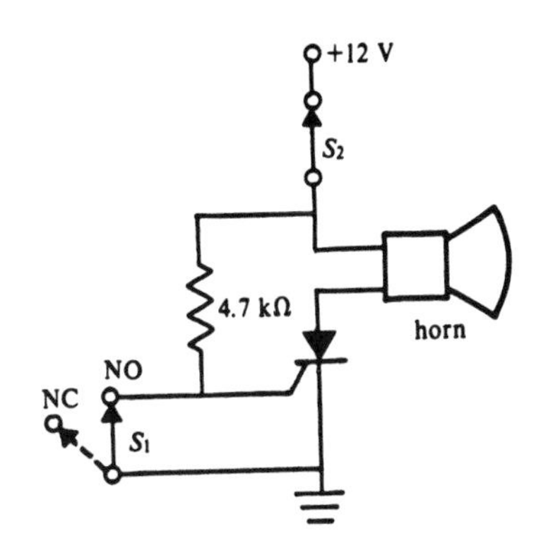

DIAGRAM A8–20.

25. A flashing circuit is desired for a conventional 100-W, 110-V_{ac} lamp with a 2-prong plug. The lamp plugs into the box that we design, and the box in turn plugs into a "mains" 110-V_{ac} socket with a 2-prong plug. The lamp is to be on for 4 seconds and off for 2 seconds in each cycle. Assume that a solid-state relay is available with a control-input requirement of 3 mA at 2 V; its triac output can switch 3 A at 125 V_{ac}. Also, a small 9-V_{dc}-encapsulated power supply is available to be incorporated in the box.
    (a) Develop a design using a 555 timer. The solid-state relay and also the power supply may be incorporated as blocks in the diagram. Design 100% over-drive current to the solid-state relay input (LED). Note that the knee of the LED diode characteristic is about 2 V.
    (b) An SCR could be considered in place of the solid-state relay in this design. However, the bulb will not shine at the 100-W brightness. Why? Why would a triac permit full brightness of the bulb during its on-time?
    (c) Use of an SCR or triac presents one point to be watched, however. The power supply output must be floating relative to ground. Explain why with the help of a sketch of an SCR design.
26. Consider a theoretical relation for g_m and r_{BE} for a transistor.
    (a) The voltage applied to the base-emitter diode in a transistor initiates a diode current that in fact almost entirely passes out the collector. For significant current levels in a transistor, we expect

$$I_c = I_R e^{\frac{|e|\, V_{BE}}{kT}}$$

Show how this relation was properly adapted from Equation A4–1.

(b) The transconductance for a transistor may be written as a derivative:

$$g_m = \frac{dI_c}{dV_{BE}}$$

Argue that this relation is properly adapted from Equation A8–1.

(c) Show the transconductance for a silicon transistor may be expected to be

$$g_m = \frac{e}{kT} I_c$$

At room temperature (300° K) show that the transconductance is given by the relation

$$g_m = (39 \text{ V}^{-1})I_c$$

(d) Justify that

$$r_{BE} = \frac{h_{fe}}{(39 \text{ V}^{-1})I_c}$$

(e) In practice, an inherent base-emitter resistance δ due to the ohmic resistance of the transistor must be added to this r_{BE} value. It may vary from a few ohms in a transistor designed for large currents to several hundred ohms for a small transistor. Why would we expect it to be more for a physically smaller transistor?

(f) Note that then

$$g_m = \frac{h_{fe}}{(h_{fe}/39\ I_c) + \delta}$$

Suppose a small transistor has an h_{fe} of 100 and a base-emitter resistance $\delta = 200\ \Omega$. Find both r_{BE} and g_m at the respective collector currents of 1 mA and 20 mA.

(g) If δ is small, the voltage gain of a common emitter amplifier is simply 39 times the number of volts across the load resistor. Justify this statement.

Manufacturers' Specifications

μA7800 Series
3-Terminal Positive Voltage Regulators
Fairchild Linear Integrated Circuits

GENERAL DESCRIPTION – The μA7800 series of monolithic 3-Terminal Positive Voltage Regulators is constructed using the Fairchild Planar* epitaxial process. These regulators employ internal current limiting, thermal shutdown and safe area compensation, making them essentially indestructible. If adequate heat sinking is provided, they can deliver over 1 A output current. They are intended as fixed voltage regulators in a wide range of applications including local (on card) regulation for elimination of distribution problems associated with single point regulation. In addition to use as fixed voltage regulators, these devices can be used with external components to obtain adjustable output voltages and currents.

- OUTPUT CURRENT IN EXCESS OF 1 A
- NO EXTERNAL COMPONENTS
- INTERNAL THERMAL OVERLOAD PROTECTION
- INTERNAL SHORT CIRCUIT CURRENT LIMITING
- OUTPUT TRANSISTOR SAFE AREA COMPENSATION
- AVAILABLE IN THE TO-220 AND THE TO-3 PACKAGE
- OUTPUT VOLTAGES OF 5, 6, 8, 8.5, 12, 15, 18, AND 24 V

ABSOLUTE MAXIMUM RATINGS

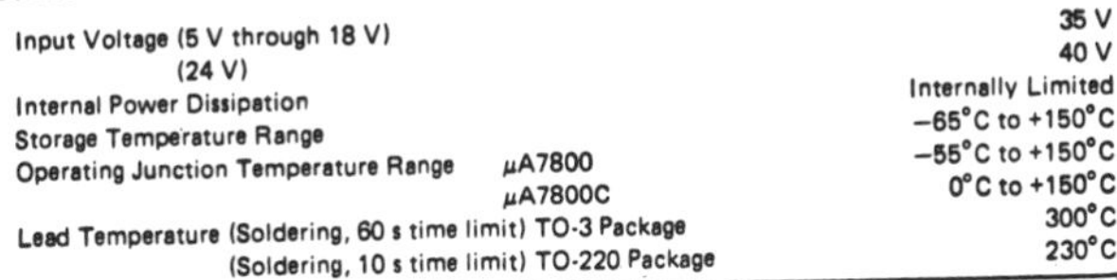

Input Voltage (5 V through 18 V)		35 V
(24 V)		40 V
Internal Power Dissipation		Internally Limited
Storage Temperature Range		−65°C to +150°C
Operating Junction Temperature Range	μA7800	−55°C to +150°C
	μA7800C	0°C to +150°C
Lead Temperature (Soldering, 60 s time limit) TO-3 Package		300°C
(Soldering, 10 s time limit) TO-220 Package		230°C

EQUIVALENT CIRCUIT

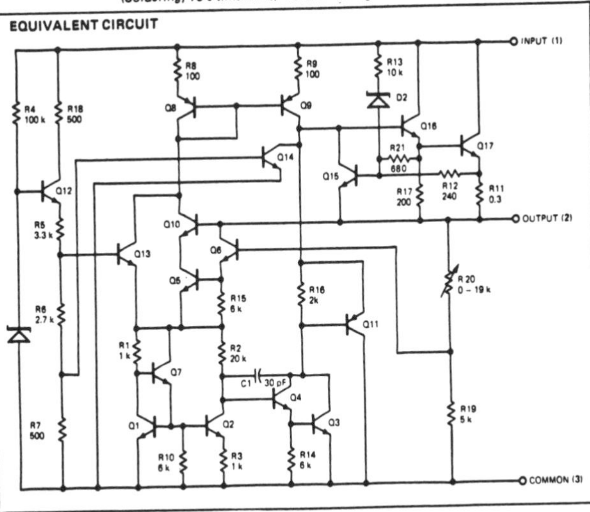

CONNECTION DIAGRAMS
TO-220 PACKAGE
(TOP VIEW)
PACKAGE OUTLINE GH
PACKAGE CODE U

OUTPUT, COMMON, INPUT; pins 2, 3, 1

ORDER INFORMATION

OUTPUT VOLTAGE	TYPE	PART NO.
5 V	μA7805C	μA7805UC
6 V	μA7806C	μA7806UC
8 V	μA7808C	μA7808UC
8.5 V	μA7885C	μA7885UC
12 V	μA7812C	μA7812UC
15 V	μA7815C	μA7815UC
18 V	μA7818C	μA7818UC
24 V	μA7824C	μA7824UC

TO-3 PACKAGE
(TOP VIEW)
PACKAGE OUTLINE GJ
PACKAGE CODE K

COMMON (3), OUTPUT, 2, 1, INPUT

ORDER INFORMATION

OUTPUT VOLTAGE	TYPE	PART NO.
5 V	μA7805	μA7805KM
6 V	μA7806	μA7806KM
8 V	μA7808	μA7808KM
8.5 V	μA7885	μA7885KM
12 V	μA7812	μA7812KM
15 V	μA7815	μA7815KM
18 V	μA7818	μA7818KM
24 V	μA7824	μA7824KM
5 V	μA7805C	μA7805KC
6 V	μA7806C	μA7806KC
8 V	μA7808C	μA7808KC
8.5 V	μA7885	μA7885KC
12 V	μA7812C	μA7812KC
15 V	μA7815C	μA7815KC
18 V	μA7818C	μA7818KC
24 V	μA7824C	μA7824KC

*Planar is a patented Fairchild process.

μA7805

ELECTRICAL CHARACTERISTICS

(V_{IN} = 10 V, I_{OUT} = 500 mA, −55°C ≤ T_J ≤ 150°C, C_{IN} = 0.33 μF, C_{OUT} = 0.1 μF, unless otherwise specified)

PARAMETER		CONDITIONS		MIN	TYP	MAX	UNITS
Output Voltage		T_J = 25°C		4.8	5.0	5.2	V
Line Regulation		T_J = 25°C	7 V ≤ V_{IN} ≤ 25 V		3	50	mV
			8 V ≤ V_{IN} ≤ 12 V		1	25	mV
Load Regulation		T_J = 25°C	5 mA ≤ I_{OUT} ≤ 1.5 A		15	50	mV
			250 mA ≤ I_{OUT} ≤ 750 mA		5	25	mV
Output Voltage		8.0 V ≤ V_{IN} ≤ 20 V 5 mA ≤ I_{OUT} ≤ 1.0 A P ≤ 15 W		4.65		5.35	V
Quiescent Current		T_J = 25°C			4.2	6.0	mA
Quiescent Current Change	with line	8 V ≤ V_{IN} ≤ 25 V				0.8	mA
	with load	5 mA ≤ I_{OUT} ≤ 1.0 A				0.5	mA
Output Noise Voltage		T_A = 25°C, 10 Hz ≤ f ≤ 100 kHz			40		μV
Ripple Rejection		f = 120 Hz, 8 V ≤ V_{IN} ≤ 18 V		68	78		dB
Dropout Voltage		I_{OUT} = 1.0 A, T_J = 25°C			2.0		V
Output Resistance		f = 1 kHz			17		mΩ
Short Circuit Current		T_J = 25°C			750		mA
Peak Output Current		T_J = 25°C			2.2		A
Average Temperature Coefficient of Output Voltage		I_{OUT} = 5 mA, 0°C ≤ T_J ≤ 150°C			−1.1		mV/°C

μA7805C

ELECTRICAL CHARACTERISTICS

(V_{IN} = 10 V, I_{OUT} = 500 mA, 0°C ≤ T_J ≤ 125°C, C_{IN} = 0.33 μF, C_{OUT} = 0.1 μF unless otherwise specified)

PARAMETER		CONDITIONS		MIN	TYP	MAX	UNITS
Output Voltage		T_J = 25°C		4.8	5.0	5.2	V
Line Regulation		T_J = 25°C	7 V ≤ V_{IN} ≤ 25 V		3	100	mV
			8 V ≤ V_{IN} ≤ 25 V		1	50	mV
Load Regulation		T_J = 25°C	5 mA ≤ I_{OUT} ≤ 1.5 A		15	100	mV
			250 mA ≤ I_{OUT} ≤ 750 mA		5	50	mV
Output Voltage		7 V ≤ V_{IN} ≤ 20 V 5 mA ≤ I_{OUT} ≤ 1.0 A P ≤ 15 W		4.75		5.25	V
Quiescent Current		T_J = 25°C			4.2	8.0	mA
Quiescent Current Change	with line	7 V ≤ V_{IN} ≤ 25 V				1.3	mA
	with load	5 mA ≤ I_{OUT} ≤ 1.0 A				0.5	mA
Output Noise Voltage		T_A = 25°C, 10 Hz ≤ f ≤ 100 kHz			40		μV
Ripple Rejection		f = 120 Hz, 8 V ≤ V_{IN} ≤ 18 V		62	78		dB
Dropout Voltage		I_{OUT} = 1.0 A, T_J = 25°C			2.0		V
Output Resistance		f = 1 kHz			17		mΩ
Short Circuit Current		T_J = 25°C			750		mA
Peak Output Current		T_J = 25°C			2.2		A
Average Temperature Coefficient of Output Voltage		I_{OUT} = 5 mA, 0°C ≤ T_J ≤ 125°C			−1.1		mV/°C

*Planar is a patented Fairchild process.

μA741
Frequency-Compensated Operational Amplifier
Fairchild Linear Integrated Circuits

GENERAL DESCRIPTION – The μA741 is a high performance monolithic Operational Amplifier constructed using the Fairchild Planar* epitaxial process. It is intended for a wide range of analog applications. High common mode voltage range and absence of "latch-up" tendencies make the μA741 ideal for use as a voltage follower. The high gain and wide range of operating voltage provides superior performance in integrator, summing amplifier, and general feedback applications.

- **NO FREQUENCY COMPENSATION REQUIRED**
- **SHORT CIRCUIT PROTECTION**
- **OFFSET VOLTAGE NULL CAPABILITY**
- **LARGE COMMON-MODE AND DIFFERENTIAL VOLTAGE RANGES**
- **LOW POWER CONSUMPTION**
- **NO LATCH UP**

ABSOLUTE MAXIMUM RATINGS

Supply Voltage	
Military (741)	±22 V
Commercial (741C)	±18 V
Internal Power Dissipation (Note 1)	
Metal Can	500 mW
DIP	670 mW
Mini DIP	310 mW
Flatpak	570 mW
Differential Input Voltage	±30 V
Input Voltage (Note 2)	±15 V

Storage Temperature Range	
Metal Can, DIP, and Flatpak	−65°C to +150°C
Mini DIP	−55°C to +125°C
Operating Temperature Range	
Military (741)	−55°C to +125°C
Commercial (741C)	0°C to +70°C
Lead Temperature (Soldering)	
Metal Can, DIP, and Flatpak (60 seconds)	300°C
Mini DIP (10 seconds)	260°C
Output Short Circuit Duration (Note 3)	Indefinite

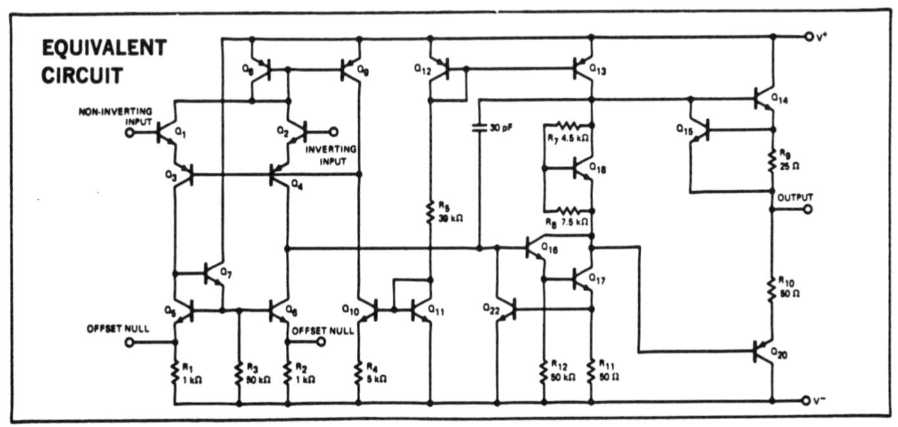

CONNECTION DIAGRAMS

8-LEAD MINI DIP
(TOP VIEW)
PACKAGE OUTLINE 9T

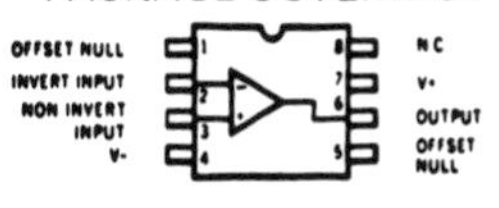

ORDER INFORMATION

TYPE	PART NO.
741C	741TC

8-LEAD METAL CAN
(TOP VIEW)
PACKAGE OUTLINE 5B

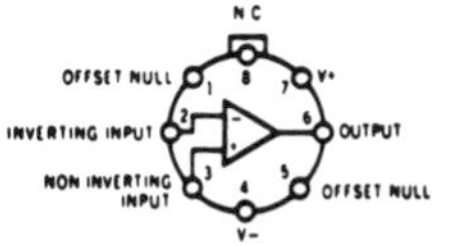

Note: Pin 4 connected to case

ORDER INFORMATION

TYPE	PART NO.
741	741HM
741C	741HC

14-LEAD DIP
(TOP VIEW)
PACKAGE OUTLINE 6A

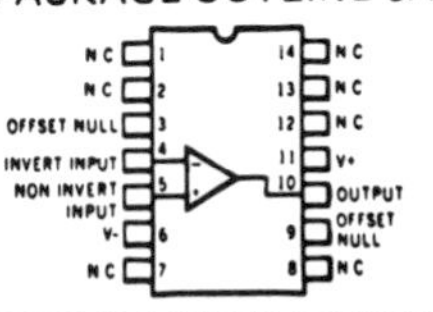

ORDER INFORMATION

TYPE	PART NO.
741	741DM
741C	741DC

741C

ELECTRICAL CHARACTERISTICS (V_S = ±15 V, T_A = 25°C unless otherwise specified)

PARAMETERS (see definitions)		CONDITIONS	MIN.	TYP.	MAX.	UNITS
Input Offset Voltage		$R_S \leq 10\ k\Omega$		2.0	6.0	mV
Input Offset Current				20	200	nA
Input Bias Current				80	500	nA
Input Resistance			0.3	2.0		MΩ
Input Capacitance				1.4		pF
Offset Voltage Adjustment Range				±15		mV
Input Voltage Range			±12	±13		V
Common Mode Rejection Ratio		$R_S \leq 10\ k\Omega$	70	90		dB
Supply Voltage Rejection Ratio		$R_S \leq 10\ k\Omega$		30	150	μV/V
Large Signal Voltage Gain		$R_L \geq 2\ k\Omega$, V_{OUT} = ±10 V	20,000	200,000		
Output Voltage Swing		$R_L \geq 10\ k\Omega$	±12	±14		V
		$R_L \geq 2\ k\Omega$	±10	±13		V
Output Resistance				75		Ω
Output Short Circuit Current				25		mA
Supply Current				1.7	2.8	mA
Power Consumption				50	85	mW
Transient Response (Unity Gain)	Risetime	V_{IN} = 20 mV, R_L = 2 kΩ, $C_L \leq 100$ pF		0.3		μs
	Overshoot			5.0		%
Slew Rate		$R_L > 2\ k\Omega$		0.5		V/μs
The following specifications apply for 0°C ≤ T_A ≤ +70°C:						
Input Offset Voltage					7.5	mV
Input Offset Current					300	nA
Input Bias Current					800	nA
Large Signal Voltage Gain		$R_L \geq 2\ k\Omega$, V_{OUT} = ±10 V	15,000			
Output Voltage Swing		$R_L \geq 2\ k\Omega$	±10	±13		V

TYPICAL PERFORMANCE CURVES FOR 741C

OUTPUT VOLTAGE SWING AS A FUNCTION OF SUPPLY VOLTAGE

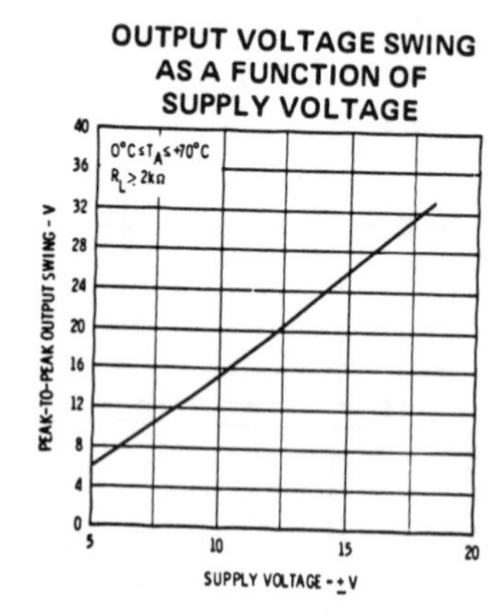

INPUT COMMON MODE VOLTAGE RANGE AS A FUNCTION OF SUPPLY VOLTAGE

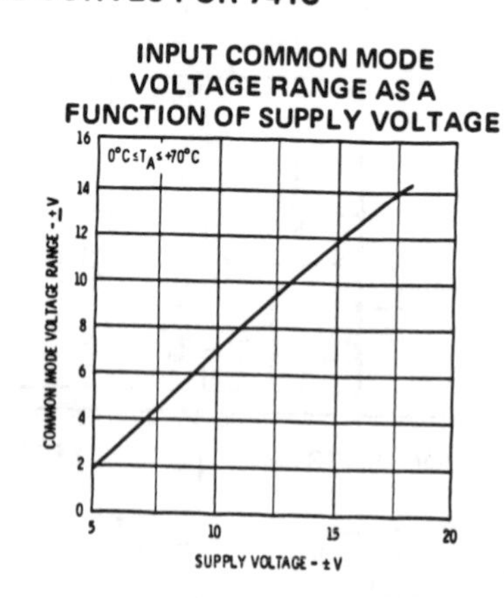

NOTES:

1. Rating applies to ambient temperatures up to 70°C. Above 70°C ambient derate linearly at 6.3 mW/°C for the Metal Can, 8.3 mW/°C for the DIP, 5.6 mW/°C for the Mini DIP and 7.1 mW/°C for the Flatpak.
2. For supply voltages less than ±15 V, the absolute maximum input voltage is equal to the supply voltage.
3. Short circuit may be to ground or either supply. Rating applies to +125°C case temperature or 75°C ambient temperature.

TYPICAL PERFORMANCE CURVES FOR 741C

INPUT BIAS CURRENT AS A FUNCTION OF AMBIENT TEMPERATURE

$V_S = \pm 15V$

INPUT BIAS CURRENT - nA

200, 160, 120, 80, 40, 0

0, 10, 20, 30, 40, 50, 60, 70

TEMPERATURE - °C

INPUT OFFSET CURRENT AS A FUNCTION OF AMBIENT TEMPERATURE

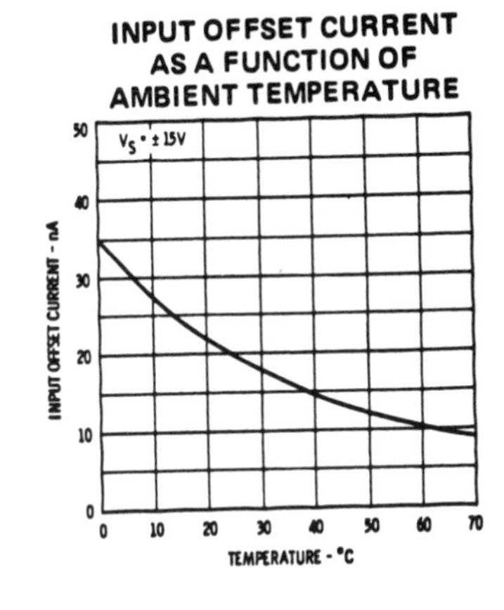

POWER CONSUMPTION AS A FUNCTION OF SUPPLY VOLTAGE

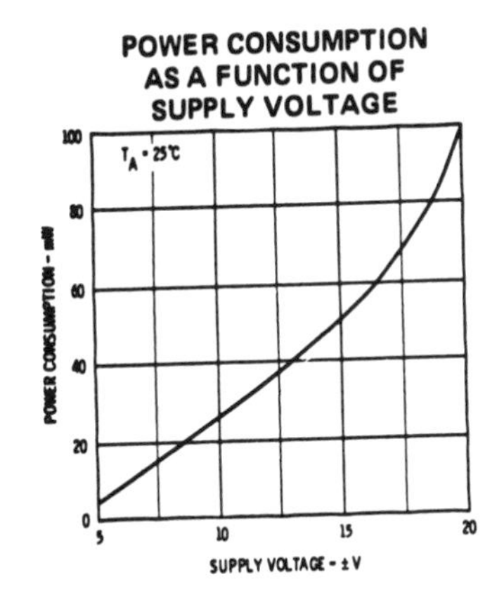

OPEN LOOP VOLTAGE GAIN AS A FUNCTION OF FREQUENCY

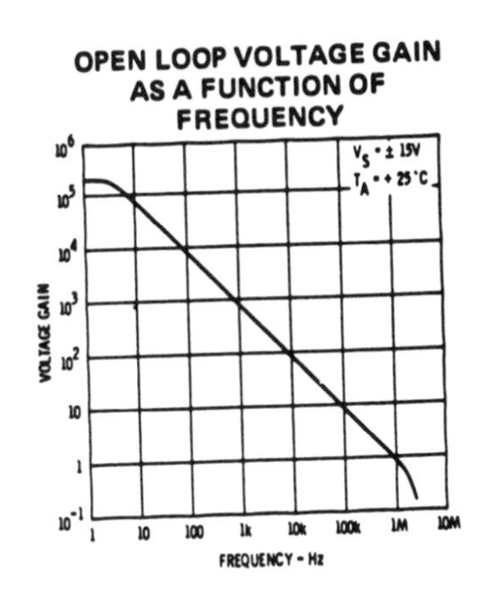

OPEN LOOP PHASE RESPONSE AS A FUNCTION OF FREQUENCY

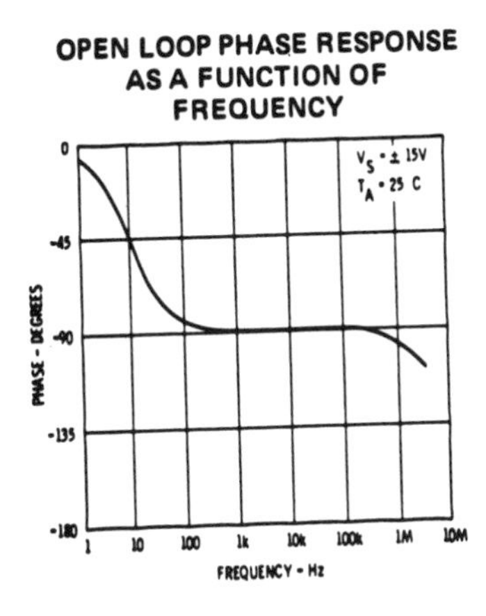

INPUT RESISTANCE AND INPUT CAPACITANCE AS A FUNCTION OF FREQUENCY

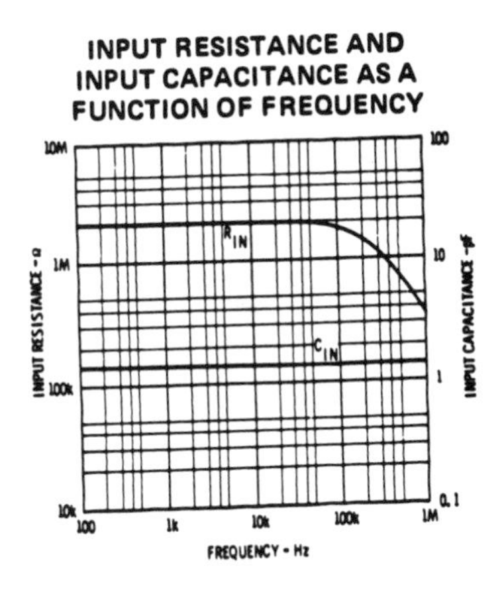

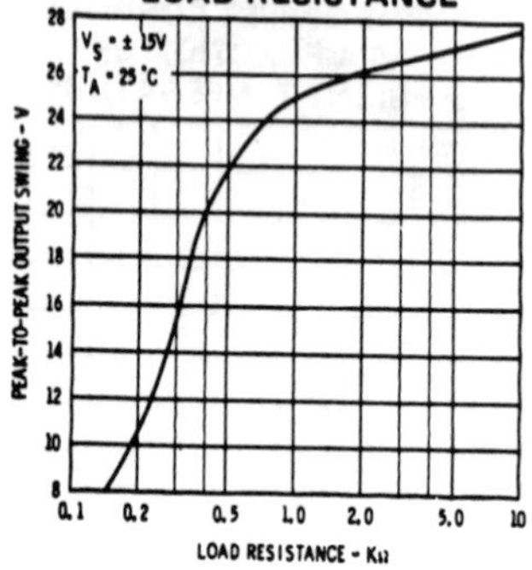

OUTPUT VOLTAGE SWING AS A FUNCTION OF LOAD RESISTANCE

V_S = ± 15V
T_A = 25°C

PEAK-TO-PEAK OUTPUT SWING - V

LOAD RESISTANCE - KΩ

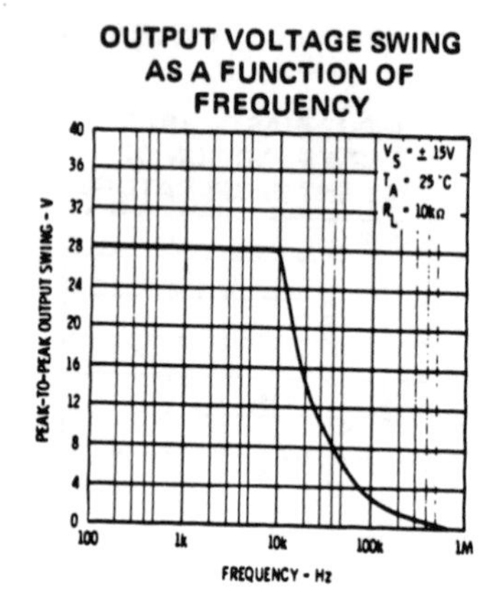

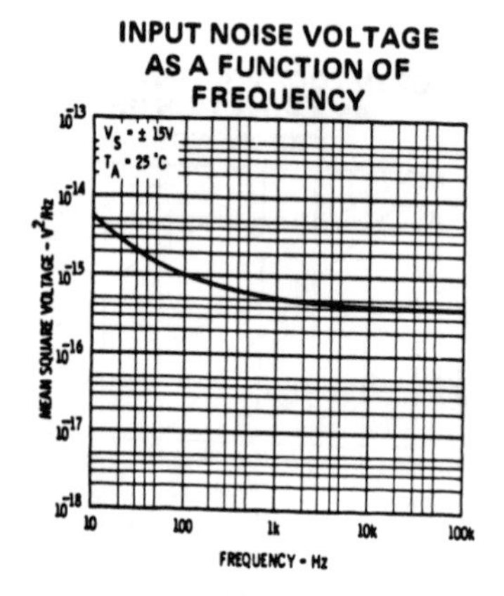

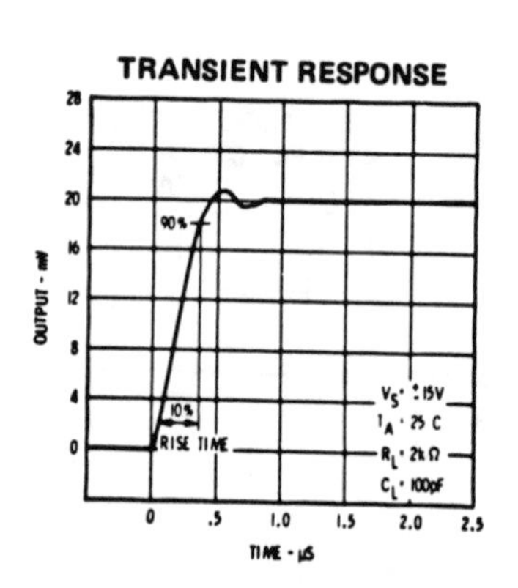

VOLTAGE OFFSET NULL CIRCUIT

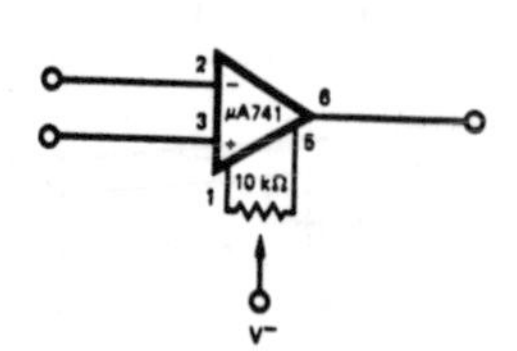

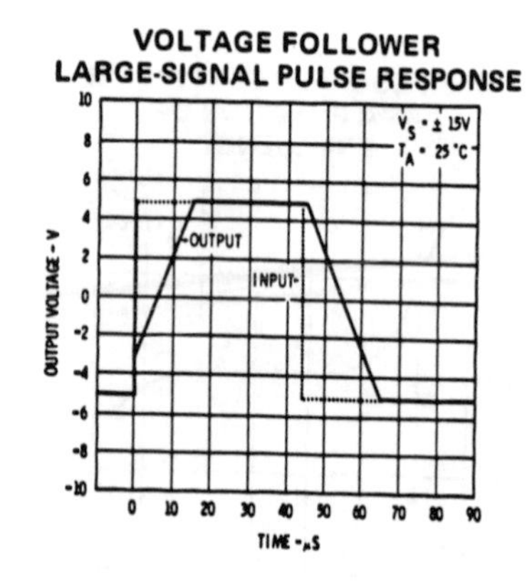

CA3140B, CA3140A, CA3140 Types
Linear Integrated Circuits
Monolithic Silicon
RCA, Solid State Division

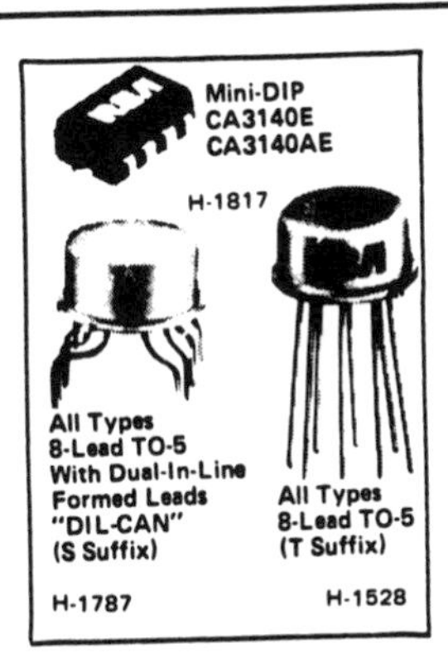

BiMOS Operational Amplifiers

With MOS/FET Input, Bipolar Output

Features:

- **MOS/FET Input Stage**
  - **(a) Very high input impedance (Z_{IN}) – 1.5 TΩ typ.**
  - **(b) Very low input current (I_I) – 10 pA typ. at ± 15 V**
  - **(c) Low input-offset voltage (V_{IO}) – to 2 mV max.**
  - **(d) Wide common-mode input-voltage range (V_{ICR}) – can be swung 0.5 volt below negative supply-voltage rail**
  - **(e) Output swing complements input common-mode range**
  - **(f) Rugged input stage – bipolar diode protected**
- **Directly replaces industry type 741 in most applications**
- **Includes numerous industry operational amplifier categories such as general-purpose, FET input, wideband (high slew rate)**
- **Operation from 4-to-44 volts Single or Dual supplies**
- **Internally compensated**
- **Characterized for ± 15-volt operation and for TTL supply systems with operation down to 4 volts**
- **Wide bandwidth – 4.5 MHz unity gain at ± 15 V or 30 V; 3.7 MHz at 5 V**
- **High voltage-follower slew rate – 9 V/μs**
- **Fast settling time – 1.4 μs typ. to 10 mV with a 10-V_{p-p} signal**
- **Output swings to within 0.2 volt of negative supply**
- **Strobable output stage**

Applications:

- **Ground-referenced single-supply amplifiers in automobile and portable instrumentation**
- **Sample and hold amplifiers**
- **Long-duration timers/multivibrators (microseconds–minutes–hours)**
- **Photocurrent instrumentation**
- **Peak detectors**
- **Active filters**
- **Comparators**
- **Interface in 5 V TTL systems & other low-supply voltage systems**
- **All standard operational amplifier applications**
- **Function generators**
- **Tone controls**
- **Power supplies**
- **Portable instruments**
- **Intrusion alarm systems**

The CA3140B, CA3140A, and CA3140 are integrated-circuit operational amplifiers that combine the advantages of high-voltage PMOS transistors with high-voltage bipolar transistors on a single monolithic chip. Because of this unique combination of technologies, this device can now provide designers, for the first time, with the special performance features of the CA3130 COS/MOS operational amplifiers and the versatility of the 741 series of industry-standard operational amplifiers.

The CA3140,CA3140A, and CA3140 BiMOS operational amplifiers feature gate-protected MOS/FET (PMOS) transistors in the input circuit to provide very-high-input impedance, very-low-input current, and high-speed performance. The CA3140B operates at supply voltages from 4 to 44 volts; the CA3140A and CA3140 from 4 to 36 volts (either single or dual supply). These operational amplifiers are internally phase-compensated to achieve stable operation in unity-gain follower operation, and, additionally, have access terminals for a supplementary external capacitor if additional frequency roll-off is desired. Terminals are also provided for use in applications requiring input offset-voltage nulling.

The use of PMOS field-effect transistors in the input stage results in common-mode input-voltage capability down to 0.5 volt below the negative-supply terminal, an important attribute for single-supply applications. The output stage uses bipolar transistors and includes built-in protection against damage from load-terminal short-circuiting to either supply-rail or to ground.

The CA3140 Series has the same 8-lead terminal pin-out used for the "741" and other industry-standard operational amplifiers. They are supplied in either the standard 8-lead TO-5 style package (T suffix), or in the 8-lead dual-in-line formed-lead TO-5 style package "DIL-CAN" (S suffix). The CA3140 is available in chip form (H suffix). The CA3140A and CA3140 are also available in an 8-lead dual-in-line plastic package (Mini-DIP-E suffix). The CA3140B is intended for operation at supply voltages ranging from 4 to 44 volts, for applications requiring premium-grade specifications and with electrical limits established for operations over the range from −55°C to +125°C. The CA3140A and CA3140 are for operation at supply voltages up to 36 volts (±18 volts). The CA3140 ages up to 36 volts (±18 volts). All types can be operated safely over the temperature range from −55°C to +125°C.

TYPICAL ELECTRICAL CHARACTERISTICS

CHARACTERISTIC		TEST CONDITIONS $V^+ = +15$ V $V^- = -15$ V $T_A = 25°C$		CA3140B (T,S)	CA3140A (T,S,E)	CA3140 (T,S,E)	UNITS
Input Offset Voltage Adjustment Resistor		Typ. Value of Resistor Between Term. 4 and 5 or 4 and 1 to Adjust Max. V_{IO}		43	18	4.7	kΩ
Input Resistance	R_1			1.5	1.5	1.5	TΩ
Input Capacitance	C_I			4	4	4	pF
Output Resistance	R_O			60	60	60	Ω
Equivalent Wideband Input Noise Voltage (See Fig. 39)	e_n	BW = 140 kHz R_S = 1 MΩ		48	48	48	μV
Equivalent Input Noise Voltage (See Fig. 10)	e_n	f = 1 kHz	R_S = 100 Ω	40	40	40	nV/√Hz
		f = 10 kHz		12	12	12	
Short-Circuit Current to Opposite Supply Source	I_{OM^+}			40	40	40	mA
Sink	I_{OM^-}			18	18	18	mA
Gain-Bandwidth Product, (See Figs. 5 & 18)	f_T			4.5	4.5	4.5	MHz
Slew Rate, (See Fig. 6)	SR			9	9	9	V/μs
Sink Current From Terminal 8 To Terminal 4 to Swing Output Low				220	220	220	μA
Transient Response: Rise Time	t_r	R_L = 2 kΩ C_L = 100 pF		0.08	0.08	0.08	μs
Overshoot (See Fig. 37)				10	10	10	%
Settling Time at 10 $V_{p\text{-}p}$, (See Fig. 17) 1 mV	t_s	R_L = 2 kΩ C_L = 100 pF Voltage Follower		4.5	4.5	4.5	μs
10 mV				1.4	1.4	1.4	

Printed in USA/3-77

Supersedes issue dated 5-76

ELECTRICAL CHARACTERISTICS FOR EQUIPMENT DESIGN
At V^+ = 15 V, V^- = 15 V, T_A = 25°C Unless Otherwise Specified

CHARACTERISTIC	LIMITS									UNITS
	CA3140B			CA3140A			CA3140			
	Min.	Typ.	Max.	Min.	Typ.	Max.	Min.	Typ.	Max.	
Input Offset Voltage, $\lvert V_{IO} \rvert$	–	0.8	2	–	2	5	–	5	15	mV
Input Offset Current, $\lvert I_{IO} \rvert$	–	0.5	10	–	0.5	20	–	0.5	30	pA
Input Current, I_I	–	10	30	–	10	40	–	10	50	pA
Large-Signal Voltage Gain, A_{OL}● (See Figs. 4,18)	50 k	100 k	–	20 k	100 k	–	20 k	100 k	–	V/V
	94	100	–	86	100	–	86	100	–	dB
Common-Mode Rejection Ratio, CMRR (See Fig.9)	–	20	50	–	32	320	–	32	320	μV/V
	86	94	–	70	90	–	70	90	–	dB
Common-Mode Input-Voltage Range, V_{ICR} (See Fig.20)	–15	–15.5 to +12.5	12	–15	–15.5 to +12.5	12	–15	–15.5 to +12.5	11	V
Power-Supply Rejection Ratio, PSRR (See Fig.11) $\Delta V_{IO}/\Delta V$	–	32	100	–	100	150	–	100	150	μV/V
	80	90	–	76	80	–	76	80	–	dB
Max. Output Voltage■ (See Figs.13,20) V_{OM}^+	+12	13	–	+12	13	–	+12	13	–	V
V_{OM}^-	–14	–14.4	–	–14	–14.4	–	–14	–14.4	–	V
Supply Current, I^+ (See Fig.7)	–	4	6	–	4	6	–	4	6	mA
Device Dissipation, P_D	–	120	180	–	120	180	–	120	180	mW
Input Current, I_I▲ (See Fig.19)	–	10	30	–	10	–	–	10	–	nA
Input Offset Voltage $\lvert V_{IO} \rvert$▲	–	1.3	3	–	3	–	–	10	–	mV
Input Offset Voltage Temp. Drift, $\Delta V_{IO}/\Delta T$	–	5	–	–	6	–	–	8	–	μV/°C
Large-Signal Voltage Gain, A_{OL}▲ (See Figs.4,18)	20 k	100 k	–	–	100 k	–	–	100 k	–	V/V
	86	100	–	–	100	–	–	100	–	dB
Max. Output Voltage,★ V_{OM}^+	+19	+19.5	–	–	–	–	–	–	–	V
V_{OM}^-	–21	–21.4	–	–	–	–	–	–	–	V
Large-Signal Voltage Gain, A_{OL}♠★	20 k	50 k	–	–	–	–	–	–	–	V/V
	86	94	–	–	–	–	–	–	–	dB

● At V_O = 26$V_{p\text{-}p}$, +12V, –14V and R_L = 2 kΩ.
■ At R_L = 2 kΩ.
▲ At T_A = –55°C to +125°C, V^+ = 15 V, V^- = 15 V, V_O = 26$V_{p\text{-}p}$, R_L = 2 kΩ.
♠ At V_O = +19 V, –21 V, and R_L = 2 kΩ.
★ At V^+ = 22 V, V^- = 22 V.

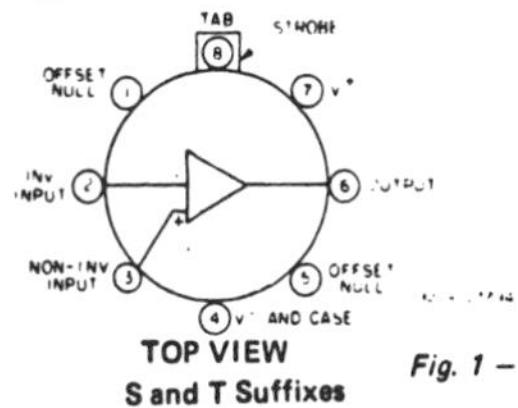

TOP VIEW
S and T Suffixes

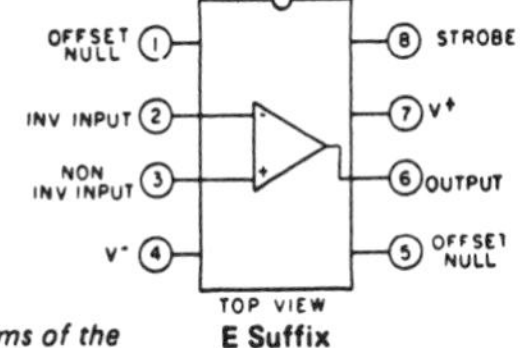

E Suffix

Fig. 1 – Functional diagrams of the CA3140 series.

MANUFACTURERS' SPECIFICATIONS, CA3140B, CA3140A, CA3140 TYPES

MAXIMUM RATINGS, *Absolute-Maximum Values:*

	CA3140, CA3140A	CA3140B
DC SUPPLY VOLTAGE (BETWEEN V^+ AND V^- TERMINALS)	36 V	44 V
DIFFERENTIAL-MODE INPUT VOLTAGE	±8 V	±8 V
COMMON-MODE DC INPUT VOLTAGE	(V^+ +8 V) to (V^- −0.5 V)	
INPUT-TERMINAL CURRENT	1 mA	
DEVICE DISSIPATION:		
WITHOUT HEAT SINK –		
UP TO 55°C	630 mW	
ABOVE 55°C	Derate linearly 6.67 mW/°C	
WITH HEAT SINK –		
Up to 55°C	1 W	
Above 55°C	Derate linearly 16.7 mW/°C	
TEMPERATURE RANGE:		
OPERATING (ALL TYPES)	−55 to + 125°C	
STORAGE (ALL TYPES)	−65 to +150°C	
OUTPUT SHORT-CIRCUIT DURATION*	INDEFINITE	
LEAD TEMPERATURE (DURING SOLDERING): AT DISTANCE 1/16 ± 1/32 INCH (1.59 ± 0.79 MM) FROM CASE FOR 10 SECONDS MAX.	+265°C	

* Short circuit may be applied to ground or to either supply.

TYPICAL ELECTRICAL CHARACTERISTICS FOR DESIGN GUIDANCE
At $V^+ = 5$ V, $V^- = 0$ V, $T_A = 25°C$

CHARACTERISTIC		CA3140B (T,S)	CA3140A (T,S,E)	CA3140 (T,S,E)	UNITS
Input Offset Voltage	$\|V_{IO}\|$	0.8	2	5	mV
Input Offset Current	$\|I_{IO}\|$	0.1	0.1	0.1	pA
Input Current	I_I	2	2	2	pA
Input Resistance		1	1	1	TΩ
Large-Signal Voltage Gain (See Figs.4,18)	A_{OL}	100 k	100 k	100 k	V/V
		100	100	100	dB
Common-Mode Rejection Ratio,	CMRR	20	32	32	μV/V
		94	90	90	dB
Common-Mode Input-Voltage Range (See Fig.20)	V_{ICR}	−0.5	−0.5	−0.5	V
		2.6	2.6	2.6	V
Power-Supply Rejection Ratio	$\Delta V_{IO}/\Delta V^+$	32	100	100	μV/V
		90	80	80	dB
Maximum Output Voltage (See Figs.13,20)	V_{OM}^+	3	3	3	V
	V_{OM}^-	0.13	0.13	0.13	V
Maximum Output Current: Source	I_{OM}^+	10	10	10	mA
Sink	I_{OM}^-	1	1	1	mA
Slew Rate (See Fig.6)		7	7	7	V/μs
Gain-Bandwidth Product (See Fig.5)	f_T	3.7	3.7	3.7	MHz
Supply Current (See Fig.7)	I^+	1.6	1.6	1.6	mA
Device Dissipation	P_D	8	8	8	mW
Sink Current from Term. 8 to Term. 4 to Swing Output Low		200	200	200	μA

Timer 555
Linear Integrated Circuits
Signetics

DESCRIPTION

The NE/SE 555 monolithic timing circuit is a highly stable controller capable of producing accurate time delays, or oscillation. Additional terminals are provided for triggering or resetting if desired. In the time delay mode of operation, the time is precisely controlled by one external resistor and capacitor. For a stable operation as an oscillator, the free running frequency and the duty cycle are both accurately controlled with two external resistors and one capacitor. The circuit may be triggered and reset on falling waveforms, and the output structure can source or sink up to 200mA or drive TTL circuits.

- TIMING FROM MICROSECONDS THROUGH HOURS
- OPERATES IN BOTH ASTABLE AND MONOSTABLE MODES
- ADJUSTABLE DUTY CYCLE
- HIGH CURRENT OUTPUT CAN SOURCE OR SINK 200mA
- OUTPUT CAN DRIVE TTL
- TEMPERATURE STABILITY OF 0.005% PER °C
- NORMALLY ON AND NORMALLY OFF OUTPUT

PIN CONFIGURATIONS (Top View)

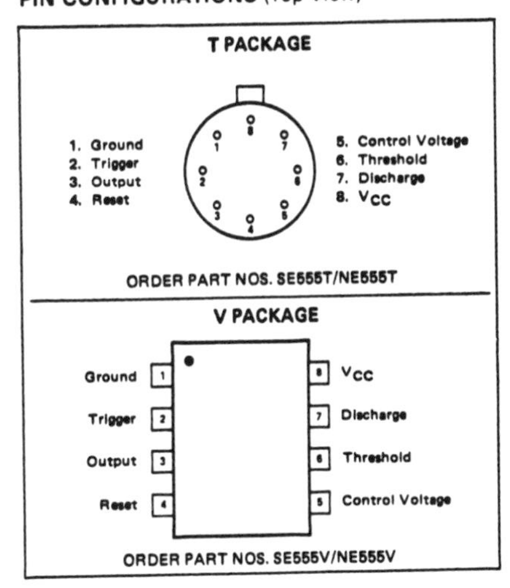

APPLICATIONS

PRECISION TIMING
PULSE GENERATION
SEQUENTIAL TIMING
TIME DELAY GENERATION
PULSE WIDTH MODULATION
PULSE POSITION MODULATION
MISSING PULSE DETECTOR

ABSOLUTE MAXIMUM RATINGS

Supply Voltage	+18V
Power Dissipation	600 mW
Operating Temperature Range	
NE555	0°C to +70°C
SE555	−55°C to +125°C
Storage Temperature Range	−65°C to +150°C
Lead Temperature (Soldering, 60 seconds)	+300°C

BLOCK DIAGRAM

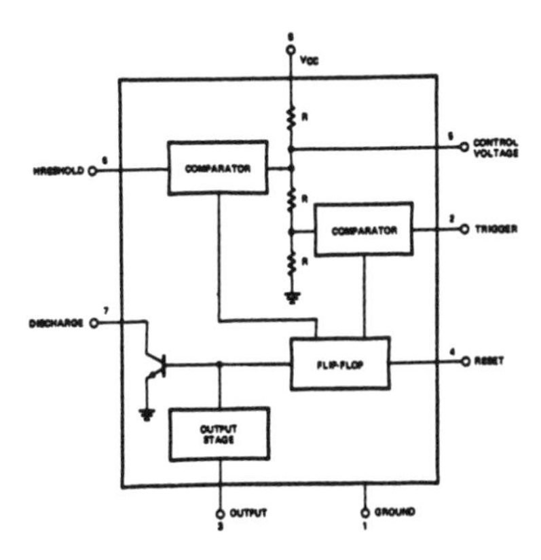

MANUFACTURERS' SPECIFICATIONS, TIMER 555

ELECTRICAL CHARACTERISTICS $T_A = 25°C$, $V_{CC} = +5V$ to $+15$ unless otherwise specified

PARAMETER	TEST CONDITIONS	SE 555			NE 555			UNITS
		MIN	TYP	MAX	MIN	TYP	MAX	
Supply Voltage		4.5		18	4.5		16	V
Supply Current	$V_{CC} = 5V$ $R_L = \infty$		3	5		3	6	mA
	$V_{CC} = 15V$ $R_L = \infty$		10	12		10	15	mA
	Low State, Note 1							
Timing Error(Monostable)	$R_A, R_B = 1K\Omega$ to $100K\Omega$							
Initial Accuracy	$C = 0.1\,\mu F$ Note 2		0.5	2		1		%
Drift with Temperature			30	100		50		ppm/°C
Drift with Supply Voltage			0.05	0.2		0.1		%/Volt
Threshold Voltage			2/3			2/3		X V_{CC}
Trigger Voltage	$V_{CC} = 15V$	4.8	5	5.2		5		V
Timing Error(Astable)	$V_{CC} = 5V$	1.45	1.67	1.9		1.67		V
Trigger Current			0.5			0.5		µA
Reset Voltage		0.4	0.7	1.0	0.4	0.7	1.0	V
Reset Current			0.1			0.1		mA
Threshold Current	Note 3		0.1	.25		0.1	.25	µA
Control Voltage Level	$V_{CC} = 15V$	9.6	10	10.4	9.0	10	11	V
	$V_{CC} = 5V$	2.9	3.33	3.8	2.6	3.33	4	V
Output Voltage (low)	$V_{CC} = 15V$							
	$I_{SINK} = 10mA$		0.1	0.15		0.1	.25	V
	$I_{SINK} = 50mA$		0.4	0.5		0.4	.75	V
	$I_{SINK} = 100mA$		2.0	2.2		2.0	2.5	V
	$I_{SINK} = 200mA$		2.5			2.5		
	$V_{CC} = 5V$							
	$I_{SINK} = 8mA$		0.1	0.25				V
	$I_{SINK} = 5mA$					.25	.35	
Output Voltage Drop (low)								
	$I_{SOURCE} = 200mA$		12.5			12.5		
	$V_{CC} = 15V$							
	$I_{SOURCE} = 100mA$							
	$V_{CC} = 15V$	13.0	13.3		12.75	13.3		V
	$V_{CC} = 5V$	3.0	3.3		2.75	3.3		V
Rise Time of Output			100			100		nsec
Fall Time of Output			100			100		nsec

NOTES

1. Supply Current when output high typically 1mA less.
2. Tested at $V_{CC} = 5V$ and $V_{CC} = 15V$
3. This will determine the maximum value of $R_A + R_B$. For 15V operation, the max total R = 20 megohm.

EQUIVALENT CIRCUIT (Shown for One Side Only)

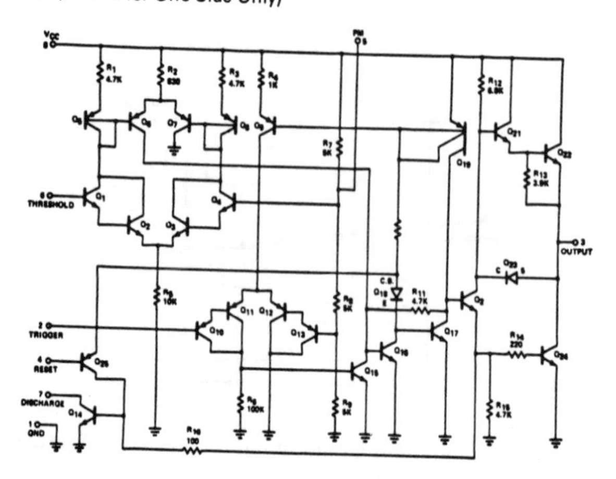

TYPICAL CHARACTERISTICS

MINIMUM PULSE WIDTH REQUIRED FOR TRIGGERING

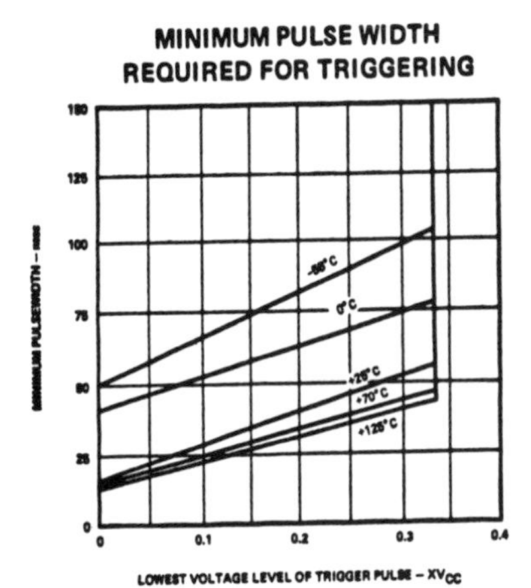

SUPPLY CURRENT vs SUPPLY VOLTAGE

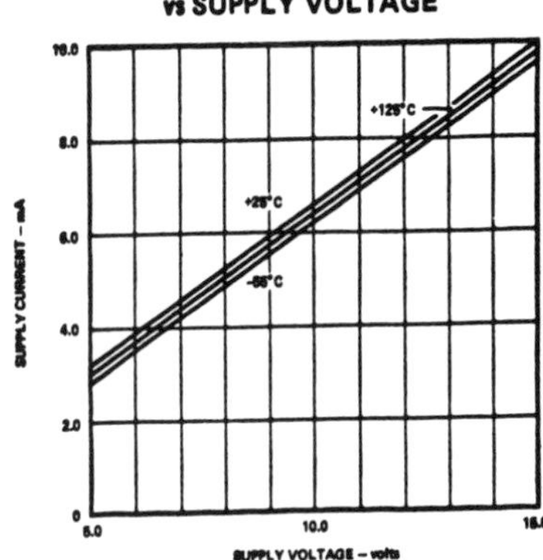

LOW OUTPUT VOLTAGE vs OUTPUT SINK CURRENT

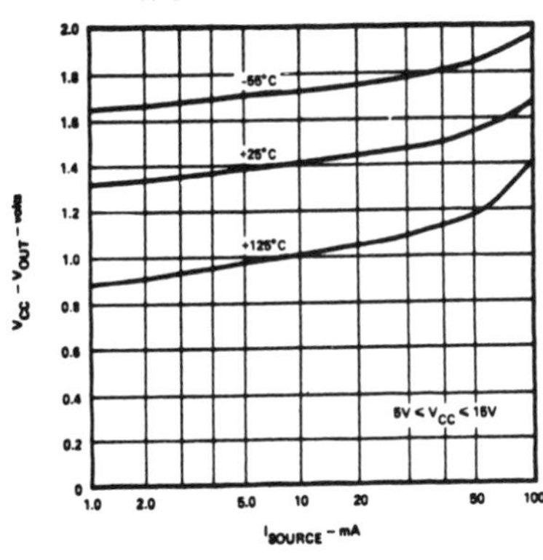

HIGH OUTPUT VOLTAGE vs OUTPUT SOURCE CURRENT

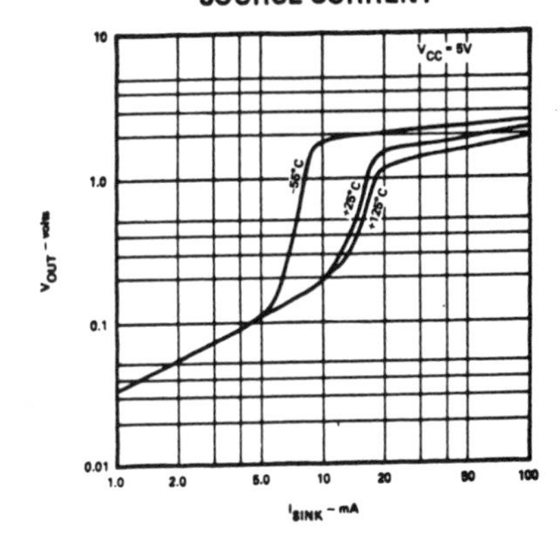

LOW OUTPUT VOLTAGE vs OUTPUT SINK CURRENT

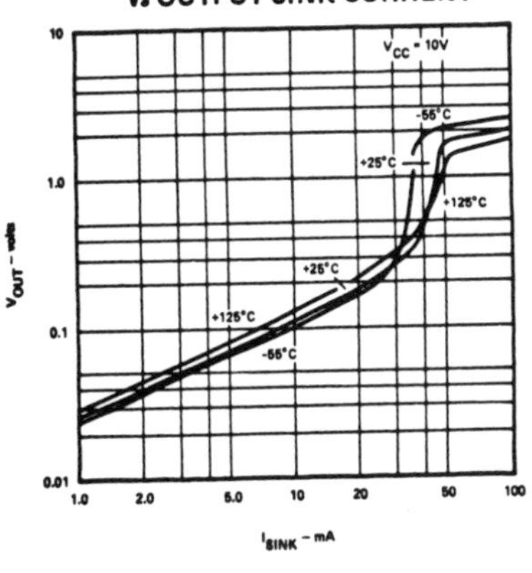

LOW OUTPUT VOLTAGE vs OUTPUT SINK CURRENT

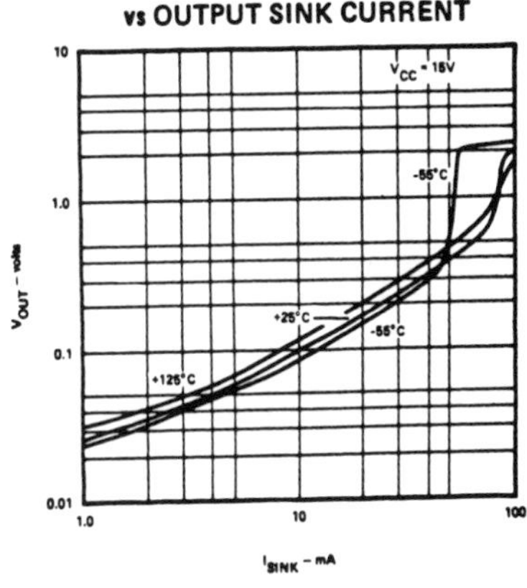

Types 2N2217 thru 2N2222, 2N2218A, 2N2219A, 2N2221A, 2N2222A NPN Silicon Transistors

DESIGNED FOR HIGH-SPEED, MEDIUM-POWER SWITCHING AND GENERAL PURPOSE AMPLIFIER APPLICATIONS

- h_{FE} . . . Guaranteed from 100 μA to 500 mA
- High f_T at 20 V, 20 mA . . . 300 MHz (2N2219A, 2N2222A)
  250 MHz (all others)
- 2N2218, 2N2221 for Complementary Use with 2N2904, 2N2906
- 2N2219, 2N2222 for Complementary Use with 2N2905, 2N2906

***mechanical data**

Device types 2N2217, 2N2218, 2N2218A, 2N2219, and 2N2219A are in JEDEC TO-5 packages.
Device types 2N2220, 2N2221, 2N2221A, 2N2222, and 2N2222A are in JEDEC TO-18 packages.

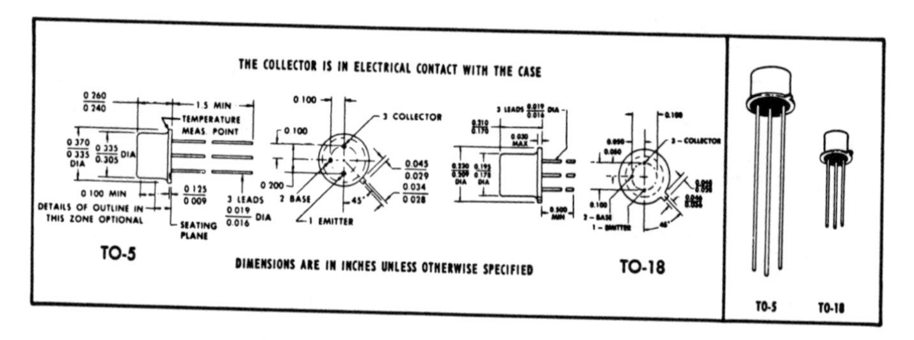

***absolute maximum ratings at 25°C free-air temperature (unless otherwise noted)**

	2N2217 2N2218 2N2219	2N2218A 2N2219A	2N2220 2N2221 2N2222	2N2221A 2N2222A	UNIT
Collector-Base Voltage	60	75	60	75	V
Collector-Emitter Voltage (See Note 1)	30	40	30	40	V
Emitter-Base Voltage	5	6	5	6	V
Continuous Collector Current	0.8	0.8	0.8	0.8	A
Continuous Device Dissipation at (or below) 25°C Free-Air Temperature (See Notes 2 and 3)	0.8	0.8	0.5	0.5	W
Continuous Device Dissipation at (or below) 25°C Case Temperature (See Notes 4 and 5)	3	3	1.8	1.8	W
Operating Collector Junction Temperature Range	−65 to 175				°C
Storage Temperature Range	−65 to 200				°C
Lead Temperature 1/16 Inch from Case for 10 Seconds	230				°C

NOTES:
1. These values apply between 0 and 500 mA collector current when the base-emitter diode is open-circuited.
2. Derate 2N2217, 2N2218, 2N2218A, 2N2219, and 2N2219A linearly to 175°C free-air temperature at the rate of 5.33 mW/°C.
3. Derate 2N2220, 2N2221, 2N2221A, 2N2222, and 2N2222A linearly to 175°C free-air temperature at the rate of 3.33 mW/°C.
4. Derate 2N2217, 2N2218, 2N2218A, 2N2219, and 2N2219A linearly to 175°C case temperature at the rate of 20.0 mW/°C.
5. Derate 2N2220, 2N2221, 2N2221A, 2N2222, and 2N2222A linearly to 175°C case temperature at the rate of 12.0 mW/°C.

USES CHIP N24

*JEDEC registered data. This data sheet contains all applicable registered data in effect at the time of publication. (Bulletin No. DL-S7311916, March 1973)

2N2217 THRU 2N2222

*electrical characteristics at 25°C free-air temperature (unless otherwise noted)

PARAMETER		TEST CONDITIONS	TO-5 → 2N2217		2N2218		2N2219		UNIT
			TO-18 → 2N2220		2N2221		2N2222		
			MIN	MAX	MIN	MAX	MIN	MAX	
$V_{(BR)CBO}$	Collector-Base Breakdown Voltage	$I_C = 10\ \mu A$, $I_E = 0$	60		60		60		V
$V_{(BR)CEO}$	Collector-Emitter Breakdown Voltage	$I_C = 10\ mA$, $I_B = 0$, See Note 6	30		30		30		V
$V_{(BR)EBO}$	Emitter-Base Breakdown Voltage	$I_E = 10\ \mu A$, $I_C = 0$	5		5		5		V
I_{CBO}	Collector Cutoff Current	$V_{CB} = 50\ V$, $I_E = 0$		10		10		10	nA
		$V_{CB} = 50\ V$, $I_E = 0$, $T_A = 150°C$		10		10		10	µA
I_{EBO}	Emitter Cutoff Current	$V_{EB} = 3\ V$, $I_C = 0$		10		10		10	nA
h_{FE}	Static Forward Current Transfer Ratio	$V_{CE} = 10\ V$, $I_C = 100\ \mu A$			20		35		
		$V_{CE} = 10\ V$, $I_C = 1\ mA$	12		25		50		
		$V_{CE} = 10\ V$, $I_C = 10\ mA$, See Note 6	17		35		75		
		$V_{CE} = 10\ V$, $I_C = 150\ mA$, See Note 6	20	60	40	120	100	300	
		$V_{CE} = 10\ V$, $I_C = 500\ mA$, See Note 6			20		30		
		$V_{CE} = 1\ V$, $I_C = 150\ mA$, See Note 6	10		20		50		
V_{BE}	Base-Emitter Voltage	$I_B = 15\ mA$, $I_C = 150\ mA$, See Note 6		1.3		1.3		1.3	V
		$I_B = 50\ mA$, $I_C = 500\ mA$, See Note 6				2.6		2.6	V
$V_{CE(sat)}$	Collector-Emitter Saturation Voltage	$I_B = 15\ mA$, $I_C = 150\ mA$, See Note 6		0.4		0.4		0.4	V
		$I_B = 50\ mA$, $I_C = 500\ mA$, See Note 6				1.6		1.6	V
$\lvert h_{fe} \rvert$	Small-Signal Common-Emitter Forward Current Transfer Ratio	$V_{CE} = 20\ V$, $I_C = 20\ mA$, $f = 100\ MHz$	2.5		2.5		2.5		
f_T	Transition Frequency	$V_{CE} = 20\ V$, $I_C = 20\ mA$, See Note 7	250		250		250		MHz
C_{obo}	Common-Base Open-Circuit Output Capacitance	$V_{CB} = 10\ V$, $I_E = 0$, $f = 1\ MHz$		8		8		8	pF
$h_{ie(real)}$	Real Part of Small-Signal Common-Emitter Input Impedance	$V_{CE} = 20\ V$, $I_C = 20\ mA$, $f = 300\ MHz$		60		60		60	Ω

NOTES: 6. These parameters must be measured using pulse techniques. $t_w = 300\ \mu s$, duty cycle ≤ 2%.
7. To obtain f_T, the $\lvert h_{fe} \rvert$ response with frequency is extrapolated at the rate of −6 dB per octave from f = 100 MHz to the frequency at which $\lvert h_{fe} \rvert = 1$.

switching characteristics at 25°C free-air temperature

PARAMETER		TEST CONDITIONS†	TYP	UNIT
t_d	Delay Time	$V_{CC} = 30\ V$, $I_C = 150\ mA$, $I_{B(1)} = 15\ mA$,	5	ns
t_r	Rise Time	$V_{BE(off)} = -0.5\ V$, See Figure 1	15	ns
t_s	Storage Time	$V_{CC} = 30\ V$, $I_C = 150\ mA$, $I_{B(1)} = 15\ mA$,	190	ns
t_f	Fall Time	$I_{B(2)} = -15\ mA$, See Figure 2	23	ns

†Voltage and current values shown are nominal; exact values vary slightly with transistor parameters.

*JEDEC registered data

Bibliography

It can be helpful to read about a topic when it is presented in a manner different from the student's own textbook. The presentation may be more elementary or more sophisticated. Furthermore, the reader may wish to study a topic in more depth, particularly if a certain project is being developed. Therefore, a short list of references follows. The first nine references are general texts on electronics for scientists and are listed in chronological order. Brief comments are included with each to help in choosing a reference. The last four references focus on topics emphasized in this book.

Malmstadt, H.V., C.G. Enke, and E.C. Toren. 1963. *Electronics for Scientists.* New York: W.A. Benjamin.

Text is one of the first books to recognize a different approach is needed for scientists. It has tube and transistor emphasis, but was among first to stress operational amplifiers. A later version of this book is *Electronic Measurements for Scientists* by H.V. Malmstadt, C.G. Enke, and S.R. Crouch. 1974. Menlo Park, Calif.: W.A. Benjamin. It is almost "encyclopedic," but a good reference. The latest version is Malmstadt, H.V., C.G. Enke, and E.C. Toren. 1981. *Electronics and Instrumentation for Scientists.* Menlo Park, Calif.: Benjamin/Cummings.

Suprynowicz, V.A. 1966. *Introduction to Electronics for Students of Biology, Chemistry, and Medicine.* Reading, Mass.: Addison-Wesley.

Stresses general principles of operation for many kinds of electronic circuits, but with tube emphasis.

Brophy, J. 1966, 1972, 1977. *Basic Electronics for Scientists.* New York: McGraw-Hill.

Good selection of material; intermediate level, with transistor emphasis.

Diefenderfer, A.J. 1972. *Principles of Electronic Instrumentation.* Philadelphia: Saunders.

Closer to the level of this book; includes digital electronics. The second edition, 1979, moved to intermediate level. Substantial transistor work.

Vassos, B., and G. Ewing. 1972. *Analog and Digital Electronics for Scientists.* New York: Wiley.

Emphasizes the operational amplifier. Uses complex impedance approach for ac circuits.

Anderson, L., and W. Beeman. 1973. *Electronic Circuits and Modern Electronics.* New York: Holt, Rinehart and Winston.

Intended for scientists, but at a moderately sophisticated level; all solid-state.

Simpson, R. 1974. *Introductory Electronics for Scientists and Engineers.* Boston: Allyn and Bacon.
Good selection of material for intermediate analog electronics course; includes digital topics.

Jones, B. 1974. *Circuit Electronics for Scientists.* Reading, Mass.: Addison-Wesley.
Quite a lot of network theory in an attempt to avoid obsolescence; also moderately sophisticated.

Weber, L., and D. McLean. 1975. *Electrical Measurement Systems for Biological and Physical Scientists.* Reading, Mass.: Addison-Wesley.
Good presentation of many topics in laboratory electronics at a very basic level; especially intended for biological scientists.

Brown, Burr. 1971. J.G. Graeme, G.E. Tobey, L.P. Huelsman, eds. *Operational Amplifiers, Design, and Applications.* New York: McGraw-Hill.
Written by a pioneering manufacturer in the field. This book is quite comprehensive and sophisticated, but may have answers when others do not.

Jung, W. 1974. *IC Op-Amp Cookbook.* Indianapolis, Ind.: Sams.
A "Sam's paperback" that is a compendium of operational amplifier techniques. Also discusses characteristics of actual important op-amp devices available.

Jung, W. 1977. *IC Timer Cookbook.* Indianapolis, Ind.: Sams.
This is another paperback; good discussion of characteristics and compendium of uses for this widely used linear integrated-circuit type (especially the 555 timer).

Sheingold, D.H., ed. 1980. "Analog Devices." *Transducer Interfacing Handbook.* Norwood, Mass.: Analog Devices, Inc.
Written by an important manufacturer of analog integrated circuits. Readable discussion of transducers and instrumentation for them. Half of the book discusses a cross section of applications.

Answers to Problems

The answers to odd-numbered problems, except for diagrams and lengthy discussions, are provided. In the case of derivations or discussion answers, only brief summary answers are given.

Chapter A1

1. 0.011 Ω
5. (a) for 47 Ω, 146 mA; for 47 kΩ, 4.6 mA; (b) for 47 Ω, 6.8 V; for 47 kΩ, 217 V
7. (b) 0.545 kΩ; (c) 11 mA; (d) for 1 kΩ, 6 mA; for 2 kΩ, 3 mA; for 3 kΩ, 2 mA
9. for 10 kΩ, 0.4 mA; for 5 kΩ, 0.695 mA; for 2 kΩ, 0.26 mA; for 3 kΩ, 0.174 mA
11. 3.36 V
13. 565 cm^2
15. plot $v = 20\ V(1 - e^{-t/20\ ms})$
17. (a) 0 V; (b) 5 V; (c) 10 V; (d) 3.54 V; (e) 0.01 Hz; (f) 100 s
21. (a) 818 turns; (b) 273 turns
23. (a.1) 0.532 mA; (a.2) 5.32 V; (a.3) 8.46 V; (a.4) 0.532; (a.5) −57.8°; (a.6) 2.83 mW; (b) 1.59 kHz
25. (a) $P_{max} = V^2/R$; (b) $P = V^2R/(R^2 + X_c^2)$; at f_o have $P = V^2/2R$
27. (a) $R = 500\ \Omega$; $C = 0.32\ \mu F$; (b) −45°
29. $I_1 = -2/19$ A; $I_3 = 12/19$ A; $I_4 = 10/19$ A
31. (a) $V_{Th} = 2.28$ V; $R_{Th} = 2.72\ \Omega$; then $I = 0.103$ A; (b) $V_{Th} = 2.5$ V; $R_{Th} = 0.75\ \Omega$; then $I = 0.526$ A
33. $I_1 = -2/19$ A; $I_3 = 12/19$ A; $I_4 = 10/19$ A
35. $I_N = V/R$; $R_N = R$
37. $V_{out} = V_{in}/[1 + (X_c/R)^2] = V_{in}/[1 + (f_o/f)^2]$; for $f_o/f >> 1$: $V_{out} = f^2(V_{in}/f_o^2)$

Chapter A2

1. (a) $R_{sh} = 1.002\ \Omega$; (b) 0.1 V
3. ±0.55°
5. (a) 1.5 kΩ; (b) 1.5 kΩ
7. $R_x = V/(10^{-3}\ A) = V(10^3\ A^{-1}) = V(k\Omega/V)$
9. 3% (worst case)
13. ΔV/V is a small fraction, requiring a high precision detector.
15. Scope alternately triggers on two different positive slope points of the waveform at a given level.
17. (a) 1/11; (b) 0.235 $V_{p\text{-}p}$; (c) 94.1 Ω
19. (a) 0.063; (b) 0.25
21. $V_{fo}/V_{pass} = 0.707 = V_2/V_1$; dB ratio = 20 log (0.707) = −3 dB

Chapter A3

1. 41.9 Ω
3. (a) 1.08 V; 0.0 V; 0.95 V; (b) B; neither; A
5. (a) 5 mV; (b) 75 mV
7. (a) A is in extension; B is in compression;

(b) $\Delta R_A \cong -\Delta R_B$ when the bar is bent. Therefore, the strain gauge pair may be put in a half-active bridge in order to double the bridge output in accordance with Problem 12(b) of Chapter A2.

(c) Support hinged plate with strain gauged bar on opposite side from hinge. Place gauges in half-active bridge and use bridge output voltages as indication of weight placed on plate.

11. (a) Ferromagnetic material placed within a coil increases its inductance; fractional insertion causes a fractional increase in inductance.

(b) Measure its inductance. However, an ac-measuring circuit is required to measure this ac parameter.

13. (a) Maximum single cell output voltage of about 0.6 V is less than voltage required across relay coil.

(b) Connect 9 cells in series.

(c) Cells must supply required current at the coil voltage; however, current capability is proportional to cell area.

Chapter A4

1. 0.575 A
3. (a) same graph as diagram A4–8a; (b) 10 V_{dc} voltage graph
5. 4.3 V_{rms}
9. (a) 10.8 V; (b) 0.02 V
13. (a) 3 mA in 1 kΩ; 1 mA in 6 kΩ; 2 mA in diode

(b) 3 mA in 1 kΩ and in 6 kΩ, 0 mA in diode

(c) As current through R_L increases, current through zener diode must decrease. But minimum zener current is zero, as obtained with $R_L = 2$ kΩ. Zener diode effectively becomes an open circuit for smaller R_L values.

Chapter A5

1. 4.0 V_{rms}
5. (b) 8.9 V_{p-p}
7. 56%
9. $v_{out}/v_{in} = [R_{in}^2/(R_{in} + R_{out})^2]A_v^3$
11. (a) for dc: 6 V; for ac: 0 V; (b) for each: 6 V_{rms}, provided the ac frequency is within the bandwidth of each.
13. (a) 1.0 V_{rms}; 10 V_{rms}; (b) 0.0793 V_{rms}; 0.793 V_{rms}; (c) 10 kΩ in amplifier 1 and 200 pF in amplifier 2; (d) 79 kHz
15. $V_1 = +0.5$ V; $V_2 = -0.5$ V; $V_3 = +1.0$ V
17. −6.0 V
19. 0.1 V_{p-p}
21. (a) plot: for $f < f$ and $f > f_2$: $A_v = 0$; for $f_1 < f < f_2$: $A_v = 1$; (b) 1.8 μV_{rms}; (c) 1.8 μV_{rms}
23. 8.5 mV_{rms} noise output; 56 μV_{rms} signal input

Chapter A6

1. $R_f = 100$ kΩ and $R = 5$ kΩ in Figure A6–2a.
3. $R_f = 9.5$ kΩ and $R = .5$ kΩ in Figure A6–5a.
5. (a) −5; 5 kΩ; (b) +6; many megohms
7. (a) $R_{icm+} \| R_{icm-}$; (c) 199.9 MΩ (for $R_{in} = 2$ MΩ, $A_{vo} = 200{,}000$)
9. (a) noninverting

(b) $v_{in} \cong [10\text{ k}\Omega/(10\text{ k}\Omega + 1000\text{ k}\Omega)][1\text{ k}\Omega/(1\text{ k}\Omega + 99\text{ k}\Omega)]v_{out} = v_{out}/1000$; then $v_{in}/v_{out} = A_v = 1000$

(c) Neither very large nor very small resistors are required.

11. Because $v_d \cong V$, voltage V appears across R; then current through R is equal to $I = V/R$. But current I flows through R_x, since there is negligible current into − input of op amp.
13. $R_1 = 3$ kΩ; $R_2 = 1.5$ kΩ; $R_3 = 1$ kΩ; $R_f = 3$ kΩ in Figure A6–7.
15. $R_f = 500$ Ω in Figure A6–11, with voltmeter at output.
17. 0.2 V
19. Turn diode in Figure A6–15 around.
21. $v_{out} = (v_B/v_A)^k$

23. (a) Descending "staircase form" with slanting descenders between horizontal steps. (b) v_{out} remains at $V-$ after reaching this level.
25. (a) $v_{out} = +0.032$ V $\cos(\omega t)$; v_{out} leads v_{in} by 90°; 0.032 V; yes
    (b) $v_{out} = -125$ V $\cos(\omega t)$; v_{out} lags v_{in} by 90°; 125 V; no
27. (a) 500 Hz; (b) 5000 Hz
29. (a) ac amplifier because of C_1 and C_2; (b) block dc offset in input signal; (c) small dc offset in op-amp output is blocked by C_2; (d) possible large dc offset at output limits output amplitude due to possible clipping at $V+$ or $V-$.
31. (a) high speed category (e.g., 318); (b) precision category (e.g., OP–07); (c) general purpose (e.g., 741); (d) high speed category (e.g., LH0032C); (e) FET input (e.g., CA3140); (f) micropower (e.g., OP–20)

Chapter A7

1. (a) −6 V; (b) −6 V; (c) −6 V; (d) −6 V; (e) +6 V
3. (a) Use $V_r = 3$ V in Figure A7–2a; have output as in Figure A7–3a, but diode D_2 is connected to +6-V battery, and diode D_1 is connected to −6-V battery (rather than ground). Also omit 1 kΩ = resistor in Figure A7–3a. (b) Use voltage divider of 9-kΩ and 3-kΩ resistors in series between $V+ = +12$ V and ground to establish $V_r = 3$ V. (One mA flows in this circuit.) Use a 3-resistor voltage divider with two 300-Ω resistors and one 600-Ω resistor connected between +12 V and −12 V to establish ±6 V clipping voltages in place of batteries in part a. (Twenty mA flows in this circuit.) Use 3-kΩ current limit resistor in op-amp circuit.
5. Upper level = 4.75 V; lower level = −0.25 V; then hysteresis = 5 V; Center = 2.25 V, not 3 V.
7. Three cascaded monostable MVs. First monostable uses $C = 1.0$ μF and switch selectable $R = 1$ MΩ, 2 MΩ, 3 MΩ; second "delay" monostable uses $R = 5$ MΩ and $C = 1.0$ μF; third monostable uses $R = 5$ MΩ and $C = 2.0$ μF.
9. (a) 5 kΩ "pulls" 555 input high to +6 V due to voltage divider action between 5 kΩ and high resistance of input. When (NO) push button is pressed, input is grounded and negative pulse is generated.
   (b) Pushing PB1 causes *RS* flip-flop to set to high volts and motor to start; pressing PB2 causes 555 to reset to low volts output and motor stops. Motor must require less than 200 mA.
   (c) Connect threshold input through 5 kΩ to ground and also through (NO) PB2 to +6 V. Positive pulse to upper comparator resets *RS* flip-flop and stops motor.
11. Three 555 timer devices in separate monostable circuits as in Figure A7–12 are required. $R = 1$ MΩ in all three, but $C = 1.8$ μF for process *A* monostable; $C = 4.5$ μF for process *B* monostable; $C = 18$ μF for process *C* monostable. A single push-button circuit like that for PB1 in Diagram A7–8 feeds both process *A* and *B* monostable trigger inputs. Process *B* monostable triggers process *C* monostable through a differentiating circuit as shown in Figure A7–14; this is required because the low voltage state from the process *B* monostable would otherwise prevent termination of the process *C* pulse.
13. Use the monostable 555 timer circuit of Problem 10. Output of this monostable connects to reset of a second 555 timer that is connected as an astable MV as in Figure A7–15. Let $R_A = 1$ kΩ; $R_B = 10$ kΩ; $C = 0.0023$ μF in this astable circuit. Operate both 555 timers from $V+ = 6$ V, and

connect output of the astable to the second transducer.

15. (b) When transistor switch S in Figure A7–15 closes, it acts as direct short between $V+$ and ground. High current flow would destroy the transistor.
17. (a) 500 Hz, 12 V; (b) 1.2 V

Chapter A8

1. (a) small resistance for A, small resistance for B; (b) large resistance for A, small resistance for B
3. 5.2 V
5. (a) For $V_{DS} < V_{\text{pinch-off}}$, I_D is proportional to V_{DS} with proportionality constant depending on V_{GS}. Then FET exhibits resistance R_{DS} between drain and source that is controlled by V_{GS}. When V_{GS} is zero, R_{DS} is a minimum.
   (b) The circuit is of the same form as the noninverting amplifier of Figure A6–5a, with R_{DS} of FET replacing resistor R. Then $A_v = (1 + R_f/R_{DS})$ is adjustable. When V_{GS} is zero, R_{DS} in minimum and voltage gain A_v is maximum.
9. $V_1 \cong 0.7$ V; $V_2 = 9$ V; $I = 3$ mA
11. (a) 100 V; $\cong 0$ V; (b) 250 kΩ; (c) $I_{C\max} >$ 20 mA; $V_{CE\max} > 100$ V
13. (a) 4.0 V; 3.3 V; 13.4 V
    (b) Since V_C does not depend on the transistor β, the operating point does not shift with changing transistor temperatures or different transistors.
    (c) The dc blocking capacitors are at the input and output.
15. I through R_L equals I_D of the FET. But $V_{GS} = 0$ here, so that $I_D = I_{DSS}$. Since I_{DSS} is approximately constant (for $V_{DS} > V_{\text{pinch-off}}$), the load current is constant. However, $V \geq V_{RL} = I_{DSS}R_L$. Then $R_L \leq V/I_{DSS}$.
17. (a) $V_{A\text{out}} = (V+) - I_{DSS}R_L < V+$; $V_{B\text{out}} = V+$; (b) no; (c) $V_{B\text{out}} = (V+) - I_C R_L < V+$; yes; (d) would forward bias the gate-channel diode; (e) negative voltage equal to pinch-off voltage
19. (a) By symmetry, ΔV of the two output terminals must be the same. Since $V_{\text{out}} = V_{\text{left}} - V_{\text{right}}$, V_{out} experiences change $\Delta V_{\text{out}} = \Delta V - \Delta V = 0$; (b) If transistors are matched, the symmetry again requires ΔV of the two output terminals must be the same. Again as in part a, $\Delta V_{\text{out}} = 0$.
21. (a) Turned on when it is bright; transistor passes current that pulls in the armature and closes the NO contact; (b) Make connection to NO switch contact rather than NC.
23. Switches S_1 and S_2 open for 0 W; only switch S_1 closed for 250 W; switch S_2 closed for 500 W. For switch S_1 closed, the SCR conducts when the instantaneous ac voltage exceeds about 10 V; however, negative half cycles are blocked by the SCR. For switch S_2 closed, the triac conducts for most of the half cycle for both positive and negative swings and 500 W is attained.
25. (a) Power the 9-V_{dc} supply from the 110 V_{ac} plug input; also connect box receptacle in series with solid-state relay output and to 110 V_{ac} input. Connect V+ and ground of 555 timer to 9-V_{dc} supply +, − outputs. Use $C = 10$ μF, $R_A = R_B = 285$ kΩ in astable connection for the 555 timer. Connect 555 timer output through 1 kΩ resistor to the solid-state relay control input (this limits input current to about 6 mA). (b) SCR permits only half-wave current flow, while triac permits full-wave current flow. (c) One terminal of the power-supply output must be connected to one of the ac lines in this approach. A short may occur if the supply output is not floating.

Index